Special Effects and German Silent Film

AF762782

Special Effects and German Silent Film

Techno-Romantic Cinema

Katharina Loew

Routledge
Taylor & Francis Group
LONDON AND NEW YORK

First published in 2021 by Amsterdam University Press Ltd.

Published 2025 by Routledge
4 Park Square, Milton Park, Abingdon, Oxon OX14 4RN
605 Third Avenue, New York, NY 10158

Routledge is an imprint of the Taylor & Francis Group, an informa business

© Katharina Loew / Taylor & Francis Group 2021

All rights reserved. No part of this book may be reprinted or reproduced or utilised in any form or by any electronic, mechanical, or other means, now known or hereafter invented, including photocopying and recording, or in any information storage or retrieval system, without permission in writing from the publishers.

Trademark notice: Product or corporate names may be trademarks or registered trademarks, and are used only for identification and explanation without intent to infringe.

ISBN: 9789463725231 (hbk)
ISBN: 9781041186526 (pbk)
ISBN: 9781003704218 (ebk)
NUR 674

Cover illustration: *Maschinenmensch* from the movie *Metropolis* (1927) by Fritz Lang, Walter Schulze-Mittendorff © / Friedrich-Wilhelm-Murnau-Stiftung and WSM Art Metropolis.

Cover design: Coördesign, Leiden

DOI: 10.5117/9789463725231

Every effort has been made to obtain permission to use all copyrighted illustrations reproduced in this book. Nonetheless, whosoever believes to have rights to this material is advised to contact the publisher.

For Product Safety Concerns and Information please contact our EU representative:
GPSR@taylorandfrancis.com
Taylor & Francis Verlag GmbH, Kaufingerstraße 24, 80331 München, Germany

Table of Contents

Acknowledgements

This book was decisively shaped by many colleagues, friends and mentors.

Before all, I offer my gratitude to Sarah Keller and Ariel Rogers, my dearest interlocutors, allies, and champions. Without their enthusiasm, perspicacious queries, and brilliant guidance, this book would have never seen the light of day.

I am also deeply indebted to Nicholas Baer, Claudia Esposito, Sergio Rigoletto, and Kathleen Karlyn for affectionately providing much-needed direction and reassurance throughout this undertaking.

Further, I thank Tessa and Marc Valabregue, Anita Chari, Anna Gunderson, Reda Cherif, Marco Aresu, Adrienne Eaton, and Ilarion Melnikov for their solidarity, love, and inspiration.

At the University of Chicago, Tom Gunning, Yuri Tsivian, David Levin, David Wellbery, and Miriam Hansen first helped conceive this project and have persistently supported it ever since. Jamie McCormack, Leigh Ann Smith-Gary, and Michelle Zimet made important suggestions for improvement in earlier stages. At the University of Oregon, Jeffrey Librett, Susan Anderson, Dorothee Ostmeier, Gantt Gurley, Sonja Boos, Ken Calhoon, Daisuke Miyao, and Matthias Vogel generously shared their expertise, and I thank them for their kindness and friendship. At the University of Massachusetts Boston, Alexander Des Forges, Wenhua Shi, Shannon McHugh, Sari Kawana, Vetri Nathan, and Dean David Terkla of the College of Liberal Arts have contributed essential advice and unfaltering support, and Marco Natoli, Yulia Mevissen, Jutta Handte, Kevin Kehl, Abdelkrim Mouhib, and Tynan Byrne have had my back at critical junctures.

Anton Kaes, Martin Lefebvre, Julie Turnock, Doron Galili, Michael Cowan, Maggie Hennefeld, Laura Lee, Christina Petersen, Tami Williams, Diane Wei Lewis, Frédéric Tabet, and Valentine Robert have offered an ear, suggestions, and encouragement. I am also grateful to Lutz Koepnick, Gerd Gemünden, Johannes von Moltke, Kristina Köhler, Franziska Heller, Scott Curtis, Jörg Schweinitz, Janet Bergstrom, Jan Olsson, Christophe Wall-Romana, Michael Raine, Helmut Asper, and Don Crafton, who provided thoughtful insights at key moments.

Maryse Elliot of Amsterdam University Press has been extraordinarily kind, knowledgeable, and supportive, and Marcel Steinlein of the Friedrich-Wilhelm Murnau Stiftung, Julia Riedel of the Deutsche Kinemathek, and Andreas Thein of the Filmmuseum Düsseldorf graciously lent a hand in the midst of the initial COVID-19 shutdown.

Finally, and most importantly, I thank Saskia Loew, Anja Loew, and David Degras, who have held me up every step of the way and whose love carries me through life. This is for you.

Introduction: Special Effects and the Techno-Romantic Paradigm

Abstract
German silent cinema is famous for its unconventional aesthetics and film-technological innovations. These characteristics were the result of efforts to reconcile the new medium's automatic reproductions of physical reality with idealist conceptions of art. Special effects played a crucial role in this endeavour. They afforded creative experiments with the cinematic apparatus and inspired filmmakers to convey ideas and emotions. Special effects embodied the "techno-romantic" project of construing technology as a means for transcending material reality. This common response to industrial modernity profoundly shaped German silent film culture. The techno-romantic paradigm formed the basis of one of the most creative periods in film history and proved instrumental in the evolution of cinematic expressivity and film art.

Keywords: special effects, techno-romantic thought, Expressionism, film art, expressivity

One of the most famous sequences in the history of cinema is the robot's anthropogenesis in *Metropolis* (Ufa, 1927, dir. Fritz Lang). The images of the metal cyborg seated like an ancient Egyptian deity on a throne enveloped in dramatic electric discharges and rings of light that glide up and down her body are as awe-inspiring as they are enigmatic and ominous. The scene has become an emblem for the unconventional aesthetics and seminal film-technological innovations of German silent cinema. It also points to a complex, even paradoxical attitude towards machine technology, one that is simultaneously characterized by fascination and apprehension. The filmmakers and intellectuals who principally shaped German silent film culture strove to reconcile their idealist conceptions of art and life with a rapidly mechanizing world. They eagerly embraced cinema as the art of

Loew, Katharina, *Special Effects and German Silent Film: Techno-Romantic Cinema*.
Taylor & Francis Group, 2021
DOI 10.5117/9789463725231_INTRO

the machine age. At the same time, however, they also insisted that it was imperative for the medium to meet key stipulations of idealist aesthetics. The leading German filmmakers were preoccupied with the creative potentials of film technology and special effects came to play a pivotal role in the endeavour to develop cinema's medium-specific expressivity. According to Sergei Eisenstein, German cinema evinced that the artistic value of special effects rivalled that of montage, which he considered the "nerve of cinema:"[1] "'The technical possibility,' foolishly called a 'trick,' is undoubtedly just as important a factor in the construction of the new film art as the new principle of montage that emerged from it."[2]

German cinema's renown for innovative uses of film technology and special effects notwithstanding, the topic has received little scholarly attention to date. Instead, studies on German silent cinema have traditionally followed the path laid out by two foundational studies, Siegfried Kracauer's *From Caligari to Hitler* (1947) and Lotte Eisner's *L'Écran démoniaque* (1952).[3] Both authors worked as film critics during the Weimar Republic and survived the Holocaust in exile. Writing in the immediate aftermath of World War II, each in their own way grappled with the question of whether the most heinous crimes in human history might have been presaged in cinema. Kracauer read recurring narrative motifs as indicative of psychological inclinations of a collective mentality during the interwar period. Eisner, for her part, traced the pictorial characteristics of Weimar cinema to the influence of nineteenth-century dark Romanticism, contemporary Expressionism, and Max Reinhardt's theatre, which she deemed symptomatic of a "German soul." Although their approaches and rhetoric differed, Kracauer and Eisner both sought to distil from Weimar films characteristics that might elucidate the rise of National Socialism. Indeed, this objective, whether pursued implicitly or explicitly, may also explain why, compared to other national contexts, socio-political history continues to play a major

1 Sergei Eisenstein, "A Dialectic Approach to Film Form [1929]," in *Film Form: Essays in Film Theory*, ed. and trans. Jay Leyda, 45-63 (San Diego, CA: Harcourt, 1949), 48.

2 Sergei Eisenstein, "The New Language of Cinematography," *Close Up* 5 (May 1929): 13. Translation amended. Many thanks to Yuri Tsivian for helping to correct it.

3 Siegfried Kracauer, *From Caligari to Hitler: A Psychological History of the German Film* [1947] (Princeton, N.J.: Princeton University Press, 2004); Lotte H. Eisner, *The Haunted Screen: Expressionism in the German Cinema and the Influence of Max Reinhardt* [1952], trans. Roger Greaves (Berkeley, CA: University of California Press, 2008).

role in German film scholarship, as for instance evidenced in the superb work of Patrice Petro and Anton Kaes, among others.[4]

For decades, studies on German cinema, particularly those of Anglophone provenance, focused on the ways film could shed light on larger socio-political issues. It was not until the 1980s and 1990s, when, as part of the paradigm shift within film studies towards "new film history," German cinema became an object of investigation in its own right. Researchers began to attach greater importance to primary sources, production processes, and the industrial contexts from which these films emerged. Careful historical investigations, initially published primarily in German-language edited collections, proceedings,[5] and monographs,[6] played an important role in shifting the focus away from Kracauer's and Eisner's grand narratives. Following Thomas Elsaesser's authoritative *Weimar Cinema and After* (2000), novel approaches that combine thorough historical analysis with theoretical questions also gained increasing prevalence in Anglophone scholarship, as for instance evinced in Noah Isenberg's and Christian Rogowsky's important edited collections.[7] This book builds

4 Patrice Petro, *Joyless Streets: Women and Melodramatic Representation in Weimar Germany* (Princeton, NJ: Princeton University Press, 1989); Anton Kaes, *Shell Shock Cinema: Weimar Culture and the Wounds of War* (Princeton, NJ: Princeton University Press, 2009).

5 Especially noteworthy are *CineGraph: Hamburgisches Centrum für Filmforschung*, which publishes annual conference proceedings (Munich: edition text + kritik, 1989-). Its important and steadily expanding encyclopaedia (since 1984) has appeared in abbreviated form as *The Concise Cinegraph: Encyclopaedia of German Cinema*, eds. Hans-Michael Bock and Tim Bergfelder (New York, NY: Berghahn, 2009). Likewise of great value are Frank Kessler, Sabine Lenk, and Martin Loiperdinger, eds., *KINtop: Jahrbuch zur Erforschung des frühen Films*, (Frankfurt am Main: Stroemfeld, 1992-2006) and Harro Seegeberg, ed., *Mediengeschichte des Films* (Munich: Wilhelm Fink Verlag, 1996-2012).

6 Important (originally) German-language monographs include: Corinna Müller, *Frühe deutsche Kinematographie: formale, wirtschaftliche und kulturelle Entwicklungen 1907-1912* (Stuttgart: Metzler-Verlag, 1994); Klaus Kreimeier, *The Ufa Story: A History of Germany's Greatest Film Company 1918-1945*, trans. Robert and Rita Kimber (Berkeley, CA: University of California Press, 1999); Michael Wedel, *Der deutsche Musikfilm: Archäologie eines Genres 1914-1945* (Munich: edition text + kritik, 2007); Chris Wahl, *Multiple Language Versions Made in Babelsberg. Ufa's International Strategy, 1929-1939* (Amsterdam: Amsterdam University Press, 2016); Philipp Stiasny, *Das Kino und der Krieg: Deutschland 1914-1929* (Munich: edition text + kritik, 2009); Tobias Nagl, *Die unheimliche Maschine: Rasse und Repräsentation im Weimarer Kino* (Munich: edition text + kritik, 2009); Joseph Garncarz, *Maßlose Unterhaltung: Zur Etablierung des Films in Deutschland 1896-1914* (Frankfurt am Main: Stroemfeld, 2010).

7 Thomas Elsaesser, *Weimar Cinema and After: Germany's Historical Imaginary* (London: Routledge, 2000); Noah Isenberg, ed., *Weimar Cinema: An Essential Guide to Classic Films of the Era* (New York, NY: Columbia University Press, 2008); Christian Rogowsky, ed., *The Many Faces of Weimar Cinema: Rediscovering Germany's Filmic Legacy* (Columbia, MD: Camden House, 2011).

on these methodologies, scrutinizing specifically correlations between filmmaking practices and pervasive ideas about the nature of cinema, art, and technology.

While German film scholarship has experienced significant methodological advances and thematic diversification in recent years, outside of specialist circles, dated concepts persist. For example, German silent cinema continues to be widely associated with the catchphrase "German Expressionism." Lotte Eisner, who later disavowed her initial far-reaching use of the term, described Expressionism as an artistic current concerned with mysticism, raw emotion, immediate experience, visions, subjectivism, and the incomprehensible.[8] While there is little doubt that these qualities figured prominently in German silent film culture, I argue that they are neither peculiar to Expressionism nor to German cinema. Many also pertained to other contemporary art movements and film cultures for instance in France, Italy, and pre-revolutionary Russia, which are not usually associated with Expressionism. What is more, the German post-war trend towards stylization went far beyond the scope of identifiable Expressionist features and few German filmmakers—certainly not those to whom it has been most commonly attributed, Fritz Lang and F. W. Murnau—accepted the Expressionist label for their work.[9] The objectives and strategies that self-identified Expressionist poets, dramatists, and painters postulated for and implemented in their art rarely found their way into the cinematic realm. As Jürgen Kasten has suggested, German films of the early 1920s adopted little more than stylistic rudiments and the sensational label from the Expressionist movement.[10] Moreover, in common parlance, "Expressionist" has not only been used as a designation for a particular early twentieth-century art movement, but also as a shorthand for any non-realistic style. As a result, the term has lost much of its specificity and meaning. Even though prominent scholars like Barry Salt, Thomas Elsaesser, and Dietrich

8 Lotte H. Eisner, "Stile und Gattungen des Films," in *Das Fischer Lexikon. Film, Rundfunk, Fernsehen*, eds. Lotte H. Eisner and Heinz Friedrich, 259-283 (Frankfurt am Main: Fischer, 1958), 264.

9 For the interwar German discourse on film style see Kristina Köhler, "Nicht der *Stilfilm* also, sondern der *Filmstil* ist wichtig!" in *Filmstil: Perspektivierungen eines Begriffs*, eds. Julian Blunk, Tina Kaiser, Dietmar Kammerer, and Chris Wahl, 91-117 (Munich: edition text + kritik, 2016). For Lang's and Murnau's attitude to the Expressionist label see Dietrich Scheunemann, "Activating the Differences: Expressionist Film and Early Weimar Cinema," in *Expressionist Film: New Perspectives*, ed. Dietrich Scheunemann, 1-31 (Rochester, NY: Camden House, 2003), 7.

10 Jürgen Kasten, "Filmstil als Markenartikel. Der expressionistische Film und das Stilexperiment Von morgens bis mitternachts," in *Die Perfektionierung des Scheins: Das Kino der Weimarer Republik im Kontext der Künste*, ed. Harro Segeberg, 37-66 (Munich: Wilhelm Fink Verlag, 2000), 41.

Scheunemann have eloquently questioned the usefulness of the Expressionist designation as a synonym for German silent cinema, many critics continue to resort to the term, thus perpetuating misleading notions about the era's stylistic and ideological orientation.[11] In this book, I intend to move past the elusive concept of Expressionism and instead scrutinize broader issues that informed German silent film culture, in particular responses to modernization and the development of cinematic modes of expressivity.

The techno-romantic paradigm

The preoccupation with German cinema's debts to Expressionism has resulted in inadequate attention to other aspects, including German film culture's pivotal and complex relationship to machine technology. The profound socio-cultural transformations in the wake of the industrial revolution—the triumph of rationalization, mechanization, and market economy—had brought about a widely perceived disregard for any spiritual, intellectual, and cultural values. Faced with massive loss of prestige and influence, intellectuals vocally deplored the "soulless" zeitgeist and the concomitant rampant "materialism."[12] Heinrich Mann observed in 1909, "[t]he hatred of intellectuals for the infamous materialism of this German Empire is considerable."[13] In response, idealist sentiment surged. It manifested in fin-de-siècle artistic movements such as symbolism, aestheticism or Art Nouveau as well as in a public discourse deeply concerned with mental processes, sensations and emotions.. As Uta Grund has argued, the decades around 1900 saw a "downright excessive use of terms such as 'mood' [*Stimmung*], 'spiritual' [*seelisch*], 'inwardly' [*innerlich*], 'sensible' [*sinnlich*], 'sensation' [*Empfindung*] and 'intimate' [*intim*]."[14] In a similar vein, Hugo Münsterberg asserted the dawn of a new idealist age:

11 Barry Salt, *Film Style and Technology: History and Analysis*. London: Starwood, 1983, 184-186; Elsaesser, *Weimar Cinema and After*, 18-60; Scheunemann, "Activating the Differences," in *Expressionist Film: New Perspectives*, 1-31.

12 While academic intellectuals and the film world only intersected in part, many of the viewpoints described by Fritz Ringer were also prevalent among film critics and practitioners. See Fritz K. Ringer, *The Decline of the German Mandarins: The German Academic Community, 1890-1933* (Cambridge, MA: Harvard University Press, 1969).

13 Heinrich Mann in a letter to René Schickele, 27 December 1909. Heinrich Mann, *Macht und Mensch: Essays*, ed. Peter-Paul Schneider (Frankfurt am Main: Fischer, 1989), 258.

14 Uta Grund, *Zwischen den Künsten: Edward Gordon Craig und das Bildertheater um 1900* (Berlin: Akademie Verlag, 2002), 32.

> The great realist wave wanes and a new idealistic one rises. Technological culture, brought about by realism, has begun to disillusion us; people everywhere begin to sense that realistic progress has not made life more valuable, better and liveable. The time has come that the accumulation of dry facts starts to leave us indifferent and everything once again moves towards a holistic worldview. [...] Realism has fulfilled its mission and its reason becomes nonsense if it is not complemented by idealistic truth."[15]

Intellectuals' persistent appeals for increased attention to the spiritual realm should however not be considered tantamount to hostility towards modernity. As Ben Singer has reminded us, "an appropriately expansive model of modernity must take into account not just dynamic sources of social and aesthetic novelty, flux, intensity and so on, but also prominent counter-forces of anti-modern sentiment that resulted from and were intertwined with, the dominant thrust forward."[16] As a matter of fact, seemingly out-dated impulses may even serve to expedite processes of modernization. As philosopher Odo Marquard argued in a different context, by offsetting modernity's dehumanizing effects, attention to the life of the mind plays a crucial role in making modernization tolerable and thus attainable in the first place.[17]

This dynamic is manifest in a sentiment towards machine technology prevalent at the turn of the twentieth century, which I describe as "techno-romantic." The term, which is not intended to characterize any systematic doctrine, refers to the inclination to construe technology as a means to evoke the imagination, emotion, and more generally the intangible or spiritual. Machines, deemed the epitome of uncreative, destructive, and dehumanizing materialism, paradoxically emerged as a safeguard of those essential human qualities under attack by the same rampant materialism. The techno-romantic mind-set allowed artists and intellectuals to affirm their fears about modernization and machine technology and simultaneously immerse themselves in the creative possibilities they afforded. The techno-romantic

15 Hugo Münsterberg, *Die Amerikaner*, vol. 1: *Das politische und wirtschaftliche Leben* (Berlin: Ernst Siegfried Mittler und Sohn, 1904), 3. Münsterberg's English-language *The Americans* differs substantially from the German version and does not include this passage.
16 Ben Singer, "The Ambimodernity of Early Cinema: Problems and Paradoxes in the Film-and-Modernity Discourse," in *Film 1900: Technology, Perception, Culture*, eds. Annemone Ligensa and Klaus Kreimeier, 38-51 (New Barnet, Herts: John Libbey Publishing, 2009), 38.
17 See Odo Marquard, "Über die Unvermeidlichkeit der Geisteswissenschaften. Hermann Lübbe zum 60. Geburtstag," *Sprache und Literatur* 57 (1986): 72-81, 76-77.

stance necessarily embodies both apprehension and enthusiasm, albeit not in a consistent ratio. Willy Ley's passionate plea for the romanticism of the space rocket, for example, constitutes a particularly technophile manifestation of this attitude: "Technology allegedly robs human kind of its spiritual assets [...] because it destroys romanticism and exoticism. The space rocket completely invalidates this accusation. [...] Modern technology does not annihilate this utmost, excessive romanticism and exoticism; it makes it possible in the first place."[18] Attention to techno-romantic responses to modernization complicates conventional views of modernity/modernism as inherently rational and progressive, views that have tended to marginalize neo-romantic perspectives as necessarily reactionary and/or anachronistic. Techno-romantic thought fostered processes of modernization by proffering meaning in the face of a rapidly modernizing world. The association of machine technology with emotional, mental or spiritual qualities corroborated the primacy of human consciousness over material reality and thus helped mitigate what was perceived as modernity's fundamental disregard for non-physical aspects of existence.

I have appropriated the term "techno-romantic" from Austrian author Karl Kraus. In his 1918 anti-war polemic "The Techno-Romantic Adventure," Kraus excoriated the simultaneity of mass extermination through industrialized warfare and archaic "knightly" values such as patriotism, honour, discipline or heroism, which for him characterized World War I.[19] In 1945, Thomas Mann made a similar observation when describing the German Empire as a "mixture of robust timeliness, efficient modernness on the one hand and dreams of the past on the other,—in a word, highly technological Romanticism."[20] Kraus and Mann diagnosed the union between technophilia and idealism as one of the era's principal characteristics. Both regarded industrialized conquest, war, and genocide, bolstered by emotional appeals to adventure, heroism, and world improvement, as the horrific outcomes of this synergy. Without minimizing this devastating legacy, I contend that efforts to construe machine technology as an agent of idealist objectives constituted one of the principal responses to the technological revolution around 1900. The wide-ranging significance of such endeavours for European intellectual and cultural history must be fathomed in areas beyond their most horrific excesses.

18 Willy Ley, *Die Möglichkeit der Weltraumfahrt* (Leipzig: Hachmeister & Thal, 1928), 340.
19 Karl Kraus, "Das technoromantische Abenteuer," *Die Fackel* 474-483 (May 1918): 41-45.
20 Thomas Mann, "Germany and the Germans," in *Thomas Mann's Addresses Delivered at the Library of Congress, 1942-1949*, 45-66 (Cabin John, MD: Wildside Press, 2008), 62.

Even more so than Kraus and Mann, Jeffrey Herf has classified the association of idealism and machine technology as a right-wing phenomenon. His concept of "reactionary modernism" describes a core feature of German right-wing and subsequently Nazi ideology, namely the combination of technophilia and irrational, illiberal, and antimodernist ideas peculiar to the German nationalist movement.[21] Techno-romantic and reactionary modernist views thus share key concerns: They embrace machine technology while disparaging modern materialism and emphasizing intangible facets of reality. However, despite prominent nationalist voices within the German film industry like Hanns Heinz Ewers, Fritz Lang, and Thea von Harbou; the right-wing roots of Ufa, Germany's largest studio; and its 1927 takeover by ultraconservative media mogul Alfred Hugenberg -- as a whole, German film culture of the 1910s and 1920s cannot be described as reactionary modernist. Concentrated in Berlin, one of the most vibrant and progressive art centres in the world, the German film industry was cosmopolitan, diverse, and involved numerous members of marginalized groups. In particular, many leading German film artists were of Jewish heritage.[22] The fact that more than two thousand filmmakers were forced into exile by the National Socialist rise to power in 1933 suggests that the identity and beliefs of a significant share of the German film industry were not compatible with National Socialism.[23]

In contrast to reactionary modernists, techno-romantic sentiments can be found across the political spectrum. What is more, a majority of German filmmakers were less concerned with any specific political agenda than with questions of art, film, and above all film art. The dominant ideology within the German film industry can thus be described as a mix of romantic inclinations, apprehensions about modernity, a pronounced patriotism, and idealist aesthetic tenets. These tendencies blended with liberal and humanist views as well as an avid commitment to pursuing novel forms of expressivity. The techno-romantic paradigm bolstered these axioms by establishing technology as a creative tool and German

21 Jeffrey Herf, *Reactionary Modernism: Technology, Culture, and Politics in Weimar and the Third Reich* (Cambridge: Cambridge University Press, 1984), 1-2.

22 Prominent Jewish filmmakers of the Weimar era include producers Paul Davidson, Jules Greenbaum, and Erich Pommer, directors Ernst Lubitsch, Joe May, Max Mack, E. A. Dupont, and Richard Oswald, screenwriters Henrik Galeen, Béla Balázs, Robert Liebmann, and Willy Haas, actors John Gottowt, Ernst Deutsch, Peter Lorre, Alexander Granach, Curt Bois, and Fritz Kortner, and cinematographers Karl Freund, Curt Courant, Helmar Lerski, and Eugen Schüfftan.

23 See Helmut G. Asper, *Filmexil in Hollywood: 'Etwas Besseres als den Tod.' Portraits, Filme, Dokumente* (Marburg: Schüren-Verlag, 2002), 20.

art cinema as a global paragon. Pervading filmmakers' aspirations and viewers' expectations, techno-romantic perspectives shaped German silent film culture and proved instrumental in cinema's eventual recognition as a full-fledged art.

Art and industry

In marked contrast to the United States, where a film's esteem was almost exclusively determined by its financial performance, in Germany it depended chiefly on perceived artistic value. That aesthetics came to play such a central role in the German film industry testifies to the persistent dominance of idealist frames of reference within the public sphere. Already in the early 1910s, at a time when the German production sector was still in its infancy, German producers turned towards more costly "artistic" films. Such prestige productions set themselves apart from the bulk of cheap, short-lived pictures, while also bolstering the medium's reputation with elite opinion leaders. Critical acclaim and increased public attention, in turn, presented the prospect of boosting sales. Following World War I, producer Erich Pommer developed this thinking into a comprehensive strategy.

Pommer was an ardent cinephile who believed in film's unifying powers and ability to overcome the antagonism and resentment left behind by a devastating world war. His strategic focus, which prioritized collaboration, aesthetic and technological innovation, and quality filmmaking, differed noticeably from the preoccupation with storytelling, box office success, and individual job performance that dominated the discourse in Hollywood. Although often blamed for Ufa's near-bankruptcy in 1926, Pommer's ambitious plan to establish German cinema as a global premium brand was in fact quite judicious. From a European perspective, Hollywood's profitable yet for the most part painfully trite pictures left plenty of room for sophisticated competition. The artistic films Pommer envisioned to that effect were not intended as elitist endeavours, but rather something akin to folk art: creative, skilled, and cultured, yet widely accessible. Pommer explained: "Today the relation between artistic and commercial films is such that an artistic film can be a gold mine, whereas a purely commercial film will almost always be inartistic."[24] Many of the most consequential works of German film history, including *Das Cabinet des Dr. Caligari* (Decla,

24 Erich Pommer, "Geschäftsfilm und künstlerischer Film," *Der Film* (10 December 1922): 1.

1920, dir. Robert Wiene), *Der letzte Mann* (Union-Film/Ufa, 1924, dir. F. W. Murnau), *Metropolis*, and *Der blaue Engel* (Ufa, 1930, dir. Josef von Sternberg), were produced under Pommer's aegis. The spirit he embodied permeated the German film industry and is evident even in films produced without Pommer's involvement, such as *Der Student von Prag* (Deutsche Bioscop, 1913, dir. Stellan Rye) or *Nosferatu* (Prana-Film, 1922, dir. F. W. Murnau).

For the most part, Germany's premium productions were created by a fairly exclusive group of filmmakers. They included directors F. W. Murnau, Fritz Lang, E. A. Dupont, Ludwig Berger, Joe May, Robert Wiene, and G. W. Pabst, scriptwriters Carl Mayer, Thea von Harbou, Henrik Galeen, Robert Liebmann, and Béla Balázs, cinematographers Fritz Arno Wagner, Karl Freund, Carl Hoffmann, Guido Seeber and Günther Rittau, Helmar Lerski, and Curt Courant as well as art directors Walther Röhrig, Robert Herlth, Hermann Warm, Erich Kettelhut, Otto Hunte, and Paul Leni, most of whom collaborated regularly in varying configurations. This creative network was initiated by Pommer, who, as Robert Herlth recounted, "through this own example, fostered idealism and an exploratory urge. Without Erich Pommer, there would not have been a 'German Era.'"[25]

Pommer saw aesthetic-technological innovation as essential for establishing German cinema's global standing. According to Herlth, he would tell his production teams to "try to invent something mad!"[26] Pommer's approach intertwined commercial considerations with aesthetic ideals by promoting medium-specific, i.e., technological forms of expressivity. Seeking to affirm idealist perspectives within the context of a capitalist mass medium, Pommer played an essential part in bolstering the techno-romantic paradigm within the German film industry.

Pommer enthusiastically advocated for the collective nature of film, the ingenuity of teamwork, and cinephilia as a creative catalyst. He declared: "Film is an art form or an art-like form, which cannot be created by an individual but only by artists in close daily collaboration and it can only be created by people obsessed with film."[27] He consequently gave his production crews far-reaching artistic freedom and encouraged perpetual experimentation. As both Erich Kettelhut and Robert Herlth recalled, under

25 Robert Herlth, "Erinnerungen (1958)," in *Filmarchitektur Robert Herlth*, 48-54 (Munich: Deutsches Institut für Film und Fernsehen, 1965), 49.
26 Robert Herlth, "With Murnau on the Set," in Lotte Eisner, *Murnau* [1964], trans. Martin Secker, 59-71 (London: Secker & Warburg, 1973), 62.
27 "Erich Pommer Interview with Radio Frankfurt [1950]," Eric Pommer Collection, Doheny Memorial Library, University of Southern California, Los Angeles, CA.

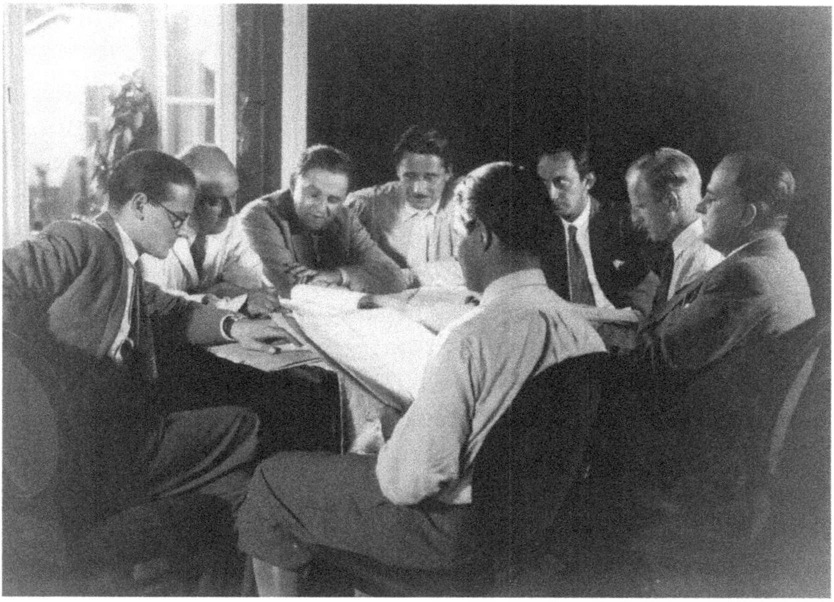

Fig. 0.1. Production meeting for *Melodie des Herzens* (1929). Left to right: composer Viktor Gertler, director Hanns Schwarz, producer Erich Pommer, cinematographer Günther Rittau, composer Richard Heymann, sound engineer Fritz Thiery, production manager Max Pfeiffer, unknown. Source: Deutsche Kinemathek

Pommer's leadership films were developed jointly in meetings of the entire production staff.[28]

To characterize the resulting atmosphere, Herlth famously invoked the collective spirit of an idealized medieval cathedral construction workshop (*Bauhütte*) that "animated everyone."[29] The idealist notion of a free community of master craftsmen and apprentices, united in their commitment to a joint artistic creation, first gained popularity around 1800 and subsequently became a central premise of Richard Wagner's total work of art, which was simultaneously conceived as a synthesis of the arts and the creation of a community of artists:

> The Art-work of the Future is an associate work, and [...] practically conceivable only in the *fellowship of every artist* [...] and for *one definite aim*, is that which forms this fellowship. This definite aim is the Drama, for which they all unite in order by their participation therein to unfold

28 Erich Kettelhut, *Der Schatten des Architekten*, ed. Werner Sudendorf (Munich: Belleville, 2009), 53, 134-135, 196; Herlth, "With Murnau on the Set," in Eisner, *Murnau*, 63-64.
29 Herlth, "Erinnerungen," 48.

> their own peculiar art to the acme of its being [...] to generate the living, breathing, moving drama.[30]

Wagner's theory of collective art production—like the *Gesamtkunstwerk* more generally—reverberated intensely with European modernists and impacted artists from Max Reinhardt, Vsevolod Meyerhold, and Konstantin Stanislavski to Walter Gropius and Bertolt Brecht. In the German film industry, the total dedication to the joint artwork along the lines of Wagner's "fellowship of artists" became a common paradigm. Fritz Lang for instance recounted, "even if we were sometimes hopping mad at each other: the work stood above everything. Here, our collective love and unconditional devotion to the idea converged."[31] The overarching objective of this concerted effort was, in Carl Hoffmann's words, "to render pictorially all of the script's thought content."[32] Special effects emerged as one of the key tools in this endeavour and consequently as an essential element in the evolution of cinematic expressivity. As I trace throughout this book, machine technology, the embodiment of "soulless" materialism, thus became a sine qua non in the pursuit of an unambiguously idealist project.

Cinema's artistic devices

Already the earliest critics identified special effects as one of film's core assets and an opportunity to emancipate the medium from actuality. When cinema emerged as a prominent mass cultural phenomenon in the early 1900s, it represented a major challenge to traditional conceptions of art. As a high-tech device for the automatic reproduction of physical reality, film was discredited as the epitome of modernity's contempt for spiritual values and the life of the mind. However, for a growing number of commentators, cinema also showed potential for expressing creativity and engaging audiences' aesthetic sensibilities. For these critics, the question of the new medium's distinctive, aesthetically relevant features was paramount. They identified two types of subject matter as cinema's proper areas of competence: scenes of nature and fantastic or mental imagery, realized by

30 Richard Wagner, "The Art-work of the Future [1850]," in *The Art-work of the Future and Other Works*, trans. William Ashton Ellis, 69-214 (Lincoln, NE: University of Nebraska Press, 1993), 196. Emphases in the original.

31 Fritz Lang, "Arbeitsgemeinschaft im Film," *Kinematograph* 887 (17 February 1924): 10.

32 Hermann Treuner, "Carl Hoffmann," in *Filmkünstler: Wir über uns selbst* (Berlin: Sibyllen-Verlag, 1928), n. pag.

means of special effects. While the originality and thus the artistic merit of nature cinematography was contentious, most critics acknowledged that trick effects embodied both creativity and the medium's characteristic technological disposition. In this context it is important to note that the German word *Technik* simultaneously refers to "technique" and "technology," a semantic inheritance from the Greek "technē." The slippage between the two concepts doubtlessly helped to assuage apprehensions about machines in the artistic realm by linguistically blurring the boundaries between artisanal craft and mechanized process and appropriating universal respect for the former in favour of the latter.

Also for aspiring film artists, filmmaking had to be about more than adeptly reproducing reality or rendering a sequence of events in a captivating way. Following the invention of photography, conceptions of the purpose of art had increasingly shifted from mimesis towards expression. Because the external world could be recreated automatically, works of art were increasingly construed as outer manifestations of inner life. Thus, like the other arts, film needed to be able to evoke concepts, sensations, and the imaginary, and elicit affective responses from the audience. Through extensive experiments with cinema's functional principles, filmmakers sought to develop the medium's ability to convey meaning beyond concrete facts and to maximize the emotional impact of filmic images or, as Herlth put it, their "intensity of impression" (*Eindringlichkeit der Wirkung*).[33] As a result, many of German silent cinema's most iconic moments result from pioneering uses of trick technology. They include the doppelganger stepping out of the mirror in *Der Student von Prag*, the Tsi-Nan-Fu hypnosis in *Dr. Mabuse, der Spieler* (Uco-Film/Decla-Bioscop, 1922, dir. Fritz Lang), the protagonist's nightclub table plummeting into an abyss *in Phantom* (Uco-Film/Decla-Bioscop, 1922, dir. F. W. Murnau), the eerie Jack-the-Ripper episode in *Das Wachsfigurenkabinett* (Neptun-Film, 1924, dirs. Paul Leni and Leo Birinski), the petrification of the dwarfs in *Die Nibelungen* (Decla-Bioscop, 1924, dir. Fritz Lang), the porter's drunken dream in *Der letzte Mann* (Union-Film/Ufa, 1924, dir. F. W. Murnau), the trapeze performances in *Varieté* (Ufa, 1925, dir. E. A. Dupont), Martin Fellman's nightmares in *Geheimnisse einer Seele* (Neumann-Film-Produktion, 1926, dir. G. W. Pabst), Gretchen's scream in *Faust* (Ufa, 1926, dir. F. W. Murnau) or the launch of the space rocket in *Frau im Mond* (Fritz Lang Film, 1929, dir. Fritz Lang). In each instance, mental states or fantastic subject matter gave occasion to imaginative applications of trick technology, produced in the context of

33 Herlth, "Erinnerungen," 51.

a "fellowship of artists." Each instance constitutes an emotional climax of the respective film and would have been impossible to achieve without the employment of trick technology. In German cinema, special effects came to epitomize the techno-romantic paradigm because it was here that machine technology fulfilled the requirements of idealist aesthetics.

In the wake of cinema's digital revolution, the transition to computer-generated imagery (CGI) in effects production, and the ascendancy of effects-heavy blockbuster aesthetics at the box office, the theory and practice of special/visual effects have come under increased scrutiny in film studies. Most important interventions, including those by Scott Bukatman, Barbara Flückinger, Lisa Purse, Kristen Whissel, and Julie Turnock, have been concerned with spectacular applications and developments in the United States since the 1970s.[34] Simultaneously, scholars like Rolf Giesen, Dan North, Réjanne Hamus Vallée, Lisa Bode, Laura Lee, and Ariel Rogers have begun to bring earlier decades and contexts outside of the United States into focus, but crucial aspects remain to be uncovered.[35] This book expands on the existing research on special effects by paying heed to the 1910s and 1920s, a formative period in their history, and to Germany's famous but insufficiently understood contributions. Partly as a result of overwhelming attention to the Hollywood model and its preoccupation with realism, storytelling, and seamlessness, two applications of special/visual effects have dominated the scholarship to date: astonishing, typically fantastic spectacles on the one hand and, to a much lesser extent, innocuous uses for practical purposes

34 See Scott Bukatman, *Matters of Gravity: Special Effects and Supermen in the 20th Century* (Durham, NC: Duke University Press, 2003); Barbara Flückinger, *Visual Effects: Filmbilder aus dem Computer* (Marburg: Schüren-Verlag, 2008); Lisa Purse, *Digital Imagining in Popular Cinema* (Edinburgh: Edinburgh University Press, 2019); Kristen Whissel, *Spectacular Digital Effects: CGI and Contemporary Cinema* (Durham, NC: Duke University Press, 2014); Julie Turnock, *Plastic Reality: Special Effects, Technology and the Emergence of 1970s Blockbuster Aesthetics* (New York, NY: Columbia University Press, 2015).

35 See Rolf Giesen, "Spezialeffekte Made in Germany," in *Künstliche Welten: Tricks, Special Effects and Computernanimation im Film von den Anfängen bis heute*, eds. Rolf Giesen and Claudia Meglin, 69-111 (Hamburg/Vienna: Europa, 2000); Dan North, *Performing Illusions: Cinema, Special Effects and the Virtual Actor* (London: Wallflower Press, 2008); Dan North, Bob Rehak, and Michael S. Duffy, eds., *Special Effects: New Histories/Theories/Contexts* (London: BFI/Palgrave, 2016); Lisa Bode, *Making Believe: Screen Performance and Special Effects in Popular Cinema* (New Brunswick, NJ: Rutgers University Press, 2017); Réjanne Hamus Vallée and Caroline Renouard, *Les effets spéciaux au cinema: 120 ans de créations en France et dans le monde* (Vanves: Armand Colin, 2018); Laura Lee, *Japanese Cinema Between Frames* (London: Palgrave Macmillan, 2017), 17-50; and Ariel Rogers, *On the Screen: Displaying the Moving Image, 1926-1942* (New York, NY: Columbia University Press, 2019), 19-57.

on the other.[36] However, a third domain must be taken into account: The use of special effects as artistic devices.

The manipulation of filmic images for the purpose of conveying sensorial qualities, emotions, attitudes or ideas played an essential role in the development of cinematic expressivity in Germany and beyond. To what extent film art was contingent on trick technology in the mind of German filmmakers becomes apparent from a 1925 conversation between Fritz Lang and journalist Ludwig Spitzer. As Lang pointed out, his goal as a filmmaker was "to prove that film is capable of exposing mental processes and thus substantiate the bare events psychologically. [...] Maybe we can then speak of true film art (whereas today, we have art films at best)." When asked about the means to convey psychology in film, Lang answered flatly: "The trick in the broadest sense."[37]

As a result of the orientation of the scholarship to date, many assumptions about special and visual effects are based on Hollywood practices since the studio era, which are often construed as the norm. However, as I show in this book, the field is much more diverse and sound-era American conditions are not universally applicable. For example, Julie Turnock has observed that in Hollywood "special effects" was above all a labour category, tracing back to the fact that certain effects were created by a specialized work force.[38] In Europe, however, film production was less departmentalized and labour therefore unsuitable as a defining criterion. Production practices in silent cinema highlight the difficulties in establishing special effects as an intrinsic (rather than a labour) category. In Germany, for instance, effects were usually created by cinematographer(s) and/or art director(s) themselves, often collaboratively and experimentally. What silent filmmakers called "tricks" can at best be described rather broadly as unusual and frequently elaborate production methods that resulted in unusual, often dramatic impressions on screen. However, because European silent films rarely relied on specialized labour, boundaries between "regular" and "unusual" shots are often fluid. Neither filmmakers nor industry observers would have classified striking visual effects resulting from extreme camera angles, unconventional camera movements (also known as "unchained camera" in 1920s diction) or innovative lighting schemes as inherently different from

36 See for instance Charlie Keil and Kristen Whissel, "Introduction," in *Editing and Special/Visual Effects*, eds. Charlie Keil and Kristen Whissel, 1-21 (New Brunswick, NJ: Rutgers University Press, 2016), 12.

37 Ludwig Spitzer, "Fritz Lang über den Film der Zukunft," *Die Filmtechnik* 2 (15 July 1925): 34-35.

38 Turnock, *Plastic Reality*, 8.

techniques such as multiple exposure composites, stop motion animation or forced perspective in set design, which in sound-era Hollywood were habitually assigned to special-effect technicians.

To complicate matters further, evolving technologies and diverse production practices render the establishment of coherent and transhistorical nomenclature exceedingly difficult. Throughout film history, a plethora of (often ill-defined) designations has been used to describe manipulated cinematic images and their various subcategories. Labels have tended to foreground either production or reception. Originally, the worldwide most common expression was "trick effects." Appropriated from stage illusions, the term stresses dexterity and legerdemain, but also carries disparaging connotations, which is why it became eventually displaced. Hollywood studio-era parlance was mainly concerned with fabrication, referring among others to mechanical, technical, photographic, optical, practical, camera or engineering effects. In addition, labels like "special photographic effects," "special process photography," "special work," and "special effects" explicitly accentuated the exceptional nature of their production. That "special effects," along with its direct translations (*effets spéciaux, Spezialeffekte, efectos especiales, effetti speciali,* Спецэффéкт, etc.), came to prevail in everyday language is likely due to the fact that that phrase seems to simultaneously refer to production and reception. It is no longer used in industry jargon, however. After repeatedly changing the name of the corresponding awards category, the Academy of Motion Pictures Arts and Sciences has been referring to "visual effects" since 1972, deemphasizing the production context in favour of reception.[39] Stephen Prince has argued that in digital cinema the term "special effects" has become obsolete since artificial imagery is not "special" anymore and not clearly distinguishable from straight cinematography.[40] As I have indicated, there were no unequivocal boundaries in silent cinema either. Nonetheless, silent effects were "special" in that they required special skills and processes and produced special impressions. Because "special effects" remains the most widely used appellation for the object under investigation, I use it here synonymously with the historical designation "trick."

This book is the first to reveal the decisive role of special effects in the evolution of cinematic expressivity. As technological devices facilitating

39 See Charlie Keil and Kristen Whissel, "Introduction," in *Editing and Special/Visual Effects*, 13.

40 Stephen Prince, *Digital Visual Effects in Cinema: The Seduction of Reality* (New Brunswick, NJ: Rutgers University Press, 2011), 3-4.

the quest for creativity and the imagination, special effects not only epitomized pervasive techno-romantic modes of thought. They also served as a conceptual and practical basis for cinema's emergence as the art of the machine age. In this book, I examine special effects from the perspective of early theoreticians, technicians as well as foreign observers, and analyse their use in three iconic films. As I show in Chapter 1, "Conceptualizing Technological Art: Early German Film Theory," special effects already figured prominently in German film culture before they were widely used in practice. Critics identified tricks as uniquely cinematic devices early on and they came to play a prominent role in the first debates about the "essence" of the filmic medium and its artistic potentials. As a form of popular mass entertainment and an apparatus for the automatic reproduction of material reality, cinema had a low social standing: artistic aspirations seemed futile. Some early commentators nonetheless asserted that the new medium could be a legitimate object of aesthetic scrutiny. In an attempt to fathom cinema's immaterial values, the first German film theorists including Gustav Melcher, Herbert Tannenbaum and Georg Lukács explored the medium's kinship with folk art, mental processes, and the fantastic. They sought to establish that film technology, specifically trick effects, could articulate ideas in a sensual form and thus provide a pathway to a spiritual dimension. Their techno-romantic lines of argument aimed at conceptualizing the new medium within established aesthetics and set the stage for the recognition of cinema as the first technological art.

Techno-romantic thought not only informed early concepts about the promise of cinema but also evolution of trick technology. Chapter 2, "Modern Magicians: Guido Seeber and Eugen Schüfftan," pays attention to two technicians who determined the development of special effects in Germany and beyond. Guido Seeber, a film pioneer and an ardent experimenter, had a penchant for methods like multiple exposure composites that allow cinematographers to simultaneously use their creativity and technical expertise. As a mentor and a publicist he was instrumental in defining the key role of technology in German cinema. Eugen Schüfftan, a painter, invented the only widely employed commercial effects technique to originate in Europe, the Schüfftan process. In contrast to Seeber's effects, which tended to foreground their craftedness and aimed at rendering abstract thought, Schüfftan's mirror-based technique sought to visualize the imagination while concealing the means by which it operated. By turning tricks into a commodity and instituting set extensions as a standard practice in Europe, Schüfftan launched a new era of effects production. In similar ways, Seeber and Schüfftan construed special effect technologies as the medium's primary creative tool, as its core means for rendering mental imagery. In accordance with the period's techno-romantic tenets, they conceived of the cinematic image as fundamentally malleable, a stance that formed the conceptual bedrock of German silent film culture.

Chapter 3, "The Uncanny Mirror: *Der Student von Prag* (1913)," examines Seeber's proudest accomplishment as a cinematographer and trick specialist. The film was co-created by actor Paul Wegener, writer Hanns Heinz Ewers and director Stellan Rye, some of Germany's most ardent early cinephiles, with the goal to demonstrate the feasibility of film art. They understood artistic filmmaking as the articulation of ideas and feelings through the original application of cinema's distinctive technological features, specifically location shots and trick effects. Accordingly, they juxtaposed images of romantic Prague with a uniquely cinematic monster, created by means of Seeber's doppelganger effects. As one of the earliest feature-length films, *Der Student von Prag* portrayed the supernatural as menacing and thus helped establish a new cinematic genre, horror. The film not only succeeded in generating unusual affective audience responses, the doppelganger motif also evoked a range of readings related to notions about identity and self, demonstrating that a silent, visual medium was in fact capable of addressing philosophical questions. Finally, the juxtaposition of trick effects and location photography called into question the ostensible verisimilitude of the photographic image and thus widespread assumptions about the nature of the cinematic medium. Contrary to the reputation of German silent cinema as studio-bound, many artistically ambitious films showcased picturesque nature scenes and confronted them with special effects. This approach also characterizes the film under investigation in the next chapter, *Nosferatu*.

Public enthusiasm for occult themes was rife in the decades around 1900 and many prominent filmmakers—including the creators of *Der Student von Prag*—were occupied with esoteric concepts. Chapter 4, "Visualizing the Occult: *Nosferatu* (1922)," examines a production that was explicitly intended to showcase the treatment of occultist ideas in the context of film art. Constitutive of occult thought is a belief in secret realities beyond our perceptual abilities and in the fundamental oneness of all there is. *Nosferatu* does not advocate for specific occultist doctrines, but many of the film's idiosyncratic aspects, particularly the appearance, behaviour, and powers of the vampire, become intelligible in novel ways when examined from an occultist perspective. The film externalizes the vampire's nature through various cinematic devices and most notably special effects. The materialization of the intangible by means of technology constitutes an essentially techno-romantic project. In *Nosferatu*, it served to consolidate objectives of occultists and cinephiles for the purpose of film art.

The massive cityscapes, the man-eating Moloch and the robot's miraculous anthropogenesis: None of the iconic imagery of Fritz Lang's *Metropolis* would exist without ground-breaking special effects, including those accomplished by means of the Schüfftan process. Chapter 5, "The Technological Sublime: *Metropolis* (1927)," examines the striking friction between the film's portrayal of technology as an agent of tyranny and dehumanization, its simultaneous flaunting of technology as spectacle, and the fact that the production itself was conceived and marketed as a marvel of film technology. The film pursued the techno-romantic project of transcending material reality by means of technology. In fact, it seeks to capture the unfathomability of technology itself. Special effects facilitate encounters with overpowering technological environments and omnipotent machines, which give rise to sentiments that are best described in terms of the "technological sublime." The sublime characterizes experiences that go beyond the earthly and finite, and thus attain a spiritual dimension. In attributing transcendent qualities to mechanical objects, the technological sublime embodies the techno-romantic paradigm.

The final chapter, "'German Technique' and Hollywood," is concerned with the reverberations of the techno-romantic mindset in the American film industry. In the mid-1920s, the innovative imagery and affective force of German productions like *Der letzte Mann* (Union-Film/Ufa, 1924, dir. F. W. Murnau), *Varieté* (Ufa, 1925, dir. E. A. Dupont) and *Metropolis* startled American critics and filmmakers. Well-known directors like F. W. Murnau, E. A. Dupont, and Paul Leni were invited to Hollywood, and their American films showcased a range of unconventional camera effects, in particular moving camera feats and extreme camera angles. What galvanized American commentators about these methods was the realization that cinematic devices could be used to visualize and augment affective content. In Hollywood, German filmmakers demonstrated that moving images could not only reveal character interiority, but also convey a mental perspective towards the events depicted. By so doing, they proffered a novel model of cinematic immersion, which strengthened the audience's absorption in the story world with figurative levels of meaning. Prompted by objectives originating in techno-romantic thought, Hollywood began to pay increased attention to the expressive potential of technical tools, with lasting effects on American filmmaking.

The conclusion considers ways in which techno-romantic thought is reflected in the vogue of speculative fiction in contemporary moving image media, which has been made possible by radical advances in digital visual effects. Computer-generated imagery has brought into reach the fully malleable photograph -- a dream that epitomizes a major triumph of the human mind over outside reality and thus an essentially techno-romantic fantasy. The same ambition also animated the German silent filmmakers who saw special effects as a key path towards shaping mechanically produced images. Their use of trick technology for conveying thoughts and emotions gives occasion to tap a new research area: special/visual effects as artistic tools.

The famous characteristics of German silent films, in particular their peculiar look and creative application of special effects, affected cinema worldwide. These idiosyncrasies trace back to efforts to establish cinema's aesthetic value. German filmmakers eagerly met this challenge and construed, as art director Walter Reimann suggested, cinema's principal task as the visual expression of ideas: "Film is work – the crazy, messy work of translating and transposing the mental via the physical into the optical. Tough handiwork that at every moment must be animated by the spirit, which, as *spiritus rector*, moves over the entire work."[41]

In early German narrative films, fantastic themes gave occasion for creative experiments with trick technology, which in turn paved the way for the emergence of special effects as one of silent cinema's principal means for conveying ideas and atmospheres, mental and affective states. Its cachet as a creative agent notwithstanding, technology never lost its demonic qualities in German silent cinema. The films examined in this book construe their protagonists—the double, the vampire, the robot, and man-eating machines—as technological creatures. They embody the threat of technology's unfathomable powers as much as they bespeak its astonishing creativity. Techno-romantic perspectives proved imperative for the process of modernization precisely because they facilitated concurrent feelings of apprehension about and enthusiasm for technology. Techno-romantic thought permeated every aspect of German silent film culture. It informed critics' expectations as well as filmmakers' goals and methods, and thus served as bedrock for one of the most innovative and influential periods in film history.

Bibliography

Helmut G. Asper, *Filmexil in Hollywood: 'Etwas Besseres als den Tod' Portraits, Filme, Dokumente* (Marburg: Schüren-Verlag, 2002).

41 Walter Reimann, "Filmarbeit," *Gebrauchsgraphik* 6 (1924): 28-32.

Hans-Michael Bock and Tim Bergfelder, eds. *The Concise Cinegraph: Encyclopaedia of German Cinema* (New York, NY: Berghahn, 2009).

Lisa Bode, *Making Believe: Screen Performance and Special Effects in Popular Cinema* (New Brunswick, NJ: Rutgers University Press, 2017).

Scott Bukatman, *Matters of Gravity: Special Effects and Supermen in the 20th Century* (Durham, NC: Duke University Press, 2003).

Lotte H. Eisner, *The Haunted Screen: Expressionism in the German Cinema and the Influence of Max Reinhardt* [1952], trans. Roger Greaves (Berkeley, CA: University of California Press, 2008).

Lotte H. Eisner, "Stile und Gattungen des Films," in *Das Fischer Lexikon. Film, Rundfunk, Fernsehen*, eds. Lotte H. Eisner and Heinz Friedrich, 259-283 (Frankfurt am Main: Fischer, 1958).

Sergei Eisenstein, "A Dialectic Approach to Film Form [1929]," in *Film Form: Essays in Film Theory*, ed. and trans. Jay Leyda, 45-63 (San Diego, CA: Harcourt, 1949).

Thomas Elsaesser, *Weimar Cinema and After: Germany's Historical Imaginary* (London: Routledge, 2000).

Barbara Flückinger, *Visual Effects: Filmbilder aus dem Computer* (Marburg: Schüren-Verlag, 2008).

Joseph Garncarz, *Maßlose Unterhaltung: Zur Etablierung des Films in Deutschland 1896-1914* (Frankfurt am Main: Stroemfeld, 2010).

Rolf Giesen, "Spezialeffekte Made in Germany," in *Künstliche Welten: Tricks, Special Effects and Computernanimation im Film von den Anfängen bis heute*, eds. Rolf Giesen and Claudia Meglin, 69-111 (Hamburg/Vienna: Europa, 2000).

Uta Grund, *Zwischen den Künsten: Edward Gordon Craig und das Bildertheater um 1900* (Berlin: Akademie Verlag, 2002).

Réjanne Hamus Vallée and Caroline Renouard, *Les Effets spéciaux au cinéma : 120 ans de créations en France et dans le monde* (Vanves: Armand Colin, 2018).

Jeffrey Herf, *Reactionary Modernism: Technology, Culture, and Politics in Weimar and the Third Reich* (Cambridge: Cambridge University Press, 1984).

Robert Herlth, "Erinnerungen (1958)," in *Filmarchitektur Robert Herlth*, 48-54 (Munich: Deutsches Institut für Film und Fernsehen, 1965).

Robert Herlth, "With Murnau on the Set," in Lotte Eisner, *Murnau* [1964], trans. Martin Secker, 59-71 (London: Secker & Warburg, 1973).

Noah Isenberg, ed., *Weimar Cinema: An Essential Guide to Classic Films of the Era* (New York, NY: Columbia University Press, 2008).

Anton Kaes, *Shell Shock Cinema: Weimar Culture and the Wounds of War* (Princeton, NJ: Princeton University Press, 2009).

Jürgen Kasten, "Filmstil als Markenartikel. Der expressionistische Film und das Stilexperiment Von morgens bis mitternachts," in *Die Perfektionierung des*

Scheins: Das Kino der Weimarer Republik im Kontext der Künste, ed. Harro Segeberg, 37-66 (Munich: Wilhelm Fink Verlag, 2000).

Charlie Keil and Kristen Whissel, "Introduction," in *Editing and Special/Visual Effects*, eds. Charlie Keil and Kristen Whissel, 1-21 (New Brunswick, NJ: Rutgers University Press, 2016).

Frank Kessler, Sabine Lenk, and Martin Loiperdinger, eds., *KINtop: Jahrbuch zur Erforschung des frühen Films*, (Frankfurt am Main: Stroemfeld, 1992-2006).

Erich Kettelhut, *Der Schatten des Architekten*, ed. Werner Sudendorf (Munich: Belleville, 2009).

Kristina Köhler, "Nicht der *Stilfilm* also, sondern der *Filmstil* ist wichtig!" in *Filmstil: Perspektivierungen eines Begriffs*, eds. Julian Blunk, Tina Kaiser, Dietmar Kammerer, and Chris Wahl, 91-117 (Munich: edition text + kritik, 2016).

Siegfried Kracauer, *From Caligari to Hitler: A Psychological History of the German Film* [1947] (Princeton, N.J.: Princeton University Press, 2004).

Karl Kraus, "Das technoromantische Abenteuer," *Die Fackel* 474-483 (May 1918): 41-45.

Klaus Kreimeier, *The Ufa Story: A History of Germany's Greatest Film Company 1918-1945*, trans. Robert and Rita Kimber (Berkeley, CA: University of California Press, 1999).

Fritz Lang, "Arbeitsgemeinschaft im Film," *Kinematograph* 887 (17 February 1924): 10.

Laura Lee, *Japanese Cinema Between Frames* (London: Palgrave Macmillan, 2017).

Willy Ley, *Die Möglichkeit der Weltraumfahrt* (Leipzig: Hachmeister & Thal, 1928).

Heinrich Mann, *Macht und Mensch: Essays*, ed. Peter-Paul Schneider (Frankfurt am Main: Fischer, 1989).

Thomas Mann, "Germany and the Germans," in *Thomas Mann's Addresses Delivered at the Library of Congress, 1942-1949*, 45-66 (Cabin John, MD: Wildside Press, 2008).

Odo Marquard, "Über die Unvermeidlichkeit der Geisteswissenschaften. Hermann Lübbe zum 60. Geburtstag," *Sprache und Literatur* 57 (1986): 72-81.

Corinna Müller, *Frühe deutsche Kinematographie: formale, wirtschaftliche und kulturelle Entwicklungen 1907-1912* (Stuttgart: Metzler-Verlag, 1994).

Hugo Münsterberg, *Die Amerikaner*, vol. 1: *Das politische und wirtschaftliche Leben* (Berlin: Ernst Siegfried Mittler und Sohn, 1904).

Tobias Nagl, *Die unheimliche Maschine: Rasse und Repräsentation im Weimarer Kino* (Munich: edition text + kritik, 2009).

Dan North, *Performing Illusions: Cinema, Special Effects and the Virtual Actor* (London: Wallflower Press, 2008).

Dan North, Bob Rehak, and Michael S. Duffy, eds., *Special Effects: New Histories/Theories/Contexts* (London: BFI/Palgrave 2016).

Patrice Petro, *Joyless Streets: Women and Melodramatic Representation in Weimar Germany* (Princeton, NJ: Princeton University Press, 1989).

Erich Pommer, "Geschäftsfilm und künstlerischer Film," *Der Film* (10 December 1922): 1.

"Erich Pommer Interview with Radio Frankfurt [1950]," Eric Pommer Collection, Doheny Memorial Library, University of Southern California, Los Angeles, CA.

Stephen Prince, *Digital Visual Effects in Cinema: The Seduction of Reality* (New Brunswick, NJ: Rutgers University Press, 2011).

Lisa Purse, *Digital Imagining in Popular Cinema* (Edinburgh: Edinburgh University Press, 2019).

Walter Reimann, "Filmarbeit," *Gebrauchsgraphik* 6 (1924): 28-32.

Fritz K. Ringer, *The Decline of the German Mandarins: The German Academic Community, 1890-1933* (Cambridge, MA: Harvard University Press, 1969).

Ariel Rogers, *On the Screen: Displaying the Moving Image, 1926-1942* (New York, NY: Columbia University Press, 2019).

Christian Rogowsky, ed., *The Many Faces of Weimar Cinema: Rediscovering Germany's Filmic Legacy* (Columbia, MD: Camden House, 2011).

Barry Salt, *Film Style and Technology: History and Analysis* (London: Starwood, 1983.)

Dietrich Scheunemann, "Activating the Differences: Expressionist Film and Early Weimar Cinema," in *Expressionist Film: New Perspectives*, ed. Dietrich Scheunemann, 1-31 (Rochester, NY: Camden House, 2003).

Harro Seegeberg, ed. *Mediengeschichte des Films* (Munich: Wilhelm Fink Verlag, 1996-2012).

Ben Singer, "The Ambimodernity of Early Cinema: Problems and Paradoxes in the Film-and-Modernity Discourse," in *Film 1900: Technology, Perception, Culture*, eds. Annemone Ligensa and Klaus Kreimeier, 38-51 (New Barnet, Herts: John Libbey Publishing, 2009).

Ludwig Spitzer, "Fritz Lang über den Film der Zukunft," *Die Filmtechnik* 2 (15 July 1925): 34-35.

Philipp Stiasny, *Das Kino und der Krieg: Deutschland 1914-1929* (Munich: edition text + kritik, 2009).

Hermann Treuner, "Carl Hoffmann," in *Filmkünstler: Wir über uns selbst* (Berlin: Sibyllen-Verlag, 1928), n. pag.

Julie Turnock, *Plastic Reality: Special Effects, Technology and the Emergence of 1970s Blockbuster Aesthetics* (New York, NY: Columbia University Press, 2015).

Richard Wagner, "The Art-work of the Future [1850]," in *The Art-work of the Future and Other Works*, trans. William Ashton Ellis, 69-214 (Lincoln, NE: University of Nebraska Press, 1993).

Chris Wahl, *Multiple Language Versions Made in Babelsberg. Ufa's International Strategy, 1929-1939* (Amsterdam: Amsterdam University Press, 2016).

Michael Wedel, *Der deutsche Musikfilm: Archäologie eines Genres 1914-1945* (Munich: edition text + kritik, 2007).

Kristen Whissel, *Spectacular Digital Effects: CGI and Contemporary Cinema* (Durham, NC: Duke University Press, 2014).

1. Imagining Technological Art: Early German Film Theory

Abstract
As a form of popular mass entertainment and an apparatus for the automatic reproduction of material reality, cinema's artistic aspirations seemed futile. Some early commentators nonetheless asserted that the new medium could be a legitimate object of aesthetic scrutiny. In an attempt to fathom cinema's immaterial values, early film theorists including Herbert Tannenbaum and Georg Lukács explored cinema's kinship with folk art, mental processes and the fantastic. They argued that film technology, specifically special effects, could articulate ideas in a sensual form and thus provide a pathway to a spiritual dimension. As this chapter shows, their techno-romantic lines of argument conceptualized the medium within established aesthetics and set the stage for the recognition of cinema as the first technological art.

Keywords: film theory, technology, film art, fantastic, trick effects

Cinema's emergence as a prominent mass culture phenomenon in the early 1900s triggered intense public discussions all over Europe. Gauging the new medium's social and aesthetic implications, commentators voiced grave concerns. Cinema was perceived as a threat to the health, tastes, and morals of mass audiences as well as a danger to established cultural institutions like the theatre. Even more importantly, cinema's technological character raised fundamental questions about the nature of art in the machine age. This chapter explores why film posed such a major aesthetic problem and how the new medium was eventually integrated with existing conceptions of art. As I argue, special effects played a key role in this endeavour.

In Germany, the first comprehensive discussion about film sprung up around 1907, at a time when cinema established itself as a public institution, permanent movie theatres became increasingly prevalent, and the first

Loew, Katharina, *Special Effects and German Silent Film: Techno-Romantic Cinema*. Taylor & Francis Group, 2021
DOI 10.5117/9789463725231_CH01

specialized film trade journals emerged. Simultaneously, the aesthetic discourse was dominated by idealist convictions that posited machine technology as diametrically opposed to art. Art generated beauty and truth through the sensuous expression of the ideal, while technology was associated with exteriority and soulless objectivity. As a mechanism that merely reproduced the appearance of the material world, film was portrayed as one-dimensional, mundane, and inherently incapable of transcending phenomena. Devoid of spiritual properties, the medium necessarily remained aesthetically inconsequential. For cinema-friendly critics, the task was therefore to ascertain film's ontology and invalidate these allegations. As I demonstrate, sympathetic commentators routinely invoked special effects as devices that—despite or because of their technological nature—allowed the medium to prevail over physical reality. I examine three prominent lines of argument. Each engaged established aesthetics in a different way, but all resorted to techno-romantic reasoning and contended that trick technology facilitated medium-specific forms of creative expressivity, which in turn substantiated cinema's artistic merit.

The first approach I consider likened film to the simple and magical world of fairy tales. Due to this ontological affinity, fairy tales also constituted preferred filmic subject matter. Although less sophisticated than forms of high art, both fairy tales and films were understood as (potential) manifestations of the collective soul and thus valuable as popular entertainment and education. The second line of reasoning detected similarities between cinema and inner psychic processes, in particular dreams. Not only the viewing situation was perceived as dream-like. Unlike any other art, film, by virtue of trick technology, could facilitate vivid impressions of mental imagery. Special effects thus established links to the immaterial and attested to the medium's spiritual depth. Third, I examine innovative theories that conceptualized cinema in the tradition of Friedrich Schiller's notion of aesthetic semblance. Focusing on essays by Gustav Melcher, Will Scheller, Herbert Tannenbaum, and Georg Lukács, I show how early commentators construed the widely assumed shallowness of film—a consequence of its technological nature—not as a liability but as a transcendent power. Inherently fantastic and, indeed, utopian, the medium offered unexpected avenues to the ideal realm. Although these approaches differed significantly, all contextualized the new medium within established aesthetics and conceived of trick technology as cinema's pathway to the imaginary and intangible and thus to film art.

Techno-romantic thought shaped film cultures all over Europe and, as I will show in Chapter 6, left marks even in the United States. In Germany, it conditioned critics' expectations and filmmakers' goals. Touching film production as well as reception, it influenced the development of the German film industry in important ways. Techno-romantic assumptions also permeate the work of leading film theorists such as Hugo Münsterberg, Jean Epstein, Rudolf Arnheim, and Béla Balázs. The early discussions examined in this chapter serve as a frame of reference for the remainder of this book. As the following chapters will reveal, the initial quest for links between trick technology and the immaterial resulted in a long-lasting preoccupation with special effects as an expressive tool. It laid the foundation for German filmmakers' persistent exploration of innovative ways to represent the supernatural and subjective. German cinema's special-effect feats and the prominent role of fairy-tale, fantastic, and horror genres trace back to the demand to integrate technology with idealist conceptions of art. Trick technology assumed such an important function in German silent cinema because it embodied the machine's triumph over its own confinement to outside appearances.

Art vs. technology

The invention of photography signified a critical juncture for art theory. Because the external world could now be recreated automatically, mimesis as art's principal task was increasingly called into question. As Gerhard Plumpe has shown, post-Hegelian aesthetic discourse was characterized—with a few prominent exceptions like Kierkegaard and Nietzsche—by remarkably consistent positions that shaped the aesthetic views of a wider public well into the twentieth century.[1] Following Hegel, nineteenth-century art theorists conceived of the purpose of art as generating beauty and truth through the sensuous expression of ideas: "In every artwork something infinite must reveal itself. The artwork obtains this infinity solely through ideas. But ideas are of extrasensory nature. Like everything that seeks to enter the material world, they must therefore assume sensory form."[2]

Art was characterized as the "beautiful fusion of the real and the ideal,"[3] while the machine served as the negative counterpart within

1 Gerhard Plumpe, *Der tote Blick: Zum Diskurs der Photographie in der Zeit des Realismus* (Munich: Wilhelm Fink Verlag, 1990), 24-25.

2 Franz Anton Nüsslein, *Lehrbuch der Ästhetik als Kunstwissenschaft* (Regensburg: Verlag von G. Joseph Manz, 1837), 83. Unless noted otherwise, all translations are my own.

3 Carl Lemcke, *Populäre Ästhetik* (Leipzig: E.A. Seemann, 1873), 244.

this dichotomous framework. Goethe had predicted as early as 1797 that "intricate mechanism, refined craftsmanship, and industrial production shall bring about the demise of art."[4] Concerns regarding industrialization, manifest in Goethe's words, and the hope that "genuine art" could provide protection against total objectivation and commercialization comprised the core of the opposition between art and machine technology forged in nineteenth-century aesthetic theory. According to the schematic terms of the discourse, art embodied life, spirit, creativity, and subjective autonomy, whereas technology was associated with materialism, heteronomy, automation, and, ultimately, death.[5] Technology was saddled with all that made modern existence seem alienating and hostile. The photographer Josef Krämer, writing under the pseudonym G. Mercator, summarized the antagonism between art and technology thus: "The task of art must always be the depiction of beauty. What we call beauty is when the extrasensory shines through the sensual, or the ideal through the real. [...] Both art and technique create. However, technical productions are dead bodies, while art inspires its products with soul; they are not dead but speak silver-tongued to a kindred spirit."[6] Like photography, which had been spurned as a "dead mirror" and contrasted negatively with the "active mirror" of the human eye,[7] cinema produced merely mechanical copies of outside reality and was therefore incapable of expressing something immaterial: "The photoplay knows nothing of God or soul. The events remain what they are: flat, figural processes without any symbolic value."[8] Many critics therefore maintained that film could never create authentic, subjective, living artworks. Prominent art historian Konrad Lange even went so far as to denounce cinema as the very antithesis of art, calling it *Unkunst*.[9] Originating in a technological apparatus that copied outside reality, cinematic representations were incompatible with established conceptions of art.

4 Johann Wolfgang Goethe, "Kunst und Handwerk [1797]," in *Gedenkausgabe der Werke, Briefe und Gespräche*, ed. Ernst Beutler, vol. 13, (Zürich: Artemis, 1971-1977), 128.

5 See Plumpe, *Der tote Blick*, 28.

6 G. Mercator (=Josef Krämer), "Ist die Photographie eine Kunst?" *Deutsche Photographen-Zeitung* 15 (1891): 193.

7 See Plumpe, *Der tote Blick*, 35.

8 F. M. Hübner, "Der Dichter und das Kino," *Allgemeine Zeitung* (Munich) 39 (27 September 1913): 629-630.

9 See Konrad Lange, "Die 'Kunst' des Lichtspieltheaters [1913]," in *Geschichte der Filmtheorie: Kunsttheoretische Texte von Méliès bis Arnheim*, ed. Helmut H. Diederichs, 75-88 (Frankfurt am Main: Suhrkamp-Verlag, 2004); Konrad Lange, "Bühne und Lichtspiel," *Deutsche Revue* (October 1913): 119-125.

The essence of cinema

All debates about cinema's artistic value during this period revolved around the question: what is the "essence" [*das Wesen*] of film? Following Lessing's assertion that any artwork must comply with the specific stylistic properties of its own medium,[10] cinema-friendly critics sought to determine cinema's innermost nature and pondered its laws, limits, and its distinctive, aesthetically relevant features. Among the established art forms cinema was considered most closely related to theatre. Taking the stage as their primary point of reference,[11] observers identified "cinematic" subject matter in areas where film had a clear advantage over the theatre. A. Günsberg articulated this view in one of the earliest German essays on film aesthetics, calling on the cinematograph to "show us precisely what the stage cannot."[12] Critics thus identified two types of subject matter as the cinema's proper areas of competence: nature cinematography and fantastic or mental imagery, realized by means of trick effects.[13] Author Friedrich Freksa for instance pointed out: "Artistically speaking, the cinematograph has two attractions: The richness of living nature, which can never appear on the stage, comes to life in all its glory on the screen. And like living ghosts, the cinema can conjure up events of a grotesque

10 Gotthold Ephraim Lessing, *Laocoön: An Essay on the Limits of Painting and Poetry* [1766], trans. Edward Allen McCormick (Baltimore: Johns Hopkins University Press, 1984).

11 Countless publications compared theatre and cinema. See for instance Herbert Tannenbaum, *Kino und Theater* (Munich: Steinebach, 1912); Willy Rath, *Kino und Bühne* (M. Gladbach: Volksvereinsverlag, 1913); Reinhard Bruck, "Kinematograph und Theater," *Der Kinematograph* 247 (20 September 1911); Friedrich Paulsen, "Theater und Kino," *Die Grenzboten* (Berlin) 45 (1913): 285-288; Hermann Koch, "Kino und Theater," *Der Strom* (Vienna) 2 (May 1914): 61-63; Ludwig Fulda, "Theater und Kinematograph," *Die Woche* (Berlin) 16 (20 April 1912): 639-642; Gustav Hartung, "Kinematographie und Theater," *Der Kinematograph* 170 (30 March 1910); Willi Warstat, "Zwischen Theater und Kino," *Die Grenzboten* 23 (1912): 483-488. Several others are reprinted in *Prolog vor dem Film: Nachdenken über ein neues Medium, 1909-1914*, ed. Jörg Schweinitz, 230-252 (Leipzig: Reclam-Verlag, 1992).

12 A. Günsberg, "Künstlerische Regie bei kinematographischen Aufnahmen und Vorführungen [1907]," in *Geschichte der Filmtheorie: Kunsttheoretische Texte von Méliès bis Arnheim*, ed. Helmut H. Diederichs, 44-47 (Frankfurt am Main: Suhrkamp-Verlag, 2004). Diederichs considers Günsberg's text "probably the earliest German-language essay on formal [film] aesthetics." See Helmut H. Diederichs, "Frühgeschichte deutscher Filmtheorie: Ihre Entstehung und Entwicklung bis zum Ersten Weltkrieg." Postdoctoral Thesis, J. W. Goethe-Universität Frankfurt am Main, 1996, 19.

13 Similar arguments can also be found in the French discourse. See for instance Georges Méliès, "Cinematographic Views [1907]," and Rémy de Gourmont, "Epilogues: Cinematograph [1907]," in *French Film Theory and Criticism: A History/Anthology 1907-1939*, ed. and trans. Richard Abel, 35-47 and 47-50 (Princeton, NJ: Princeton University Press, 1988).

and magical kind, which never take place in life and could never appear on the stage."[14]

Although commentators habitually referred to the stage to explain their attention to images of nature and tricks, they rarely explained their reasoning in detail. One of the few critics who made explicit what others just implied was cinema reformer Hermann Häfker, whose articles (since 1907) and monograph *Kino und Kunst* (1913) are among the most sophisticated early reflections on cinema's cultural significance and aesthetic potential. As Häfker set forth, images of nature and tricks were genuinely cinematic (*kinoecht*) because they were based on the medium's "technological essence."[15] The implication was that the essence of the filmic medium manifested itself in subject matter that was made possible by film technology in the first place. Cinema's validation as an art was thus posited as dependent on its technological medium specificity, while technology and art were simultaneously seen as mutually exclusive. As I will show, it was techno-romantic lines of argument that helped resolve this apparent paradox.

Prior to World War I, Häfker was the only German critic to argue that cinema's most significant artistic property was the reproduction of reality. Häfker specifically referred to the calm, enigmatic, and timeless stirrings of nature: "Nature is ready-made poetry, we simply did not have the tools to reproduce it in a genuine way."[16] For Häfker, film technology had the unique ability to capture the mysticism of nature. However, his espousal of cinematic realism came at least one decade too early. In the pre-war period images of nature—although widely acknowledged as genuinely cinematic— were considered of little artistic value.

While there was widespread consensus about cinema's "technological essence" and what constituted genuinely cinematic subject matter, the question of whether the medium could actually produce full-fledged works of art remained a fiercely disputed issue. In order to be granted the status of an art form, cinema had to simultaneously heed its "technological essence"

14 Friedrich Freksa, "Theater, Pantomime and Cinema [1916]," in *The Promise of Cinema: German Film Theory 1907-1933*, eds. Anton Kaes, Nicholas Baer, and Michael Cowan, trans. Alex H. Bush, 111-114 (Berkeley, CA: University of California Press, 2016), 114.

15 See Hermann Häfker, "Atlantis [1914]," in *Prolog vor dem Film: Nachdenken über ein neues Medium, 1909-1914*, ed. Jörg Schweinitz, 405-406 (Leipzig: Reclam-Verlag, 1992) and Hermann Häfker, "Hauptmanns Atlantis," *Bild & Film* III:6 (1913/1914): 139-140. Incidentally, what Häfker and others deemed *kinoecht* corresponds to what Siegfried Kracauer would half a century later describe as cinema's inherent formative and realistic tendencies. Siegfried Kracauer, *Theory of Film: The Redemption of Physical Reality* [1960] (Princeton, NJ: Princeton University Press, 1997), 30-37.

16 Hermann Häfker, *Kino und Kunst* (M. Gladbach: Volksvereinsverlag, 1913), 36.

and live up to Schiller's postulation: "The ultimate purpose of art is to depict what transcends the realm of the senses."[17] Critics therefore explored how film technology could provide access to a sphere beyond what its mechanical trappings suggested.

Techno-romantic perspectives

Film scholars, particularly since the 1980s, have routinely associated cinema with concepts like novelty, speed, technology, youth, and urbanity. However, as Ben Singer has shown, the discourse on film and modernity during the silent era portrays the new medium not only in terms of the "novel and dynamic dimensions of twentieth century life" but also explores "cinema's potential as an instrument of metaphysical revelation and spiritual transcendence."[18] Indeed, uncovering affinities between machine technology and the immaterial was a matter of widespread concern. It reflects the strong urge to reaffirm affective and spiritual aspects of existence in an age that seemed defined by unbridled, stifling and destructive reification and rationalization. The reaction against the prevalent ideology of objectivity also found expression in various artistic movements that emphasized subjectivity, mysticism, exoticism, and the imagination, including symbolism, art nouveau, and neo-romanticism. It might be tempting to characterize such impulses simply as backward-looking. However, far from rejecting modernity, commentators like philosopher Richard von Coudenhove-Kalergi, in his *Apologia of Technology* (1922), instead emphasized the pertinence of ideal or spiritual aspects in the modern world: "Technology is the body, ethics is the soul of culture. [...] All technological progress becomes harmful and worthless if the human being, while conquering the world, loses his soul: in that case it would have been better if he had stayed animal."[19] In a similar vein, Hugo Münsterberg, referencing a dictum from Goethe's *Faust*, called for a reconciliation of

17 Friedrich Schiller, "On the Pathetic" [1793], in *Friedrich Schiller Essays*, eds. Walter Hinderer and Daniel O. Dahlstrom, trans. Daniel O. Dahlstrom, 45-69 (New York, NY: Continuum, 1993), 45.

18 Ben Singer, "The Ambimodernity of Early Cinema: Problems and Paradoxes in the Film-and-Modernity Discourse," in *Film 1900: Technology, Perception, Culture*, eds. Annemone Ligensa and Klaus Kreimeier, 38-51 (New Barnet, Herts: John Libbey Publishing, 2009), 45.

19 Richard Nikolaus Coudenhove-Kalergi, *Apologie der Technik* (Leipzig: Verlag der neue Geist Dr. Peter Reinhold, 1922), 68.

modern materialism and idealism: "Realism has fulfilled its mission and its 'reason becomes unreason' if it is not complemented by idealist truth."[20]

Efforts to defend the imagination, feelings, and more generally the immaterial against the perceived onslaught of materialism and rationalism were rife in the decades around 1900. To designate the conceptual antithesis to "modern" attitudes, contemporaries often resorted to the term "romantic." "Romantic" described a penchant for nostalgia, exoticism, fantasy, sentiment, and introspection. It was also interpreted as an inclination "to flee this-worldliness" and retreat from distressing living conditions.[21]

Particularly leftists like Ernst Bloch perceived the coincidence of romantic tendencies and the highly industrialized societies in which they flourished as deeply troubling. Bloch diagnosed a synchronicity of different historical modalities within modernity, a phenomenon he described as "contemporaneity of the non-contemporaneous."[22] He believed that ideas and ways of life that challenged and resisted the present moment constituted a key feature of modernity. At the same time, Bloch asserted, backward-looking thought provided a breeding ground for radical political ideologies and thus paved the way for National Socialism. For example, a powerful influence on the development of National Socialism was the *völkisch* movement, whose orientation was distinctly anti-modern. *Völkisch* thinkers construed the Germanic peoples as a bio-mystical and primordial community that had been corrupted by foreign, particularly Jewish elements. They advocated for an ethno-national rebirth based on the spiritual essence of the Nordic race and an adherence to "eternal" laws and ways of life.

There is little doubt that the political right effectively capitalized on the widespread longing for spirit and emotion, whereas the left tended to emphasize reason and objectivity, which may have, as Oskar Negt has suggested, ultimately benefited its opponents.[23] The frictions between "old-fashioned" sentiment and idealism and "modern" detachment and indifference constituted a dominant theme in the public discourse. Romantic escapism should therefore not necessarily be interpreted as a right-wing phenomenon or an outright rejection of modernity. After all, by rendering

20 Hugo Münsterberg, *Die Amerikaner* (Berlin: Ernst Siegfried Mittler und Sohn, 1904), 3.

21 Rudolf Kayser, "Amerikanismus," *Vossische Zeitung* 458 (27 September 1925).

22 Ernst Bloch, *The Heritage of Our Times* [1935], trans. Neville and Stephen Plaice (Berkeley, CA: University of California Press, 1990), 104-116.

23 Oskar Negt, "Erbschaft aus Ungleichzeitigkeit und das Problem der Propaganda," in *Es muss nicht immer Marmor sein. Ernst Bloch zum 90. Geburtstag*, ed. Detlef Horster (Berlin: Wagenbach, 1975), 12.

modern existence less intolerable, romantic impulses also contributed to the advancement and greater acceptance of modernization processes.

Efforts to safeguard the ideal realm against the dehumanizing materialism that for many characterized modern existence were pervasive in the decades around 1900. In this book, I am specifically concerned with seemingly paradoxical attempts to construe machine technology itself as a means to attain the intangible and spiritual. Such techno-romantic lines of argument were prevalent in dealings with the new cinematic medium, which—despite or precisely because of its nature as a technological mass medium—was construed as imaginative refuge from the harsh realities of industrial society. Author Kurt Pinthus explained:

> The people of our times, trapped in their never-ending routine of daily work, transform themselves, during their leisure time, into romantics. They wish to see not simply realistic things, but also realistic things raised into a more ideal, more fantastic sphere. The world should be garnished (like a Sunday roast) with adventures and oddities; a more plausible logic should prevail, and things should be shed of their weightiness and causality. All of this they find in the cinema.[24]

In fact, cinema was widely perceived as one of the "sanctuaries of modern romanticism."[25] According to commentators like Coudenhove-Kalergi and Hanns-Heinz Ewers, the new medium afforded viewers the opportunity to temporarily remove themselves from their mundane existence and offered a reprieve from the alienating and dehumanizing reality of modern life. Ewers for instance maintained that by facilitating an imaginative engagement with a "strange, mysterious world," cinema allows us to "dream in the dark."[26] In this context it is important to emphasize that already the earliest film theorists explicitly attributed cinema's ability to transport audiences to imaginary spheres to its technological nature. They identified the medium as a site wherein the ideal could reveal itself through technology.

To be sure, the conception of technology as means of transcending objective reality was not limited to the cinematic realm. Also in other areas critics

24 Kurt Pinthus, "The Photoplay: A Serious Introduction for Those Who Think Ahead and Reflect [1913]," in *The Promise of Cinema: German Film Theory 1907-1933*, eds. Anton Kaes, Nicholas Baer, and Michael Cowan, trans. Don Reneau, 199-203 (Berkeley, CA: University of California Press, 2016), 201.

25 Coudenhove-Kalergi, *Apologie der Technik*, 69.

26 Hanns Heinz Ewers, "Vom Kinema [1910]," in *Kein Tag ohne Kino: Schriftsteller über den Stummfilm*, ed. Fritz Güttinger, 20-23 (Frankfurt am Main: Deutsches Filmmuseum, 1984), 21.

resorted to techno-romantic lines of argument, which allowed them to embrace technology while simultaneously insisting on the pre-eminence of spirit. However, techno-romantic thought played a particularly important role in the evolution of the cinematic medium. Cinema, as journalist Heinz Michaelis wrote in 1923, was seen as an object lesson for harnessing technology for the benefit of the immaterial:

> It is thus first necessary to disrupt the absolutism of technology and bend it to the will of mind and soul. [...] The most basic motivation behind the invention of cinematography may have been the desire to bring technology into the realms that were almost entirely reserved for art. In a sense, the apparatus was crying out to be endowed with a soul.[27]

Emblematic of the views of many intellectuals, Michaelis's comments reflect a deep apprehension about the decreasing importance of immaterial aspects of life, which he attributes to technology as a catalyst of ubiquitous reification. At the same time, Michaelis invokes cinema as the locus in which technology, and consequently the modern world, might be spiritually sublimated.

Techno-romantic conceptions of cinema were not a uniquely German phenomenon. Many French film theorists, for instance, likewise explored cinema's ability to grant access to a spiritual dimension, which they also considered a prerequisite for film art. Most famously, discussions about *photogénie* during the 1920s assumed that film technology possessed mysterious, impenetrable, quasi-supernatural powers to transform material reality and reveal its hitherto hidden facets.[28] For critics like Louis Delluc, Jean Epstein, and René Clair *photogénie* embodied cinema's technology-based spiritual essence that manifested itself in fleeting moments of pure visual poetry.

Even though, as we have seen, responses to the technological revolution had somewhat different inclinations on the political right and left, techno-romantic thought can be found across the political spectrum. From Alfred Baeumler, one of Nazi Germany's few influential philosophers and nationalists such as Hanns Heinz Ewers and Will Scheller to leftists like

27 Heinz Michaelis, "Art and Technology in Film [1923]," in *The Promise of Cinema: German Film Theory 1907-1933*, eds. Anton Kaes, Nicholas Baer, and Michael Cowan, trans. Alex H. Bush, 581-582 (Berkeley, CA: University of California Press, 2016), 581.

28 For a detailed discussion of the concept of *photogénie* see Richard Abel, "Photogénie and Company," in *French Film Theory and Criticism: A History/Anthology 1907-1939*, ed. and trans. Richard Abel, 95-124 (Princeton, NJ: Princeton University Press, 1988).

Kurt Pinthus, Hermann Häfker, and Georg Lukács critics of all political shades embraced cinema as the site wherein something infinite revealed itself through technology. For example, Baeumler ascribed transcendent qualities to the cinematic experience: "Everything real flows and vanishes. Only appearance [*Schein*] is eternal because it has no reality. This moment of eternity, which a reader might take from a contemporary poet, is offered to naïve people when they see their lives reflected in film."[29] As Baeumler suggested, cinema was capable of making accessible the eternal realm of ideas. Although, in his view, the medium remained aesthetically inferior, its spiritual impact was comparable to that of high art. Like Baeumler, Lukács affirmed cinema's ability to transcend what is perceptible to the senses. In contrast to Baeumler, however, he did not conceptualize the medium as a lesser version of the traditional arts. Instead, he saw it as a novel way to reach beyond the limits of ordinary experience. Lukács wrote, film "is so strongly unmetaphysical, so exclusively empirically alive, that through this sheer extremity of its nature another entirely different metaphysics arises."[30] As Lukács explained, because cinema's replicas of life exist outside of human time, they obtain a spiritual quality. Although their arguments differ drastically, Baeumler and Lukács both concluded that the cinematic medium was capable of providing access to the infinite. At first sight, critics' contention that a machine could capture something in the ideal realm might be interpreted as a radical break with traditional aesthetics. In actual fact, however, techno-romantic lines of argument simultaneously upheld and broadened those very principles. When pointing to the ways technology pertained to non-physical aspects of existence, commentators made the oxymoronic notion of "technological art" first conceivable.

Transcending physical reality

The primary obstacle in establishing cinema's artistic potential was that many observers considered it questionable that photographic images could be anything but direct imprints of physical reality. The cinematic apparatus

29 Alfred A. Baeumler, "The Effects of the Film Theater: An Attempt at an Apologia for the Cinematographic Theater [1912]," in *The Promise of Cinema: German Film Theory 1907-1933*, eds. Anton Kaes, Nicholas Baer, and Michael Cowan, trans. Alex H. Bush, 381-384 (Berkeley, CA: University of California Press, 2016), 383.

30 Georg Lukács, "Thoughts Toward an Aesthetic of the Cinema [1911/1913]," in *The Promise of Cinema: German Film Theory 1907-1933*, eds. Anton Kaes, Nicholas Baer, and Michael Cowan, trans. Janelle Blankenship, 377-381 (Berkeley, CA: University of California Press, 2016), 379.

was seen as somewhat similar to the gramophone, which could reproduce but not create music. Author Paul Ernst, for example, emphatically denied cinema all artistic aspirations because "it is simply a fact that art first of all requires spirit, and spirit is not to be found in actuality."[31] Konrad Lange likewise took what Tom Gunning has called photography's "truth claim"[32] at face value: "what we see in the cinema is not art, but reality."[33] Since for many commentators film conveyed nothing but objective facts, the medium was perceived as epitomizing the ultimate triumph of materialism and was thus fiercely attacked.

The sole area that provided evidence for cinema's aptitude to rise above the automatic production of monochrome and silent copies of outside reality was special effects. Incidentally, as we have seen, tricks were also considered one of the medium's core areas of competence, areas that reflected its "technological essence." Special effects substantiated that film technology was capable of *creating* images rather than simply reproducing them. Many commentators noted that they enabled film to render the immaterial. Representations of dreams, visions, and supernatural occurrences helped to establish technology as a creative agent, which in turn corroborated cinema's artistic aspirations. In 1918, a review of Robert Reinert's film *Opium* expressed this thought with exemplary precision: "Gesellius's opium fantasies conjure up images that only film—on account of its trick technology—is not merely able to reproduce, but also to create! Film's creativity thus provided the potential to achieve the extraordinary."[34]

Although observers asserted the originality of film technology in general terms early on, its practical application as stylistic device rarely attracted attention during the pre-war years. Their efforts to overcome theoretical frictions between film, technology, and art notwithstanding, early critics were reluctant to deliberately engage with cinematic technique. As Helmut H. Diederichs has noted, they almost never discussed formal aspects like camera angles or editing.[35] It was not until April 1916 that a prominent figure proclaimed the camera—rather than for instance the actors' performance—as cinema's principal means of expression. Actor-director Paul

31 Paul Ernst, "Möglichkeiten einer Kinokunst [1913]," in *Kein Tag ohne Kino: Schriftsteller über den Stummfilm*, ed. Fritz Güttinger, 69-73 (Frankfurt am Main: Deutsches Filmmuseum, 1984), 73.

32 Tom Gunning, "What's the Point of an Index? Or, Faking Photographs," *Nordicom Review* 1/2 (September 2004): 39.

33 Konrad Lange, "Die 'Kunst' des Lichtspieltheaters," in *Geschichte der Filmtheorie*, ed. Diederichs, 82.

34 C. B., "Opium," *Der Film* 7 (15 February 1919).

35 Diederichs, "Frühgeschichte deutscher Filmtheorie," 129.

Wegener, who had earned great respect for his merits in advancing German artistic filmmaking, declared:

> The true poet of the film must be the *camera*. The possibilities to make large things small and small things large, to blend and superimpose photographic images; the possibility of a constantly changing standpoint for the viewer; the endless tricks rendered by split-screen, reflection, and so forth—in short, the *technology* of film must become significant for the choice of content.[36]

This explicit equation of camera and poet, of apparatus and artist, constituted a radical shift in perspective and resonated strongly within the German cultural sphere. The fact that most techniques Wegener cites can be considered special effects testifies to their essential role in raising awareness for cinema's technological means of expression. Due to their ability to transcend physical reality, they became the starting point for fervent experimentation for the purpose of conveying ideas or emotions cinematically. This ambition, as I will show in subsequent chapters, shaped the evolution and character of German silent cinema.

Although Wegener was the first personality to assert the camera's creativity and expressivity, pre-war commentators had already zeroed in on "cinematic" subject matter. At once reacting to what they saw in movie theatres and envisioning the medium's future direction, early critics highlighted two types of content: fairy tales on the one hand and representations of dreams and other mental processes on the other. Both were considered ideally suited for substantiating cinema's specificity and artistic potential. Relying on cinema's "technological essence," i.e., trick technology, they demonstrated that cinema could transcend the mere reproduction of outside reality and present vividly what the traditional arts could only approximate. In addition, both fairy tales and dreams were attributed ontological affinities with the new medium. Georg Lukács articulated this perspective when he wrote that "there arises in the cinema a new homogeneous and harmonious, coherent and changing world, one that corresponds to the fairy tale and dream in literature and life."[37]

36 Paul Wegener, "On the Artistic Possibilities of the Motion Picture [1916]," in *The Promise of Cinema: German Film Theory 1907-1933*, eds. Anton Kaes, Nicholas Baer, and Michael Cowan, trans. Alex H. Bush, 206-208 (Berkeley, CA: University of California Press, 2016), 207. Emphases in the original.

37 Lukács, "Thoughts Toward an Aesthetic of the Cinema," in *The Promise of Cinema*, eds. Kaes et al., 379.

Cinema's perceived likeness to folk culture and fairy tales in particular can be traced to several attributes. Over the course of the nineteenth century, folk culture attracted increasing attention from cultural elites. The interest in folklore reflected a desire to secure community traditions before they were lost to industrialization and to preserve a "genuine" collective identity. On account of their simple, direct, and "naïve" character of representation, films seemed to display strong ontological affinities to cultural practices of the common people. Commentators frequently likened the medium to puppet theatres, the epic poems and chapbooks.[38] However, the folklore genre most commonly on critics' minds was the fairy tale, which, in the decades around 1900, was experiencing a revival of its own. Widespread interest in subjectivity, mysticism, exoticism and the imagination inspired a vogue of both literary and folk fairy tales. Authors like Oscar Wilde, Hugo von Hofmannsthal, Manfred Kyber, Robert Walser and Hermann Hesse each produced a considerable corpus of fairy tales, while, in 1912, the prominent Diederichs-Verlag began publishing an immensely successful series of international fairy tales, *Märchen der Weltliteratur,* which by 1940 encompassed forty volumes. To many observers, cinema's ability to conjure up the magic of a make-believe world and to mesmerize its childlike audiences was evocative of folk stories. Author Hans Land for instance claimed that cinema "is really nothing but an oriental teller of fairy tales, transported here and rendered technologically. We are being told a story. Our eyes are listening and the ears get a little music as consolation for the surroundings."[39]

Both as a result of the public enthusiasm for folk stories and cinema's presumed ontological similarities, fairy tales became a fixture in German film culture early on. Messter's *Hänsel und Gretel* (1897) and *Rapunzel* (1897) were among the first fictional films produced in Germany and a sizable body of fairy-tale films based on both German and Middle Eastern sources emerged in the following decades. Well-known examples include Paul

38 For examples see Rudolf Rotheit, "Kasperle und Kinematograph," *Der Kinematograph* 5 (3 February 1907); Emil Perlmann, *Das Kino als modernes Volkstheater: Eine Entgegnung auf unberechtigte Angriffe* (Berlin: Agitations-Komitee der kinematographischen Fachpresse zur Förderung der Lichtbildkunst, 1912); Hermann Duenschmann, "Cinematograph and Crowd Psychology: A Sociopolitical Study [1912]," in *The Promise of Cinema: German Film Theory 1907-1933*, eds. Anton Kaes, Nicholas Baer, and Michael Cowan, trans. Eric Ames, 256-258 (Berkeley, CA: University of California Press, 2016); Ulrich Rauscher, "Das Kintop-Epos [1913]," in *Kein Tag ohne Kino: Schriftsteller über den Stummfilm*, ed. Fritz Güttinger, 137-140 (Frankfurt am Main: Deutsches Filmmuseum, 1984).

39 Hans Land, "Lichtspiele [1910]," in *Prolog vor dem Film: Nachdenken über ein neues Medium, 1909-1914*, ed. Jörg Schweinitz, 18-19 (Leipzig: Reclam-Verlag, 1992), 18.

Wegener's *Rübezahls Hochzeit* (PAGU, 1916), Paul Leni's *Dornröschen* (PAGU, 1917), Ernst Lubitsch's *Sumurun* (PAGU, 1920), Ludwig Berger's *Der verlorene Schuh* (Decla-Bioscop, 1923), Lotte Reiniger's animated feature *Die Abenteuer des Prinzen Achmed* (Comenius-Film, 1926) and Alexandre Volkoff's French-German co-production *Geheimnisse des Orients/Shéhérazade* (Ufa/Ciné Alliance, 1928). Audiences, filmmakers, and critics valued fairy-tale films because they accommodated a range of economic, ideological, and aesthetic interests. They attracted diverse groups of viewers: adults as well as children, common people as well as members of the intelligentsia, and were widely seen as innocuous amusement. Paul Wegener, creator of numerous fantastic films based on folk motifs, for instance praised the genre for facilitating the production of "films that satisfy the legitimate requirements of entertainment and visual pleasure without leaving behind a regrettable aftertaste. In *The Student of Prague* [...], in *The Golem*, and most recently in *Rübezahl*, I have attempted to tackle subjects that intrinsically suit the technical aspects of the motion picture."[40] As Wegener suggests, his fanciful films concurrently sought to cater to audiences, appease critics, and make full use of the medium's technological means. Fairy-tale films reflected the techno-romantic paradigm as they relied on special-effect technologies in order to visualize the imagination and thus invoke both individual childhood memories and the collective heritage. Through technology, cinematic folk stories made tangible forms of immaterial culture and sustained them in the industrial age.

Since fairy-tale films were generally considered a wholesome form of entertainment, tasteful and suitable also for susceptible patrons, few critics found fault with the genre as such. Even the most culturally conservative pundits, many of whom otherwise objected to all fictional subjects on film, deemed fairy tales a valuable pedagogic tool. They "cultivated" the masses without appropriating (and thus devaluing) high works of literature. Furthermore, though ostensibly apolitical, fairy-tale films also cultivated a sense of national identity and patriotism. By fostering collective nostalgia and perpetuating group-defining traditions, they held out the prospect of improving social cohesion and generate uplift, as for instance author Fedor von Zobeltitz hoped: "All of our fairy tales and myths would celebrate a vivid comeback in the cinema; the poetic power of our old folk tales could permeate society anew."[41] For many advocates of the cinematic fairy tale the

40 Wegener, "On the Artistic Possibilities of the Motion Picture," in *The Promise of Cinema*, eds. Kaes et al., 207.

41 Fedor von Zobeltitz, "Film-Literatur," *Das literarische Echo* 13:15 (1 May 1911): 1103.

underlying hope was the modern medium of film might help repair some of the socio-cultural havoc that modernization had wrought.

One of the principal arguments in support of ontological affinities between silent cinema and folk narratives involved their comparable lack of character psychology. Neither, critics observed, was concerned with or fully capable of exploring mental states and processes. Instead, protagonists' qualities and inner life directly expressed themselves outwardly through action. In traditional fairy tales, as folklorist Maria Tatar has noted,

> A character's physical and mental traits serve not only to define him, but also to elicit an action—they are at once descriptive and causal. Those traits, furthermore, nearly always have a single predictable consequence, rather than opening multiple possibilities for the course of the tale. [...] Odd as it sounds, it appears that we must look to the plot in order to read the minds of fairy-tale characters."[42]

For example, the envy of the evil queen in Snow White not only constitutes her foremost trait; it dictates the plot line. All events result from the queen's attempts to inflict misfortune on Snow White and they in turn allow readers and listeners to fathom her mental traits.

A similar relationship between external action and characters' inner life also distinguishes silent cinema. Kurt Pinthus observed, "cinema presents only action, effects, and visible phenomena, while the theatre aspires to convey subtle psychology. On the stage, words are more important than visible events. The cinema must do without words, and consequently without everything that words reveal."[43] Many early observers commented on the medium's lack of psychological complexity compared to works of high literature. They perceived cinema as a quintessentially surface-oriented medium—an aspect that I will revisit in more detail below.

Simultaneously, however, cinema's alleged superficiality was also seen as an exciting challenge: Filmmakers began to explore uniquely cinematic ways to reveal characters' mental states, which in turn also affected the medium's technological evolution, as cinematographer Karl Freund indicated: "The future development of cinematography will be shaped by the ever new

42 Maria Tatar, *The Hard Facts of the Grimms' Fairy Tales* (Princeton, NJ: Princeton University Press, 2003), 79.

43 Kurt Pinthus, "Quo Vadis, Cinema? On the Opening of the Königspavillon-Theater [1913]," in *The Promise of Cinema: German Film Theory 1907-1933*, eds. Anton Kaes, Nicholas Baer, and Michael Cowan, trans. Don Reneau, 186-188 (Berkeley, CA: University of California Press, 2016), 186.

psychological tasks it faces."[44] The ambition to visualize ideas and sensations shaped German film culture throughout the silent era and resulted in some of German cinema's most memorable sequences including the veterans' posttraumatic hallucinations in *Nerven* (Monumental-Filmwerke, 1919, dir. Robert Reinert), Kriemhild's falcon dream in *Die Nibelungen* (Decla-Bioscop, 1924, dir. Fritz Lang), the hypnotized shadow play in *Schatten* (Pan-Film, 1923, dir. Arthur Robison) the doorman's encounter with the spiteful neighbour women in *Der letzte Mann* (Union Film/Ufa, 1924, dir. F. W. Murnau), Auguste's jealous anguish in *Dirnentragödie* (Pantomim-Film, 1927, dir. Bruno Rahn) or Freders fever dreams in *Metropolis* (Ufa, 1927, dir. Fritz Lang), several of which will be discussed in subsequent chapters.

Although, for many observers, film was characterized by its lack of complexity and depth, others engaged in explicitly techno-romantic lines of argument and claimed that cinema's trick technological capacities enabled the medium to render the life of the mind. Already in 1914, in one of the earliest essays to analyse film in a scholarly context, psychologist Otto Rank asserted "the uniqueness of film technology in visibly portraying psychological events."[45] In sequences depicting mental images, subjective experiences or emotional states film technology could serve as a mirror of the psyche rather than as one of physical reality. By thus representing the ideal, cinema could meet the most crucial criterion for art.

Although the links between mental processes and film technology would prove foundational for German film culture, the theorist who formulated them most cogently was at no point a part of the German discourse. German-American psychologist Hugo Münsterberg first became interested in cinema in 1915,[46] at a point when, on account of the war, the intellectual ties between Germany and the United States had largely been cut. His film theoretical work *The Photoplay: A Psychological Study* (1916) thus remained unknown in Germany and Rudolf Arnheim, for instance, reportedly first heard about it in the 1970s.[47] In *The Photoplay*, Münsterberg analogized cinema and the human mind and described tools such as close-ups and

44 "Meinungen von Karl Freund," *Filmtechnik* 1 (5 January 1926): 4-5, 27-28.

45 Otto Rank, *The Double: A Psychoanalytic Study* [1914], ed. and trans. Harry Tucker, Jr. (Chapel Hill, NC: The University of North Carolina Press, 1971), 7. Translation amended.

46 See Margaret Münsterberg, *Hugo Münsterberg: His Life and Work* (New York, NY: D. Appleton and Company, 1922), 281.

47 See Rudolf Arnheim, "Zum Geleit [für Hugo Münsterberg: Das Lichtspiel]," montage A/V 2 (2000): 55-57. Dudley Andrew likewise concludes that Münsterberg had little effect on subsequent film theory. See J. Dudley Andrew, *The Major Film Theories: An Introduction* (London: Oxford University Press, 1976), 27.

flashbacks as correlating to mental mechanisms like attention and memory. For Münsterberg cinema replaced physical reality with a subjective rendering of experience: "*the photoplay tells us the human story by overcoming the forms of the outer world, namely, space, time, and causality, and by adjusting the events to the forms of the inner world, namely, attention, memory, imagination, and emotion.*"[48] Münsterberg proceeded on the same techno-romantic assumptions as his colleagues in Germany and likewise hoped that technology could overcome its material substance and facilitate a new form of consciousness:

> For the first time the psychologist can observe the starting of an entirely new esthetic development, a new form of true beauty in the turmoil of a technical age, created by its very technique and yet more than any other art destined to overcome outer nature by the free and joyful play of the mind.[49]

While Münsterberg was undoubtedly the most sophisticated early thinker concerned with the affinities between the psychic and the cinematic apparatus, he was not the only one. Above all, critics identified similarities between the cinema and dreams. Author Theodor Heinrich Mayer claimed that cinema shared "so many common features with dream-apparitions that the feeling of dreaming must be taken as one of the cinematograph's underlying principles."[50] Two areas of kinship were often highlighted. On the one hand, the films themselves were perceived as resembling oneiric images. Alfred Polgar observed that for both films and dreams "reality is without waste products. For both the laws of nature have been suspended, the force of gravity has ceased to exist and existence is without conditionality. And for both the enigmatic word of Gurnemanz [in Richard Wagner's *Parsifal*] is valid and meaningful: 'You see, my son, here time turns into space.'"[51] Films resemble dreams in that they present a succession of vivid, phantasmagoric images that are not bound by the parameters of the physical world. On the other

48 Hugo Münsterberg, *The Photoplay: A Psychological Study* (New York, NY: D. Appleton and Company, 1916), 173. Emphasis in the original.

49 Münsterberg, *The Photoplay*, 232-233.

50 Theodor Heinrich Mayer, "Lebende Photographien [1912]," in *Kein Tag ohne Kino: Schriftsteller über den Stummfilm*, ed. Fritz Güttinger, 119-129 (Frankfurt am Main: Deutsches Filmmuseum, 1984), 120.

51 Alfred Polgar, "Das Drama im Kinematographen [1911]," in *Prolog vor dem Film: Nachdenken über ein neues Medium, 1909-1914*, ed. Jörg Schweinitz, 159-164 (Leipzig: Reclam-Verlag, 1992), 163.

hand, film reception, i.e., the affective engagement with overpowering images in the darkened movie theatre, was also likened to dreaming. In the French context, author Jules Romains wrote: "The group dream now begins. They sleep; their eyes no longer see. They are no longer conscious of their bodies. Instead there are only passing images, a gliding and rustling of dreams.[52]

The similarities between film and dreams prompted some critics to engage in techno-romantic fantasies of cinema as an external manifestation of the soul. Hanns Heinz Ewers described film reception in 1910 as a "remote encounter" with one's own dreams: "In the cinema [...] I can dream. I live in the world of wonder, and yet this world comes alive only through my dreams."[53] For Ewers, the filmic universe engulfs the spectator and is simultaneously the product of his or her imagination. The spectator's mental life and the cinematic apparatus become indistinguishable. Such visions were not necessarily judged favourably, however. In 1911, R. Seligmann described cinema as "the external projection of dreams, and their materialization into external objects. It is the bold attempt to replace our poetic and dreaming imagination with an artificial mechanism."[54] Rather than embracing cinema as the realization of our inner life that Ewers envisioned, Seligmann's reaction reveals apprehension and a sense of loss. While differing in their attitudes, Ewers and Seligmann both characterize cinema as a technological extension of the spectator's mental life, which puts their ideas in proximity to Münsterberg's film theory a few years later.

Compared to Münsterberg's detailed examination of the relationship between film technology and mental processes, the thinking of early European critics remained rather approximate. They conjectured that the cinema, precisely as a consequence of its "technological essence," had an affinity to mental processes and emotional states. At the same time, however, they refrained from examining this connection, specifically with regard to the creative potentials of film technology or to formal aspects like camera work or editing in any detail. As a result, prior to World War I the links between camera and mind were not expanded into a full-fledged theory in Europe.

52 Jules Romains, "The Crowd at the Cinematograph [1911]," in *French Film Theory and Criticism: A History/Anthology 1907-1939*, ed. and trans. Richard Abel, 53-54 (Princeton, NJ: Princeton University Press, 1988), 53.

53 Ewers, "Vom Kinema," in *Kein Tag ohne Kino,* ed. Güttinger, 21.

54 R. Seligmann, "Kinematograph und Traum," *Frankfurter Zeitung und Handelsblatt* 289 (18 October 1911): 2.

Surface art

The attention to the relationship between cinema and psyche also aimed at invalidating the charge that film was inherently superficial or, as author Joseph August Lux framed it, nothing but "surface art."[55] The medium was deemed shallow in two respects. First, it literally lacked depth. Along with silence and monochromaticity, two-dimensionality was consistently cited as cinema's chief formal feature.[56] Second, film's literal flatness also corroborated a metaphorical absence of profundity. Critic Carl Forch asserted that "cinema is indeed a product of its time, which has raised the superficial to infinite heights and runs the risk of losing sight of the inner value of things."[57] Cinema seemed shallow in its slavish reproduction of outward appearances and in the absence of spoken words without a discernible pathway to a spiritual dimension. Thus, author Hermann Kienzl concluded, "the cinematograph will not convince anyone to create poetry of the soul. It cannot go beyond situation, movement, and bare action."[58]

Remarkably, however, cinema's shallowness was not only invoked to denounce the medium, but also to make a case for its privileged access to the ideal realm. In fact, some early critics argued that it was precisely cinema's superficiality that allowed it to exceed the limits of ordinary experience. The cinematic universe was, according to Herbert Tannenbaum, "one that exists entirely beyond our life, light as air (psychologically and visually) and two-dimensional, a world that has absolutely nothing to do with the anthropocentric, intellectual world of the theatre but is in its own way completely balanced, unified and homogeneous."[59] Trick technology allowed cinema to surpass the this-worldly sphere. Relieved from the burdens of the

55 Joseph August Lux, "Die Muse des Films," *Velhagen & Klasings Monatshefte* 11 (July 1913): 429.

56 See for instance Egon Friedell, "Prologue before the Film [1913]," and Herbert Tannenbaum, "Film Advertising and the Advertising Film [1920]," in *The Promise of Cinema: German Film Theory 1907-1933*, eds. Anton Kaes, Nicholas Baer, and Michael Cowan, trans. Don Reneau; Tara Hottman, 168-169; 179-182 (Berkeley, CA: University of California Press, 2016).

57 Carl Forch, "Thrills in Film Drama and Elsewhere [1913]," in *The Promise of Cinema: German Film Theory 1907-1933*, eds. Anton Kaes, Nicholas Baer, and Michael Cowan, trans. Sara Hall, 35-38 (Berkeley, CA: University of California Press, 2016), 38. Translation amended.

58 Hermann Kienzl, "Theater und Kinematograph," *Der Strom* (Vienna) 8 (November 1911): 219-221. An abbreviated version of this essay appears in *The Promise of Cinema: German Film Theory 1907-1933*, eds. Anton Kaes, Nicholas Baer, and Michael Cowan, trans. Michael Cowan, 30-31 (Berkeley, CA: University of California Press, 2016).

59 Herbert Tannenbaum, "Problems of the Film Drama [1913]," in *The Promise of Cinema: German Film Theory 1907-1933*, eds. Anton Kaes, Nicholas Baer, and Michael Cowan, trans. Alex H. Bush, 192-196 (Berkeley, CA: University of California Press, 2016), 194.

human condition, the filmic realm was dynamic, boundless, anarchistic and creative. The association of fantasy, play, freedom, and art traces back to Friedrich Schiller, who understood playful aesthetic semblance (as distinct from actual, deceitful semblance) as the basis of art. Inherently unreal, semblance liberates humanity from purpose, compulsion, and worry and can therefore, in Schiller's words, "set man free both morally and physically."[60] Focussing on essays by Gustav Melcher, Will Scheller, Herbert Tannenbaum, and Georg Lukács, I show how some early theorists radically redefined the common assumptions about the relationship between materialism and artistic substance based on Schiller's concept of semblance. By so doing, they both substantiated cinema's essence and artistic merit from a techno-romantic perspective and reframed nineteenth-century aesthetics in ways that anticipate the classical film theories of the 1920s and beyond.

As early as 1909, the painter Gustav Melcher, co-founder of the "Society for the Promotion of Motion Picture Art" in Düsseldorf, published an extraordinary accolade of the new medium titled "On Living Photography and the Photoplay." Melcher's essay anticipates the work of Jean Epstein, Siegfried Kracauer, Dziga Vertov, and Béla Balázs by identifying cinematic vision as a transcendent force. As Malcom Turvey has argued, these later critics conceive of the revelatory capacities of cinema as "an awesome, even miraculous power that, rather than extending the power of the human eye, escapes its limitations and thereby has the potential to bring about a fundamental change for the better in human existence."[61] In a similar vein, Melcher characterizes the cinematograph as a "new visual organ."[62] Devices like the microscope and the telescope exceed the visual acuity of our biological eyes and have brought about a hybrid, techno-biological form of vision: "The primitive visual tools given to us by nature are but the substrate for the eyes of today and no one who relies upon our inherited eyes alone can truly *see* in the contemporary sense of the word."[63] For Melcher, technology has revolutionized our concept of seeing and unmediated vision has become

60 Friedrich Schiller, *On the Aesthetic Education of Man* [1795], trans. Reginald Snell (Mineola, NY: Dover Publications, 1992), 74.

61 Malcom Turvey, *Doubting Vision: Film and the Revelationist Tradition* (Oxford: Oxford University Press, 2008), 6.

62 Gustav Melcher, "On Living Photography and the Film Drama [1909]," in *The Promise of Cinema: German Film Theory 1907-1933*, eds. Anton Kaes, Nicholas Baer, and Michael Cowan, 17-20, trans. Alex H. Bush and Jon Cho-Polizzi, 17-20 (Berkeley, CA: University of California Press, 2016), 18.

63 Melcher, "On Living Photography," in *The Promise of Cinema*, eds. Kaes et al., 18. Emphasis in the original.

a thing of the past. Far more so than any other device, the cinematograph increases our visual acuity in ways that appear preternatural.

The powers of cinematic vision transcend space, time, and even physical reality. First, according to Melcher, film is not bound by any spatial restrictions: "With the aid of this international visual organ, the cinematograph, we see around the very surface of the earth [...]. We expand our optical horizons, or perhaps abolish them entirely to gaze unhindered upon the light of the world."[64] Film technology holds out the promise of unlimited freedom. The religious overtones of Melcher's phrasing (Jesus uses the expression "light of the world" to describe himself in the New Testament)[65] emphasize the spiritual dimension that total vision entails. Second, Melcher deems cinematic vision transcendent in its ability to observe and display anything imaginable: "The fly has more than ten thousand eyes. The flounder has eyes that can wander across its body. But twentieth-century man has the cinematograph. He sees more than the visual world: he sees what he wants to see."[66] The eye of the cinematograph is superior not only to man-made optical devices, but also to the most fascinating visual organs in nature: surpassing physical reality, it turns vision into a matter of free choice.[67] Lastly, film exceeds the limits of time. Unaffected by human finitude, the cinematograph occupies a liminal space between death and life:

> The modern eye [the cinematograph] sees beyond the grave, peers into the past, stretches across the ravages of time. It looks both backwards and forwards, sees living death and deathly life. The Kingdom of Death no longer provides the imminent frontier to life. Through these eyes, Leo XIII need not have ceased to project his benevolent smile, to pray for us and bless us with this papal grace. The Kingdom of Death stops at the border of the cinematograph and the gramophone.[68]

Cinema's conquest of death liberates us from the limitations of the human condition. The medium opens up a realm of pure surface appearance that

64 Melcher, "On Living Photography," in *The Promise of Cinema*, eds. Kaes et al., 19.

65 John 8:12.

66 Melcher, "On Living Photography," in *The Promise of Cinema*, eds. Kaes et al., 19. Translation amended.

67 Blaise Cendrars came back to the notion a decade later, describing cinema as "an eye more marvellous than the multifaceted eye of the fly." Blaise Cendrars, "The Modern: A New Art, the Cinema [1919]," in *French Film Theory and Criticism: A History/Anthology 1907-1939*, ed. and trans. Richard Abel, 182-183 (Princeton, NJ: Princeton University Press, 1988), 182.

68 Melcher, "On Living Photography," in *The Promise of Cinema*, eds. Kaes et al., 19.

is simultaneously one of endless possibility: "Cinema leads us into the free realm of semblance [*Schein*]. It shows us figures that can triumph over death, figures that only in our imagination move exactly like this."[69] Here, semblance does not serve as a negative foil for essence, as was generally the case in post-Hegelian aesthetics. Instead, Melcher construes semblance in the tradition of Schiller as a powerful liberating force, a free space for the imagination that overcomes even death.

Semblance and film's ability to transcend the material reality of the here and now play also a key role in author Will Scheller's 1914 essay "The New Illusion." Compared to Melcher, Scheller's views are more conventional, but he emphatically embraces cinema as art. Scheller sees film as a coping mechanism that sublimates recipients' consciousness and provokes powerful emotions comparable to traditional arts. Cinema is the ultimate modern art form as modern subjects no longer command the stamina or heightened mental capacities to face the abstract, strenuous stimuli of traditional arts. At the same time, however, modern subjects have a deep-seated yearning for aesthetic experiences, an "instinctual desire to get away from material things from time to time."[70] Cinema compensates for the dire realities of modern existence in a simple, immediate and effortless manner. Trick technology affords cinema with "unlimited possibilities," allowing "the most unexpected fantasies to run wild."[71] It allows us to leave behind the everyday and the modern predicament that the medium itself epitomizes. Freedom, unreality, and immediacy are constitutive of cinema's transcendent powers: "*This* is the long-desired, powerful illusion, which transcends the small concerns of daily life and enters into the realms of greater strains of existence [*Daseinsspannungen*]; this is a pathos that is not dependent upon conceptual interpretations, but rather emerges unambiguously from the ethical passions of events themselves."[72] Most commentators at the time sought to justify the medium's artistic merits by dissecting its innermost essence. Scheller in contrast embraces cinema as a surface phenomenon that owes its spiritual dimension to its ability to transport its viewers away from materiality. Its transcendent qualities valorise cinema as art.

69 Gustav Melcher, "Die künstlerischen Vorzüge der Kinematographie," *Der Kinematograph* 116 (17 March 1909).

70 Will Scheller, "The New Illusion [1914]," in *The Promise of Cinema: German Film Theory 1907-1933*, eds. Anton Kaes, Nicholas Baer, and Michael Cowan, trans. Alex H. Bush, 196-199 (Berkeley, CA: University of California Press, 2016), 197.

71 Scheller, "The New Illusion," in *The Promise of Cinema*, eds. Kaes et al., 199.

72 Scheller, "The New Illusion," in *The Promise of Cinema*, eds. Kaes et al., 198.

Scheller's ideas are in part reminiscent of Herbert Tannenbaum's essay "Problems of the Photoplay," which was published in the same journal six months earlier, in December 1913. Like Scheller, Tannenbaum contrasts a troublesome modern existence with the carefree and unconventional world of cinema and emphasizes freedom and unreality as the medium's core attractions: "The film drama leads us out of ourselves into an exhilarating world, which knows nothing of us and our pain. As long as we sit before the cinema screen, we forget our grievances and live happy moments in a realm without motives or consequences."[73] By providing relief from the constraints of everyday life, film represents a powerful liberating force.

Like many early critics, Tannenbaum explores the affinity between filmic images and incorporeal and soulless spectres, which had first been highlighted in Maxim Gorky's famous 1896 "Kingdom of Shadows" review:[74] "Though the lack of any spiritual depth, film characters completely lose their earthly grounding: they become peculiar, instinctually living, moving phantoms; they become uncanny, fantastic."[75] Cinema's soundless, achromatic and incorporeal images represent a simple, two-dimensional, and strangely homogenous world:

> The essential shadowiness of human beings in the cinema creates total uniformity between them and all things in the visible world. Based on inner meaning, evaluated solely as exciting moments, there is no difference between the man Bumke, the dog that follows him, the dairy maid that he bowls over, the flower trough that he falls into, and the tree he finally climbs."[76]

Later critics concurred with Tannenbaum's view that film blurs the distinction between humans and the object world. Béla Balázs, for example, noted that in silent cinema "objects [...] *share with human beings a quality of silence*

73 Tannenbaum, "Problems of the Film Drama," in *The Promise of Cinema*, eds. Kaes et al., 195.

74 Maxim Gorky, "A Review of the Lumière Programme at the Nizhni-Novgorod Fair [1896]," in Jay Leyda, *Kino: A History of the Russian and Soviet Film*, trans. Leda Swan, 407-410 (London: George Allen and Unwin, 1960), 407. See also Willy Rath, "Künstlerische Möglichkeiten des Lichtspiels [1913]," in *Geschichte der Filmtheorie: Kunsttheoretische Texte von Méliès bis Arnheim*, ed. Helmut H. Diederichs, 121-131 (Frankfurt am Main: Suhrkamp-Verlag, 2004), 122; Malwine Rennert, "Heureka [1913]," in *Geschichte der Filmtheorie: Kunsttheoretische Texte von Méliès bis Arnheim*, ed. Helmut H. Diederichs, 143-146 (Frankfurt am Main: Suhrkamp-Verlag, 2004), 145.

75 Tannenbaum, "Problems of the Film Drama," in *The Promise of Cinema*, eds. Kaes et al., 193.

76 Tannenbaum, "Problems of the Film Drama," in *The Promise of Cinema*, eds. Kaes et al., 193.

that makes the two almost homogeneous."[77] Similarly, as Miriam Hansen has shown, Kracauer repeatedly referred to the film actor as "a thing among things."[78] For Tannenbaum, the reification of human beings on film is the result of their lack of formal and spiritual depth. He refutes the assertion that motion pictures are mere mechanical copies of reality by positing two-dimensionality and unreality as the essence of cinema. Even though the "drama of pure visuality"[79] resembles our world at first glance, on account of their two-dimensionality in form and substance filmic images are principally unreal. Cinema's technological spectres epitomize the medium's distinctive stylistic properties that, if observed, can bring forth film art.

Tannenbaum's conception of cinema as inherently otherworldly is closely related to Georg Lukács's "Thoughts Toward an Aesthetic of the Cinema" (1911/1913), which had been reprinted in the prominent *Frankfurter Zeitung* just a few months before Tannenbaum's article. Lukács's essay constitutes the most far-reaching early attempt to theorize cinema's superficiality in terms of a utopian beyond. Lukács lays out that cinema, as a matter of principle, lacks any sense of the ultimate truth of life, namely that all that exists will inevitably perish. Because every moment captured on film is endlessly repeatable, cinema's essential feature is the absence of any consciousness of mortality and life's transience. The theatre, in contrast, is characterized by total spatial and temporal presence: the physical presence of the actor shaping the present moment that subsequently will be lost forever. The theatre's mystical presence according to Lukács is comparable to Benjamin's notion of aura, which is based on an immediate, interactive relationship between perceiver and perceived. Like Lukács, Benjamin links the decline of the aura directly to the process of mechanical reproduction, which positions an apparatus between perceiver and perceived and thus breaks off their "magical" exchange of immediate gazes.[80] On account of cinema's "absence of presence," as Lukács calls it, filmic images merely replicate the external

77 Béla Balázs, *Early Film Theory: Visible Man and The Spirit of Film*, ed. Erica Carter; trans. Rodney Livingstone (New York, NY/Oxford: Berghahn Books, 2010), 23. Emphasis in the original.
78 Miriam Bratu Hansen, *Cinema and Experience: Siegfried Kracauer, Walter Benjamin, and Theodor W. Adorno* (Berkeley, CA: University of California Press, 2011), 29; Miriam Bratu Hansen, "Introduction," in Siegfried Kracauer, *Theory of Film: The Redemption of Physical Reality*, vii-xlv (Princeton, NJ: Princeton University Press, 1997), xvii.
79 Tannenbaum, "Problems of the Film Drama," in *The Promise of Cinema*, eds. Kaes et al., 193.
80 See Lukács, "Thoughts Toward an Aesthetic of the Cinema," in *The Promise of Cinema*, eds. Kaes et al., 378; Walter Benjamin, "The Work of Art in the Age of Its Technological Reproducibility [second version, 1935/1936]," in *Walter Benjamin Selected Writings, vol. 3: 1935-1938*, eds. Howard Eiland and Michael W. Jennings, trans. Edmund Jephcott and Harry Zohn, 101-133 (Cambridge, MA: Harvard University Press, 2002), 112; Walter Benjamin, "On Some Motifs in Baudelaire

appearance of life, i.e., the movements and actions of human beings. The medium itself lacks "presence" and therefore any spiritual depth; its images are but "uncannily lifelike"[81] facades. As a result, cinema is an inherently unreal, yet creative daydream that both consists in and transcends the technological surface from which it originates: "Not only in their technique, but also in their effect cinematic images [...] maintain a life of a completely different kind. In a word, they become *fantastic*. The fantastical element is not a contrast of living life, however, but is only a new aspect of the same: a life [...] without soul, of pure surface."[82] Lukács maintains that life as represented on film does not end, it merely changes; it knows no doom, only possibility: "'*Everything is possible*': that is the worldview of the 'cinema.'"[83]

Since cinema is not affected by death and finitude, there are no inevitable outcomes—all options remain open at any given moment. Film's ability to represent both outside reality *and*, by means of trick technology, any possibility imaginable constitutes the medium's dual disposition: cinema is simultaneously empirical *and* creative. Lukács does not understand these aspects (which correspond to what Kracauer later described as the formative and realistic tendencies of the filmic medium[84]) to be oppositional. Rather, they are constitutive of cinema's utopian potential. Because film technology renders both empirical reality and the most excessive figments of our imagination, cinema becomes a visionary tool. Without simulating the existence of a third dimension below its surface, film can establish a connection to a utopian realm. What Lukács describes as cinema's intrinsically fantastic nature is the result of the unity of reality and possibility at the core of the medium.

Each in their own way, Melcher, Scheller, Tannenbaum and Lukács contested the assertion that, being mere mechanical copies of empirical reality, filmic images were devoid of spiritual substance and could therefore never be art. Instead, pursuing techno-romantic lines of argument, they maintained that film technology and particularly special effects could provide a pathway to an immaterial dimension. Melcher conceived of cinema

(1939)," in *Selected Writings*, vol. 4, eds. Howard Eiland and Michael W. Jennings, trans. Harry Zohn, 313-355 (Cambridge, MA: Harvard University Press, 2003), 338.

81 Lukács, "Thoughts Toward an Aesthetic of the Cinema," in *The Promise of Cinema*, eds. Kaes et al., 378

82 Lukács, "Thoughts Toward an Aesthetic of the Cinema," in *The Promise of Cinema*, eds. Kaes et al., 378. Emphasis in the original.

83 Lukács, "Thoughts Toward an Aesthetic of the Cinema," in *The Promise of Cinema*, eds. Kaes et al., 379. Emphasis in the original.

84 Kracauer, *Theory of Film*, 30-37.

as a new visual organ that transcended time, space and empirical reality and freed the human condition from all limitations. Scheller understood the medium as a simultaneously shallow and emotionally effective coping mechanism that incited aesthetic ecstasy and thus offset the constraints of modern existence. For Tannenbaum, cinema's spiritual and pictorial two-dimensionality accounted for its liberating unreality in essence and effect. Lukács argued that cinema resembled life without being subjected to its limitations and thus possessed unprecedented creative and utopian power. They all believed that cinema, both on account of its superficiality and immense (trick-) technological capacities, was the antithesis of reality. Echoing Schiller, for whom both art and human freedom originated in playful aesthetic semblance, they saw cinema's technology-induced fantasies as a powerful emancipating force.

This techno-romantic conception of cinema anticipates later theories of film in interesting ways. It prefigures Benjamin's emphasis on film technology's liberating energies, which according to Benjamin blast away the oppressive realities of modern existence: "Then came film and exploded this prison-world with the dynamite of the split second, so that now we can set off calmly on journeys of adventure among its far-flung debris."[85] Like Melcher, Scheller, Tannenbaum, and Lukács, Benjamin contends that film actualizes its utopian potential by opening up new spaces for play and virtual action (*Spielraum* in Benjamin's words).[86] Melcher, Scheller, Tannenbaum, and Lukács also anticipate some of Kracauer's views. As Miriam Hansen has shown, Kracauer's early writings on film undermine his reputation as a straightforward realist.[87] In fact, Kracauer was initially much concerned with the essence of cinema, which he described, like Melcher, Scheller, Tannenbaum, and Lukács, as inherently fantastic. In a 1923 review, for example, Kracauer asserts that "improbability, which runs contrary to any kind of naturalism, corresponds perfectly to the essence of film. If film wants to realize its own substance, it has to thoroughly shatter the natural contexts of our life."[88] Kracauer also embraces surface as cinema's

85 Benjamin, "The Work of Art," in *Selected Writings*, 3, eds. Eiland and Jennings, 117. A similar phrasing can also be found in Walter Benjamin, "Reply to Oskar A. H. Schmitz [1927]," in *Selected Writings*, 2, eds. Howard Eiland and Michael W. Jennings, trans. Rodney Livingstone, 16-19 (Cambridge, MA: Harvard University Press, 1999), 17.

86 For an in-depth analysis of Benjamin's concept of *Spiel* (which translates variously as play, gamble, game, acting), see Hansen, *Cinema and Experience*, 183-204.

87 See Hansen, Cinema and Experience, 5.

88 Siegfried Kracauer, "Zwischen Flammen und Bestien [1923]," in Siegfried Kracauer, *Kleine Schriften zum Film*, vol. 6.1., *1921-1927*, ed. Inka Mülder-Bach, 35-36 (Frankfurt am Main:

essential characteristic.[89] For him, however, surface is not coupled with a spiritual realm beyond, prompting him to call on cinema to "renounce the portrayal of spiritual substance in favour of a film-specific rendering of semblance-like [*scheinhaft*] surface life."[90]

As a form of popular mass entertainment cinema had initially a low social standing. What is more, as a result of the conceptual opposition between machines and the ideal, its technological nature thwarted any artistic aspirations. In an effort to establish cinema as an object of aesthetic scrutiny, cinema-friendly critics sought to demonstrate that cinema was capable of exceeding the mere reproduction of material reality and simultaneously remain true to its medium specificity, its "technological essence." Exploring cinema's affinities with folk art, mental processes and the fantastic, cinema's supporters claimed that film technology could in fact facilitate connections to the ideal realm. Special effects served as key evidence in support of cinema's ability to transcend this-worldliness and give expression to feelings and thoughts. Techno-romantic reasoning proved vital in the effort to conceptualize the nature of art in the machine age and to reconcile cinema with existing aesthetics. It cleared the path for cinema's emergence as a new, technological art.

Bibliography

Richard Abel, "Photogénie and Company," in *French Film Theory and Criticism: A History/Anthology 1907-1939*, ed. and trans. Richard Abel, 95-124 (Princeton, NJ: Princeton University Press, 1988).

J. Dudley Andrew, *The Major Film Theories: An Introduction* (London: Oxford University Press, 1976).

Rudolf Arnheim, "Zum Geleit [für Hugo Münsterberg: Das Lichtspiel]," *montage A/V* 2 (2000): 55-57.

Suhrkamp-Verlag, 2004), 36.

89 For in-depth analyses of the surface metaphor in Kracauer's early work, see Inka Mülder-Bach, "Der Umschlag der Negativität. Zur Verschränkung von Phänomenologie, Geschichtsphilosophie und Filmästhetik in Siegfried Kracauers Metaphorik der 'Oberfläche,'" *Deutsche Vierteljahrsschrift für Literaturwissenschaft und Geistesgeschichte* 61:2 (June 1987): 359-373; Heide Schlüpmann, "Phenomenology of Film: On Siegfried Kracauer's Writings of the 1920s," *New German Critique* 40 (Winter 1987): 97-114; Miriam Hansen, "Decentric Perspectives: Kracauer's Early Writings on Film and Mass Culture," *New German Critique* 54 (Autumn 1991): 47-76.

90 Siegfried Kracauer, "Hochstaplerfilme [1923]," in Siegfried Kracauer, *Kleine Schriften zum Film*, vol. 6.1., *1921-1927*, ed. Inka Mülder-Bach, 37 (Frankfurt am Main: Suhrkamp-Verlag, 2004).

Alfred A. Baeumler, "The Effects of the Film Theater: An Attempt at an Apologia for the Cinematographic Theater [1912]," in *The Promise of Cinema: German Film Theory 1907-1933*, eds. Anton Kaes, Nicholas Baer, and Michael Cowan, trans. Alex H. Bush, 381-384 (Berkeley, CA: University of California Press, 2016).

Béla Balázs, *Early Film Theory: Visible Man and The Spirit of Film*, ed. Erica Carter; trans. Rodney Livingstone (New York, NY/Oxford: Berghahn Books, 2010).

Walter Benjamin, "On Some Motifs in Baudelaire [1939]," in *Selected Writings,* vol. 4: *1938-1940*, eds. Howard Eiland and Michael W. Jennings, trans. Harry Zohn, 313-355 (Cambridge, MA: Harvard University Press, 2003).

Walter Benjamin, "Reply to Oskar A. H. Schmitz [1927]," in *Selected Writings*, vol. 2: *1927-1934*, eds. Howard Eiland and Michael W. Jennings, trans. Rodney Livingstone, 16-19 (Cambridge, MA: Harvard University Press, 1999).

Walter Benjamin, "The Work of Art in the Age of Its Technological Reproducibility [second version, 1935/1936]," in *Walter Benjamin: Selected Writings,* vol. 3: *1935-1938*, eds. Howard Eiland and Michael W. Jennings, trans. Edmund Jephcott and Harry Zohn, 101-133 (Cambridge, MA: Harvard University Press, 2002).

Ernst Bloch, *The Heritage of Our Times* [1935], trans. Neville and Stephen Plaice (Berkeley, CA: University of California Press, 1990).

Reinhard Bruck, "Kinematograph und Theater," *Der Kinematograph* 247 (20 September 1911).

C. B., "Opium," *Der Film* 7 (15 February 1919).

Blaise Cendrars, "The Modern: A New Art, the Cinema [1919]," in *French Film Theory and Criticism: A History/Anthology 1907-1939*, ed. and trans. Richard Abel, 182-183 (Princeton, NJ: Princeton University Press, 1988).

Richard Nikolaus Coudenhove-Kalergi, *Apologie der Technik* (Leipzig: Verlag der neue Geist Dr. Peter Reinhold, 1922).

Helmut H. Diederichs, "Frühgeschichte deutscher Filmtheorie: Ihre Entstehung und Entwicklung bis zum Ersten Weltkrieg." Postdoctoral Thesis, J. W. Goethe-Universität Frankfurt am Main, 1996.

Hermann Duenschmann, "Cinematograph and Crowd Psychology: A Sociopolitical Study [1912]," in *The Promise of Cinema: German Film Theory 1907-1933*, eds. Anton Kaes, Nicholas Baer, and Michael Cowan, trans. Eric Ames, 256-258 (Berkeley, CA: University of California Press, 2016).

Paul Ernst, "Möglichkeiten einer Kinokunst [1913]," in *Kein Tag ohne Kino: Schriftsteller über den Stummfilm*, ed. Fritz Güttinger, 69-73 (Frankfurt am Main: Deutsches Filmmuseum, 1984).

Hanns Heinz Ewers, "Vom Kinema [1910]," in *Kein Tag ohne Kino: Schriftsteller über den Stummfilm*, ed. Fritz Güttinger, 20-23 (Frankfurt am Main: Deutsches Filmmuseum, 1984).

Carl Forch, "Thrills in Film Drama and Elsewhere [1913]," in *The Promise of Cinema: German Film Theory 1907-1933*, eds. Anton Kaes, Nicholas Baer, and Michael Cowan, trans. Sara Hall, 35-38 (Berkeley, CA: University of California Press, 2016).

Friedrich Freksa, "Theater, Pantomime and Cinema [1916]," in *The Promise of Cinema: German Film Theory 1907-1933*, eds. Anton Kaes, Nicholas Baer, and Michael Cowan, trans. Alex H. Bush, 111-114 (Berkeley, CA: University of California Press, 2016).

Egon Friedell, "Prologue before the Film [1913]," in *The Promise of Cinema: German Film Theory 1907-1933*, eds. Anton Kaes, Nicholas Baer, and Michael Cowan, trans. Don Reneau, 168-169 (Berkeley, CA: University of California Press, 2016).

Ludwig Fulda, "Theater und Kinematograph," *Die Woche* (Berlin) 16 (20 April 1912): 639-642.

Johann Wolfgang Goethe, "Kunst und Handwerk [1797]," in *Gedenkausgabe der Werke, Briefe und Gespräche*, ed. Ernst Beutler, vol. 13, (Zurich: Artemis, 1971-1977).

Maxim Gorky, "A Review of the Lumière Programme at the Nizhni-Novgorod Fair [1896]," in Jay Leyda, *Kino: A History of the Russian and Soviet Film*, trans. Leda Swan, 407-410 (London: George Allen and Unwin, 1960).

Rémy de Gourmont, "Epilogues: Cinematograph [1907]," in *French Film Theory and Criticism: A History/Anthology 1907-1939*, ed. and trans. Richard Abel, 47-50 (Princeton, NJ: Princeton University Press, 1988).

A. Günsberg, "Künstlerische Regie bei kinematographischen Aufnahmen und Vorführungen [1907]," in *Geschichte der Filmtheorie: Kunsttheoretische Texte von Méliès bis Arnheim,* ed. Helmut H. Diederichs, 44-47 (Frankfurt am Main: Suhrkamp-Verlag, 2004).

Tom Gunning, "What's the Point of an Index? Or, Faking Photographs," *Nordicom Review* 1/2 (September 2004).

Hermann Häfker, "Atlantis [1914]," in *Prolog vor dem Film: Nachdenken über ein neues Medium, 1909-1914*, ed. Jörg Schweinitz, 405-406 (Leipzig: Reclam-Verlag, 1992).

Hermann Häfker, "Hauptmanns Atlantis," *Bild & Film* III:6 (1913/1914): 139-140.

Hermann Häfker, *Kino und Kunst* (M. Gladbach: Volksvereinsverlag, 1913).

Miriam Bratu Hansen, *Cinema and Experience: Siegfried Kracauer, Walter Benjamin, and Theodor W. Adorno* (Berkeley, CA: University of California Press, 2011).

Miriam Hansen, "Decentric Perspectives: Kracauer's Early Writings on Film and Mass Culture," *New German Critique* 54 (Autumn 1991): 47-76.

Miriam Bratu Hansen, "Introduction," in Siegfried Kracauer, *Theory of Film: The Redemption of Physical Reality*, vii-xlv (Princeton, NJ: Princeton University Press, 1997).

Gustav Hartung, "Kinematographie und Theater," *Der Kinematograph* 170 (30 March 1910).

F. M. Hübner, "Der Dichter und das Kino," *Allgemeine Zeitung* (Munich) 39 (27 September 1913): 629-630.

Rudolf Kayser, "Amerikanismus," *Vossische Zeitung* 458 (27 September 1925).

Hermann Kienzl, "Theater und Kinematograph," *Der Strom* (Vienna) 8 (November 1911): 219-221.

Hermann Koch, "Kino und Theater," *Der Strom* (Vienna) 2 (May 1914): 61-63.

Siegfried Kracauer, "Hochstaplerfilme [1923]," in Siegfried Kracauer, *Kleine Schriften zum Film*, vol. 6.1., *1921-1927*, ed. Inka Mülder-Bach, 37 (Frankfurt am Main: Suhrkamp-Verlag, 2004).

Siegfried Kracauer, *Theory of Film: The Redemption of Physical Reality* [1960] (Princeton, NJ: Princeton University Press, 1997).

Siegfried Kracauer, "Zwischen Flammen und Bestien [1923]," in Siegfried Kracauer, *Kleine Schriften zum Film*, vol. 6.1., *1921-1927*, ed. Inka Mülder-Bach, 35-36 (Frankfurt am Main: Suhrkamp-Verlag, 2004).

Hans Land, "Lichtspiele [1910]," in *Prolog vor dem Film: Nachdenken über ein neues Medium, 1909-1914*, ed. Jörg Schweinitz, 18-19 (Leipzig: Reclam-Verlag, 1992).

Konrad Lange, "Bühne und Lichtspiel," *Deutsche Revue* (October 1913): 119-125.

Konrad Lange, "Die 'Kunst' des Lichtspieltheaters [1913]," in *Geschichte der Filmtheorie: Kunsttheoretische Texte von Méliès bis Arnheim*, ed. Helmut H. Diederichs, 75-88 (Frankfurt am Main: Suhrkamp-Verlag, 2004).

Carl Lemcke, *Populäre Ästhetik* (Leipzig: E. A. Seemann, 1873).

Gotthold Ephraim Lessing, *Laocoön: An Essay on the Limits of Painting and Poetry* [1766], trans. Edward Allen McCormick (Baltimore: Johns Hopkins University Press, 1984).

Georg Lukács, "Thoughts Toward an Aesthetic of the Cinema [1911/1913]," in *The Promise of Cinema: German Film Theory 1907-1933*, eds. Anton Kaes, Nicholas Baer, and Michael Cowan, trans. Janelle Blankenship, 377-381 (Berkeley, CA: University of California Press, 2016).

Joseph August Lux, "Die Muse des Films," *Velhagen & Klasings Monatshefte* 11 (July 1913): 429.

Theodor Heinrich Mayer, "Lebende Photographien [1912]," in *Kein Tag ohne Kino: Schriftsteller über den Stummfilm*, ed. Fritz Güttinger, 119-129 (Frankfurt am Main: Deutsches Filmmuseum, 1984).

"Meinungen von Karl Freund," *Filmtechnik* 1 (5 January 1926): 4-5, 27-28.

Gustav Melcher, "Die künstlerischen Vorzüge der Kinematographie," *Der Kinematograph* 116 (17 March 1909).

Gustav Melcher, "On Living Photography and the Film Drama [1909]," in *The Promise of Cinema: German Film Theory 1907-1933*, eds. Anton Kaes, Nicholas Baer, and Michael Cowan, 17-20, trans. Alex H. Bush and Jon Cho-Polizzi, 17-20 (Berkeley, CA: University of California Press, 2016).

Georges Méliès, "Cinematographic Views [1907]," in *French Film Theory and Criticism: A History/Anthology 1907-1939*, ed. and trans. Richard Abel, 35-47 (Princeton, NJ: Princeton University Press, 1988).

G. Mercator (=Josef Krämer), "Ist die Photographie eine Kunst?" *Deutsche Photographen-Zeitung* 15 (1891).

Heinz Michaelis, "Art and Technology in Film [1923]," in *The Promise of Cinema: German Film Theory 1907-1933*, eds. Anton Kaes, Nicholas Baer, and Michael Cowan, trans. Alex H. Bush, 581-582 (Berkeley, CA: University of California Press, 2016).

Inka Mülder-Bach, "Der Umschlag der Negativität. Zur Verschränkung von Phänomenologie, Geschichtsphilosophie und Filmästhetik in Siegfried Kracauers Metaphorik der 'Oberfläche,'" *Deutsche Vierteljahrsschrift für Literaturwissenschaft und Geistesgeschichte* 61:2 (June 1987): 359-373.

Hugo Münsterberg, *Die Amerikaner* (Berlin: Ernst Siegfried Mittler und Sohn, 1904).

Hugo Münsterberg, *The Photoplay: A Psychological Study* (New York, NY: D. Appleton and Company, 1916).

Margaret Münsterberg, *Hugo Münsterberg: His Life and Work* (New York, NY: D. Appleton and Company, 1922).

Oskar Negt, "Erbschaft aus Ungleichzeitigkeit und das Problem der Propaganda," in *Es muss nicht immer Marmor sein. Ernst Bloch zum 90. Geburtstag*, ed. Detlef Horster (Berlin: Wagenbach, 1975).

Franz Anton Nüsslein, *Lehrbuch der Ästhetik als Kunstwissenschaft* (Regensburg: Verlag von G. Joseph Manz, 1837).

Friedrich Paulsen, "Theater und Kino," *Die Grenzboten* (Berlin) 45 (1913): 285-288.

Emil Perlmann, *Das Kino als modernes Volkstheater: Eine Entgegnung auf unberechtigte Angriffe* (Berlin: Agitations-Komitee der kinematographischen Fachpresse zur Förderung der Lichtbildkunst, 1912).

Kurt Pinthus, "Quo Vadis, Cinema? On the Opening of the Königspavillon-Theater [1913]," in *The Promise of Cinema: German Film Theory 1907-1933*, eds. Anton Kaes, Nicholas Baer, and Michael Cowan, trans. Don Reneau, 186-188 (Berkeley, CA: University of California Press, 2016).

Kurt Pinthus, "The Photoplay: A Serious Introduction for Those Who Think Ahead and Reflect [1913]," in *The Promise of Cinema: German Film Theory 1907-1933*, eds. Anton Kaes, Nicholas Baer, and Michael Cowan, trans. Don Reneau, 199-203 (Berkeley, CA: University of California Press, 2016).

Gerhard Plumpe, *Der tote Blick: Zum Diskurs der Photographie in der Zeit des Realismus* (Munich: Wilhelm Fink Verlag, 1990).

Alfred Polgar, "Das Drama im Kinematographen [1911]," in *Prolog vor dem Film: Nachdenken über ein neues Medium, 1909-1914*, ed. Jörg Schweinitz, 159-164 (Leipzig: Reclam-Verlag, 1992).

Otto Rank, *The Double: A Psychoanalytic Study* [1914], ed. and trans. Harry Tucker, Jr. (Chapel Hill, NC: The University of North Carolina Press, 1971).

Willy Rath, *Kino und Bühne* (M. Gladbach: Volksvereinsverlag, 1913).

Willy Rath, "Künstlerische Möglichkeiten des Lichtspiels [1913]," in *Geschichte der Filmtheorie: Kunsttheoretische Texte von Méliès bis Arnheim,* ed. Helmut H. Diederichs, 121-131 (Frankfurt am Main: Suhrkamp-Verlag, 2004).

Ulrich Rauscher, "Das Kintop-Epos [1913]," in *Kein Tag ohne Kino: Schriftsteller über den Stummfilm,* ed. Fritz Güttinger, 137-140 (Frankfurt am Main: Deutsches Filmmuseum, 1984).

Malwine Rennert, "Heureka [1913]," in *Geschichte der Filmtheorie: Kunsttheoretische Texte von Méliès bis Arnheim,* ed. Helmut H. Diederichs, 143-146 (Frankfurt am Main: Suhrkamp-Verlag, 2004).

Jules Romains, "The Crowd at the Cinematograph [1911]," in *French Film Theory and Criticism: A History/Anthology 1907-1939*, ed. and trans. Richard Abel, 53-54 (Princeton, NJ: Princeton University Press, 1988).

Rudolf Rotheit, "Kasperle und Kinematograph," *Der Kinematograph* 5 (3 February 1907).

Will Scheller, "The New Illusion [1914]," in *The Promise of Cinema: German Film Theory 1907-1933*, eds. Anton Kaes, Nicholas Baer, and Michael Cowan, trans. Alex H. Bush, 196-199 (Berkeley, CA: University of California Press, 2016).

Friedrich Schiller, *On the Aesthetic Education of Man* [1795], trans. Reginald Snell (Mineola, NY: Dover Publications, 1992).

Friedrich Schiller, "On the Pathetic [1793]," in *Friedrich Schiller Essays*, eds. Walter Hinderer and Daniel O. Dahlstrom, trans. Daniel O. Dahlstrom, 45-69 (New York, NY: Continuum, 1993).

Heide Schlüpmann, "Phenomenology of Film: On Siegfried Kracauer's Writings of the 1920s," *New German Critique* 40 (Winter 1987): 97-114.

Jörg Schweinitz, ed., *Prolog vor dem Film: Nachdenken über ein neues Medium, 1909-1914* (Leipzig: Reclam-Verlag, 1992).

R. Seligmann, "Kinematograph und Traum," *Frankfurter Zeitung und Handelsblatt* 289 (18 October 1911): 2.

Ben Singer, "The Ambimodernity of Early Cinema: Problems and Paradoxes in the Film-and-Modernity Discourse," in *Film 1900: Technology, Perception, Culture*, eds. Annemone Ligensa and Klaus Kreimeier, 38-51 (New Barnet, Herts: John Libbey Publishing, 2009).

Herbert Tannenbaum, "Film Advertising and the Advertising Film [1920]," in *The Promise of Cinema: German Film Theory 1907-1933*, eds. Anton Kaes, Nicholas Baer, and Michael Cowan, trans. Tara Hottman, 179-182 (Berkeley, CA: University of California Press, 2016).

Herbert Tannenbaum, *Kino und Theater* (Munich: Steinebach, 1912).

Herbert Tannenbaum, "Problems of the Film Drama [1913]," in *The Promise of Cinema: German Film Theory 1907-1933*, eds. Anton Kaes, Nicholas Baer, and Michael Cowan, trans. Alex H. Bush, 192-196 (Berkeley, CA: University of California Press, 2016).

Maria Tatar, *The Hard Facts of the Grimms' Fairy Tales* (Princeton, NJ: Princeton University Press, 2003).

Malcom Turvey, *Doubting Vision: Film and the Revelationist Tradition* (Oxford: Oxford University Press, 2008).

Willi Warstat, "Zwischen Theater und Kino," *Die Grenzboten* 23 (1912): 483-488.

Paul Wegener, "On the Artistic Possibilities of the Motion Picture [1916]," in *The Promise of Cinema: German Film Theory 1907-1933*, eds. Anton Kaes, Nicholas Baer, and Michael Cowan, trans. Alex H. Bush, 206-208 (Berkeley, CA: University of California Press, 2016).

Fedor von Zobeltitz, "Film-Literatur," *Das literarische Echo* 13:15 (1 May 1911): 1103.

2. Modern Magicians: Guido Seeber and Eugen Schüfftan

Abstract
Two technicians had a particularly formative impact on the evolution of special effects in Germany. Film pioneer Guido Seeber favoured methods like multiple exposure composites, which allow the cinematographer to excel both technically and creatively. Aiming at forging convincing composite spaces on screen, Eugen Schüfftan invented the only widely used commercial special-effects technique originating in Europe, the Schüfftan process. In similar ways, Seeber's photographic and Schüfftan's perceptual effects construe technology as cinema's core creative tool and the cinematic image as fundamentally malleable. Both shared techno-romantic views, which is apparent from their devotion to the goal of film art and commitment to devising medium-specific means for transcending material reality and expressing emotions and ideas.

Keywords: Guido Seeber, Eugen Schüfftan, film technicians, multiple exposure composites, Schüfftan process

All major German cinematographers of the silent era were masters of trick technology. Internationally the best known was Karl Freund, but the artistry of Carl Hoffmann, whom Freund considered the greatest of all cinematographers, Fritz Arno Wagner, and Günther Rittau equalled Freund's.[1] Even though German cameramen, particularly in the early 1920s, often complained about being undervalued, by international standards

1 Herbert G. Luft, "Der Mann hinter der Kamera: Karl Freund – schon Filmgeschichte," *Der Filmkreis* (July 1963): 16-19; in the ASC Collection, Margaret Herrick Library, Los Angeles, CA. While filmmakers like Leni Riefenstahl, Lotte Reiniger, and Lola Kreutzberg certainly shot (parts of) their own films, I am not aware of any camerawomen working in a German studio until after World War II. However, further research is needed to conclusively substantiate this assumption.

Loew, Katharina, *Special Effects and German Silent Film: Techno-Romantic Cinema*.
Taylor & Francis Group, 2021
DOI 10.5117/9789463725231_CH02

they were highly respected. Ubiquitous techno-romantic views elevated the cinematographer's prestige. They rendered conceivable the ostensibly oxymoronic notion of technician as artist. Because medium-specific creativity, i.e., the imaginative use of film technology to shape cinematic images, was deemed a prerequisite for film art, film technicians became regarded as co-creators of the filmic artwork early on in Germany. The ability to create striking trick effects was therefore not merely a hallmark of cinematographers' professional merit, but fundamental to their status as artists.

Tricks, as Guido Seeber defined them in his 1927 book-length study on the subject, are "skilful cinematographic devices" [*Kunstgriffe aufnahmetechnischer Art*], specifically of the "photographic, optical, physical, and chemical" kind.[2] Many of the devices that Seeber examines would still be considered special effects today. They include techniques that manipulate the photographic image itself, like mattes, miniatures, rear projection, and multiple exposure composites as well as methods that manipulate the illusion of movement like stop, reverse, slow, and fast motion. At the same time, however, several of the techniques covered in Seeber's book, such as extreme camera angles, capturing certain photographically challenging objects like the moon, or the use of particular film stock, were later no longer regarded as "tricks." This suggests that "tricks" or "special effects" are not inherent categories but determined by their deviation from the conventional production or reception contexts in which they are situated.

Whether or not the methods discussed in Seeber's book would have been deemed "special effects" at other historical junctures, in the context of 1920s Germany, all were "special" in three regards: First, they afforded images beyond the scope of customary straight photography, second, they relied on production processes that were perceived as unusual, and third, they required extraordinary skill on the part of their creators. In 1927, these features pertained to handheld cameras as much as to multiple-exposure composites. Even if, as Stephen Prince has argued, the term "special effects" has become obsolete today, in many ways these characteristics also apply to digital visual effects.[3]

As I have shown in Chapter 1, special effects paved the way for cinema's validation as an art by adhering to cinema's technological essence,

2 Guido Seeber, *Der praktische Kameramann*, vol. 2, *Der Trickfilm in seinen grundsätzlichen Möglichkeiten* [1927] (Frankfurt am Main: Deutsches Filmmuseum, 1979), III; 13.

3 Stephen Prince, *Digital Visual Effects in Cinema: The Seduction of Reality* (New Brunswick, NJ: Rutgers University Press, 2011), 3-4.

transcending physical reality, and demonstrating the medium's creativity. On top of this, their thrice-extraordinary nature also helped foster the cinematographer's renown, elevating the camera operator to an artist. Critic Lotte Eisner proclaimed: "The productive cameraman, who transposes the assumptions of author and director based on his own intuition into the visual realm, is the obvious precondition for the artistic film."[4] As a creator of fictional images who combined exceptional technical skills with creative ingenuity, the cinematographer gained acceptance as key co-creator of the filmic artwork. Günther Rittau asserted:

> Just as the poet can only through complete command of language fully realize his will of expression, just as the musician can only through technical mastery of his instrument make his emotional world resound in the listener, the cinematographer must through mastery and incessant exploration of film technology merge with the camera, which is often degraded to a mere copying tool, into one artistic-creative consubstantiality.[5]

The cameraman embodied the connection between the material and the immaterial that characterized the techno-romantic outlook. Virtuoso command of film technology was considered the precondition for realizing an artistic vision, which in turn allowed the machine to transcend the physical realm.

While many German camera artists made major trick technological contributions, two individuals stand out. Like few others, Guido Seeber (1879-1940) and Eugen Schüfftan (1886-1977) shaped the discourse about and the production of trick effects in the German film industry.[6] Seeber, a film pioneer, cameraman, inventor, and historian of film technology, was the first German to concern himself with effects production. He had immense influence on the development of trick technology in Germany, which is at least in part attributable to his efforts as a publicist. In deviation from the standard practice of keeping trick techniques a secret, Seeber publically shared his know-how. Although he did not single-handedly transform the

4 Lotte Eisner, "Der produktive Kameramann wird zum Mitschöpfer," *Film-Kurier* 272 (26 November 1931).

5 Günther Rittau, "Günther Rittau über Filmkunst und Filmtechnik," *Filmtechnik* 3 (5 February 1926): 50.

6 Schüfftan's date of birth is often given as 1893, yet his autographed Curriculum Vitae in his Paul Kohner agency file at the Deutsche Kinemathek Berlin shows the date of birth as July 21, 1886.

secretiveness of the trick trade, Seeber initiated a national conversation that helped shift attitudes about film technicians and film technology and fostered a consensus about technology as cinema's principal creative tool. Perhaps most importantly, the ready availability of knowledge about trick production elevated levels of competence among cinematographers and encouraged young talent, which was instrumental in establishing the world fame of German cinematography.

In contrast to Seeber, who directed his efforts mainly towards the German context, Eugen Schüfftan acquired international renown, first as the inventor of the mirror-based Schüfftan process and subsequently as one of the leading European cinematographers of the twentieth century. Schüfftan challenged the established effects paradigm in several important ways. Unlike Seeber, Schüfftan did not create effects in a capacity as a cinematographer, but initially operated as an external consultant administering only one technique, his own invention. This degree of specialization did not become prevalent until later decades and Schüfftan remained an anomaly in 1920s Europe. His heavily patented invention lent itself to the creation of imperceptible, economizing special effects. Although matte and glass paintings gained currency in European cinema roughly around the same time, it was above all the Schüfftan process that established set extensions as a regular production technique. A standardized method that was commercially available through a specialized company and promoted itself as a means of cost reduction, the Schüfftan process epitomizes a crucial modernization step in the use and production of trick effects in European cinema.

Guido Seeber and Eugen Schüfftan each had a lasting impact on the development of special effects. In similar ways, they construed trick technologies as cinema's primary creative tool, as its core means for rendering mental images. As film technicians, they aspired to provide innovative tools and superior expertise that allowed for the cinematic representation of anything imaginable. Their categorical insistence on the inherent malleability of the cinematic image is testament to their techno-romantic convictions. Affirming the primacy of the ideal, they simultaneously touted technology as means to attain it. Both provoked major transformations in special effects production, yet they remained committed to a professional ethos characteristic for most leading German cinematographers and art directors: Seeber and Schüfftan always understood themselves as master craftsmen charged with blending creative sensibility and technological excellence in the service of film art.

Early German film technology

In the earliest years of cinema, Germany quickly emerged as an international leader in the field of cinematic hardware, but was slow to embrace film production. Economic parameters explain this striking discrepancy in part, but equally important was the conceptual opposition between art and technology, which limited the appeal of silent narrative films in the eyes of producers and critics. German filmmakers started to show interest in film art and special effects around the same time as intellectuals embraced techno-romantic perspectives on cinema.

Before approximately 1907, Germany produced hardly any silent narrative films and trick films were not part of its early output. In contrast to France, Britain, and the United States, the country therefore had also no share in the initial evolution of special effects. Instead of narrative films, producers prioritized either comparatively cheap nonfiction subjects, or expensive, yet more lucrative sound pictures (*Tonbilder*), which relied on various sound-on-disk systems.[7] Low prices on the film market and strong foreign competition made the domestic production of narrative films unprofitable. Corinna Müller has described the (silent) film trade in Germany before 1910/1911 as an environment of cutthroat competition. Inexpensive ticket prices prompted cinema owners to buy the cheapest prints available, which they screened relentlessly. Struggling to turn a profit, producers churned out low-quality films at a maximum pace, which further aggravated the problem of oversupply and depressed prices. In this environment, experimentation or aesthetic innovation were dispensable. Given that the creation of trick films is particularly time intensive, it is little surprising that the genre played no role in early German cinema.

In addition to the adverse economic situation in the film market, producers also grappled with cinema's low social standing. Whereas actuality and scenic films were accredited with educational value and *Tonbilder* reproduced stage performances, fiction films were initially considered one of the lowest forms of cultural expression. In 1907, the trade journal *Der Kinematograph* explained the reasons for the German reluctance to engage in narrative film production in its section "For Our English Readers" in the following way: "In Germany, the land of the most earnest scientists, the taking of kinematographic photographs, of

7 Corinna Müller, *Frühe deutsche Kinematographie: formale, wirtschaftliche und kulturelle Entwicklungen 1907-1912* (Stuttgart: Metzler-Verlag, 1994), 81. For a contemporary assessment, see for instance A. Berein, "Kinematographisches," *Der Komet* 1109 (23 June 1906): 6.

regulated theatrical representations, was regarded as something very like playing; that sort of photographic picture-making was more suited to the temperament of the Frenchman than to that of the fundamentally earnest-thinking German."[8]

Instead of film production, German technicians initially engaged in film engineering. Capitalizing on the public veneration of the applied sciences and on the engineer's status as a modern national hero, Germans devised cameras, shutters, and projectors as well as natural colour, sound, and stereoscopic film systems. Chronophotographer Ottomar Anschütz acquired renown for his Electrotachyscope (1887), a viewing device for moving images. Max Skladanowsky, inventor of the Bioskop double projector, was the first to exhibit film projections to a paying audience, some two months before the public debut of the Lumière Brothers' technically superior Cinématographe. Inventor, producer, distributor, and exhibitor Oskar Messter, a pivotal figure for early German cinema, devised the four-slotted Maltese Cross construction (patented in 1900), which became an essential part of analogue film projectors. His Biophon process launched the first era of sound film in Germany (1903-1913). While German film engineering prospered early on, filmmaking technique remained inadequate. In 1912, film technician K. W. Wolf-Czapek lamented in the trade journal *Lichtbild-Bühne*:

> For decades, Germany has produced the best photographic-optical instruments. For several years, Germany has produced the best film cameras and projectors. Most recently, Germany started producing raw stock of excellent quality, which will soon acquire a leading position on the world market. Germany spearheads the production of safety film. Why does Germany of all countries produce the most abysmal cinematographic pictures on the entire globe? Why does almost every German film stand out (in addition to the abovementioned lack in taste that cannot be discussed here) because the shots are out of frame, the image is entirely out of focus and the lighting is badly chosen? Additional characteristics by which anyone can recognize a German film after only a few seconds are its lack of focus due to insufficient shutter speed, its lack of image detail due to hard lighting as well as its splotchy shadows resulting from

8 "The German Kinematograph and Foreign Countries," *Der Kinematograph* 38 (18 September 1907).

inadequate exposure, faulty processing and possibly the use of too high contrast positive stock.[9]

The discrepancy between first-rate hardware and inferior technique, which Wolf-Czapek criticized, traces back to the assumed incompatibility of art and technology. As a scientific venture, the engineering of filmmaking devices was highly valued, but the application of technology to aesthetic purposes was not taken seriously. Techno-romantic perspectives played an essential role in reconciling the two, ostensibly mutually exclusive phenomena. In 1910, a new distribution system known as "monopoly film" (*Monopolfilm*) was introduced and made high-quality productions profitable for the first time. As I will examine in more detail in Chapter 3, a majority of monopoly films were touted as "artistic" and consequently not only aspired to better workmanship but also to meeting critics' aesthetic expectations. As a result, special effects as the embodiment of cinema's medium-specific creativity came into focus for the first time. Economic zeal and artistic ambitions spurred each other: "artistic" filmmaking was seen as a path to greater profitability, while improved economic conditions were a crucial precondition for the realization of any artistic aspirations. Following the much-respected Asta Nielsen series (1911-1914), the German art film movement of 1913/1914 was built on techno-romantic tenets. It was here, for the first time in German film history, that artistic pursuit and technological excellence were construed as interdependent, a perspective that would characterize quality film production in Germany for decades to come. Centrally involved in this was Guido Seeber. The importance of his leadership in asserting technology as cinema's key artistic device cannot be overestimated.

Guido Seeber: The camera as musical instrument

Trained as a photographer, Seeber began in the film industry in 1897 as the exhibitor of "Seeber's Giant Living Photographs." After having obtained his first camera from Oskar Messter, he developed his own "Seeberograph" projector, which he combined in 1904 with a sound-on-disk

9 K. W. Wolf-Czapek, "Deutsche Films – technisch bewertet," *Lichtbild-Bühne* 21 (25 May 1912) 22-23. The substandard technical quality of German films was also noted internationally. See for instance W. Stephen Bush, "'The Life of Richard Wagner' Made by Messter & Co. in Four Reels and Imported by Klaw and Erlanger," *Moving Picture World* 8-13 (1913): 995.

Tonbild system under the name "Seeberophon." Between 1909 and 1914, he created several animated and live-action trick films, including *Prosit Neujahr 1910!* (Deutsche Bioscop, 1909), *Die geheimnisvolle Streichholzdose* (Deutsche Bioscop, 1910) and *Das verkehrte Berlin* (Deutsche Bioscop, 1911). Although these films were surely not exceptional by any international standards, they are noteworthy for constituting the earliest German trick films.

As technical director of Deutsche Bioscop, Seeber established the production facilities in Babelsberg near Berlin in 1911/1912, which later became the nucleus of Ufa's studio lot and today's Babelsberg Film Studios. When Deutsche Bioscop signed the promising Danish actress Asta Nielsen and her director Urban Gad, Guido Seeber photographed their thirteen first pictures and Asta Nielsen became Germany's first film star. In 1913/1914, Seeber collaborated with actor Paul Wegener, best-selling author Hanns Heinz Ewers, and Danish director Stellan Rye on a series of artistically ambitious films for Deutsche Bioscop. Most, including *Ein Sommernachtstraum in unserer Zeit*, *Die Eisbraut*, *Die Augen des Ole Brandis*, and *Erlkönigs Töchter*, dealt with supernatural subject matter and thus flaunted numerous trick effects. Solely *Der Student von Prag* is known to survive, but it seems likely that all of these productions adopted a similarly techno-romantic approach when relying heavily on special effects in pursuit of film art.

Following his wartime service, Seeber worked as a freelance cinematographer. He made significant contributions to mobilizing the camera. As I will discuss in more detail in Chapter 6, his use of camera dollies in the four-part *Fridericus Rex* series (Cserépy Film, 1922/1923, dir. Arsen von Cserépy) and *Sylvester* (Rex-Film, 1924, dir. Lupu Pick) prefigure the "unchained camera" that acquired international renown in Friedrich Wilhelm Murnau's *Der letzte Mann* (Union-Film/Ufa, 1924). Films like *Das wandernde Bild* (May Film, 1920, dir. Fritz Lang), *Die freudlose Gasse* (Sofar-Film-Produktion, 1925, dir. G. W. Pabst) and *Dirnentragödie* (Pantomim-Film, 1927, dir. Bruno Rahn) showcase Seeber's innovative lighting schemes and his artistry as a portraitist of both faces and milieus. However, as Thomas Brandlmeier explains,

> Almost curiously, Seeber's artistic role is that of a precursor and trailblazer. Be it chiaroscuro lighting, the Fanck school [of mountain photography], the unchained camera or New Objectivity—others always seized his impulses and completed his outlines. The major revolutions, the milestones of film history are not Seeber's, but they undoubtedly emerged from his

> periphery. Seeber is some kind of central star, without which the planetary system of German camera art would not cohere.[10]

Seeber's merits as a vanguard cinematographer derive to a large extent from his ardour for experimentation. "In my films," he declared in an interview, "technological innovations have always particularly appealed to me."[11] His persistent quest for novel film-technological means of expression culminated in a number of experimental projects like a talking picture *Der sprechende Film* (Tri-Ergon-Film, 1923, dir. W. Doerry), a promotional short for the 1925 industrial Cinema and Photo Exhibition (Kipho) in Berlin, and a series of eight rebus films, animated crossword puzzles, which he created with Paul Leni and Hans Brennert in 1925/1926.[12] Some of these films bear resemblance to the work of avant-garde artists like Viking Eggeling, Germaine Dulac, Dziga Vertov, Hans Richter, Walter Ruttmann, and Oskar Fischinger. However, although Seeber shared their faith in the unlimited expressive possibilities of film technology and its central role in shaping innovative film art, he did not aspire to radical social or aesthetic change. He therefore remained, as Malte Hagener put it, a "marginal figure" of the avant-garde.[13]

By the 1920s, Seeber was commonly referred to as the "doyen" (*Altmeister*) in the German film industry. This renown was to no small extent based on countless publications. With the goal to promote exchange between film technicians, he co-founded and co-edited two technical trade journals, *Kinotechnik* (1919-1943) and *Die Filmtechnik* (1925-1943). In 1927 Seeber authored two book-length studies that introduced aspiring cinematographers to the basics of the trade. The first volume *Der praktische Kameramann* (*The Practical Cameraman*) deals with camera and lighting equipment and presents an important source for the state of European filmmaking equipment in the mid-1920s.[14] The second volume *Der Trickfilm in seinen grundsätzlichen Möglichkeiten* (*The Trick Film in its Basic Possibilities*) is the first-ever

10 Thomas Brandlmeier, *Kameraautoren: Technik und Ästhetik* (Marburg: Schüren Verlag, 2008), 369.

11 Alexander Kossowsky, "Die Männer der Kurbel VI: Guido Seeber," *Film-Kurier* (22 August 1925).

12 For an in-depth discussion of these "rebus" films see Michael Cowan, "Moving Picture Puzzles: Training Urban Perception in the Weimar 'Rebus Films,'" *Screen* 51:3 (Autumn 2010): 197-218.

13 Malte Hagener, *Moving Forward, Looking Back: The European Avant-Garde and the Invention of Film Culture 1919–1939* (Amsterdam: Amsterdam University Press, 2007), 139.

14 Guido Seeber, *Der praktische Kameramann* [1927] (Frankfurt am Main: Deutsches Filmmuseum, 1980).

book-length study on special effects.[15] It provides an overview over the evolution and current capabilities of trick technology and also gives an idea of the considerable challenges associated with creating even the most basic effects. Reputedly at the instigation of Sergei Eisenstein a Russian-language version of *Der Trickfilm* appeared in 1929 under the title *Technika kino-trjuka*. Eduard Tissé's camera assistant Vladimir Nilsen was in charge of the translation. In their dedication to Seeber's personal copy of *Technika kino-trjuka*, Eisenstein and Tissé expressed their deep admiration for his work: "To the grand master of the trick film Guido Seeber in deepest reverence. S. Eisenstein. E. Tissé."[16]

The art of multiple exposure composites

As a historian, inventor, mentor, and advocate, Seeber had lasting effect on the discourse on film technology in the German film industry. His writings not only elevated standards by making knowhow about trick technology widely available, they also raised awareness about technology's creative potentials and demanded respect for the work of film technicians. His remark "one must play the camera like a musical instrument" reveals a techno-romantic conception of the cinematographer as an artist, a creative master of camera technique.[17] In keeping with this professional ethos, Seeber favoured trick techniques that require cinematographers to make full use of their instrument and photographic skills. Above all, he focused his attention on multiple exposure composites, which are achieved by exposing a single negative to multiple images before the film is processed.

In the silent era, producing multiple exposure composites was a tedious endeavour. The main scene was usually recorded first and "apparitions" added later. During the first exposure, the cameraman, cranking the camera at the usual two or three turns per second, counted at the top of his voice, so that he could be heard over the noisy silent movie sets. Assistants monitored his count and carefully recorded where the second exposure was scheduled to begin and end. After the take, the negative was rewound and the "apparition" positioned in front of a black velvet backdrop. Unable to actually see the camera image, the cameraman had to resort to aids like screens with

15 Guido Seeber, *Der praktische Kameramann*, vol. 2, *Der Trickfilm in seinen grundsätzlichen Möglichkeiten* [1927] (Frankfurt am Main: Deutsches Filmmuseum, 1979).

16 The dedication is reproduced in Stiftung Deutsche Kinemathek, ed. *Das wandernde Bild: Der Filmpionier Guido Seeber* (Berlin: Stiftung Deutsche Kinemathek, 1979), 151.

17 Alexander Kossowsky, "Die Männer der Kurbel VI: Guido Seeber," *Film-Kurier* (22 July 1925).

gridlines to frame the second exposure. The film was then advanced to the count where the apparition was to materialize. The cameraman would slowly open the shutter, expose the film for the second time and gradually close the shutter at the designated count. Any mistake would ruin the negative and the entire procedure had to start from the beginning.

The main challenge regarding multiple exposure composites was that the slightest disruption to the film registration results in a shaky image against a steady background. At the time, neither film perforation nor intermittent mechanisms were accurate enough to ensure identical film registration for every take.[18] In addition, it was difficult to keep cameras absolutely steady: Tripods were insufficiently sturdy and hand-cranked cameras made a vibration-free environment almost a physical impossibility. In fact, as Seeber explained, more so than any other piece of film equipment, tripods were were the biggest obstacle to steady composites.[19]

During the silent era, cinematographers were usually expected to provide their private camera equipment for principal photography. However, at approximately $6,000 the American Bell & Howell and Mitchell cameras, which featured advanced intermittent mechanisms with registration pins and thus afforded particularly steady composites, were out of reach for most European cinematographers. It was not until 1924 that Ufa acquired two Mitchell cameras specifically with the trick work for *Metropolis* (1927, dir. Fritz Lang) in mind. The most commonly used studio camera in Europe during the silent period and also Seeber's camera of choice was the French Debrie Parvo. Although it lacked pilot-pin registration, Seeber considered the Debrie Parvo just as suitable for complex trick work as the American models, assuming, as he emphasized, that it was handled by a skilled cinematographer.[20]

To combine two or more image components by means of multiple exposures *without* superimposition effects, parts of the film must be protected from light during each exposure. This can be achieved either by masking the film in proximity to the lens or by covering parts of the scenery in black velvet. Such "non-actinic" backgrounds, which cause little reaction in the photographic emulsion, can facilitate a range of illusions from multiplications to changes in size. An obvious drawback, however, is the conspicuous black area that inevitably dominates the frame. Sets with non-actinic areas,

18 Intermittent mechanisms transport the film and hold it in place during exposure.

19 See for instance, Alexander Kossowsky, "Die Männer der Kurbel II: Reimar Kuntze," *Film-Kurier* (14 May 1925); Alexander Kossowsky, "Die Männer der Kurbel III: Werner Brandes," *Film-Kurier* (19 May 1925).

20 Seeber, *Der Trickfilm*, 59-60.

such as dark doors, picture frames, or fireplaces, were common early on. In fact, non-actinic sets were so omnipresent in trick films that Georges Méliès, in *Le Portrait spirite* (Star Film, 1903), self-ironically touted on screen "dissolving effect obtained without black background. GREAT NOVELTY."

In order to attain multiple exposure composites without black areas or superimposition effects, the frame must be partitioned and mattes and counter-mattes placed either inside the camera or directly in front of the lens. The resulting split screen composites replaced parts of the frame like in *Le Portrait spirite* or the stationmaster's office window in *The Great Train Robbery* (Edison Manufacturing Company, 1903, dir. Edwin S. Porter) or facilitated stunts like falls or encounters with wild animals. Seeber was particularly fond of the doppelganger illusion, which relies on split screen mattes and was already popular in nineteenth century still photography. I will examine this effect in the context of *Der Student of Prag* in Chapter 3.

Seeber's passion for research and experimentation is epitomized in a travelling matte process that he unearthed for *Lebende Buddhas* (Paul Wegener Film, 1923/1925, dir. Paul Wegener). Sadly, only a few fragments survive of this film, which depicted the clash between modern European rationalism and the avenging gods of ancient Tibet. Seeber considered it his trick technological *chef-d'œuvre. Lebende Buddhas* not only thematically alluded to the conflict underlying the techno-romantic paradigm, but also actualized it by visualizing orientalist spirituality by means of an abundance of pioneering special effects.

In one climactic shot, which effectively functions as an icon of techno-romantic thought, the gaze of a powerful Grand Lama is shown to control the course of an ocean liner (Figure 2.1.). As the steamer enters the frame, the Grand Lama's monumental head ascends from behind the waterline until it is fully visible and fills three quarters of the frame. According to Seeber, "the head following the action was framed in close-up and thus appeared vis-à-vis the steamer exceedingly uncanny and impressive."[21] While passing, parts of the ship's superstructure cover the Lama's face without generating substantial superimposition effects—a feat that at the time only the Williams process rendered possible.[22] In order to achieve this, Seeber applied a somewhat obscure contact-printing method described by

21 Seeber, *Der Trickfilm*, 97.

22 For the Williams process see Julie Turnock, "Patient Research on the Slapstick Lots: From Trick Men to Special Effects Artists in Silent Hollywood," *Early Popular Visual Culture* 2 (May 2015): 152-173.

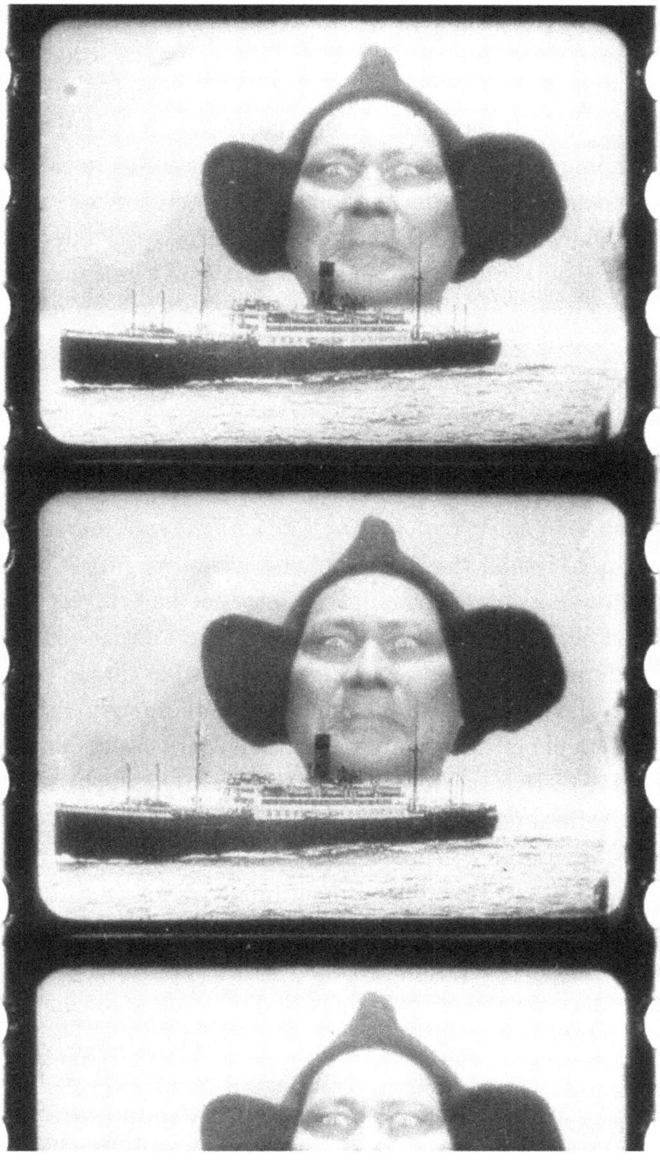

Fig. 2.1. Travelling matte process by Guido Seeber for *Lebende Buddhas* (1923/1925). Source: Deutsche Kinemathek.

pioneer film historian Émile Kress in 1912.[23] Wegener's head was positioned in between two white flats and Seeber bipacked two reels into his camera,

23 Reprinted as E. Kress, "Trucs et illusions. Applications de l'optique et de la mécanique au cinématographe (c. 1912)," *1895. Revue d'histoire du cinéma* 27 (September 1999): 7-20.

a pre-exposed and developed film of an ocean liner together with regular raw stock. During the exposure, the ocean liner contact-printed onto the unexposed stock along with the images of Wegener's head from the camera lens, resulting in a fairly convincing composite. Two aspects aid the transition between the image components. The hazy dividing line between image components coincides with the horizon, which allows it to be rationalized as a weather phenomenon. This and the fact that the vessel moves visibly *in front of* the face facilitate a spatial coherence that is instrumental in turning, as a contemporary reviewer of *Lebende Buddhas* put it, "the impossible into an optical experience."[24]

Montage shots

The composites in *Le Portrait spirite, Der Student von Prag*, or *Lebende Buddhas* seek to render the transitions between image components as imperceptible as possible and simulate a consistent diegetic space. Simultaneously, however, multiple exposure composites were also used in contexts where they enunciate their patchwork nature. As Jan Olsson has shown, between 1906 and 1916 split screen composites were the standard device for representing telephone conversations, particularly on European screens.[25] Typically, three vertical panels depicted different spaces, i.e., two sides of a phone conversation and a mid-panel that provided additional information of various kinds. Tracing back to nineteenth-century postcards, advertisements, and lanternslides, this type of composition favoured simultaneity over continuity and constituted an alternative to parallel editing. Split screen compositions that were recognizable as such also appear in non-telephonic shots to introduce a pertinent elsewhere that is linked to the master space, for instance to illustrate characters' accounts, recollections, or dreams.[26]

As Olsson argues, by the late-1910s, split screen composites became rare in the context of telephone conversations. However, juxtapositions of separately recorded images in different parts of the frame were employed frequently throughout the 1920s. Three-panel compositions, horizontal ones like those depicting the Foxtrot epidemic in *Die Austernprinzessin* (PAGU, 1919, dir.

24 –s, "Kritik der Leinwand," *Die Filmwoche* 22 (1925): 519.

25 Jan Olsson, "Framing Silent Calls: Coming to Cinematographic Terms with Telephony," in *Allegories of Communication: Intermedial Concerns from Cinema to the Digital*, eds. John Fullerton and Jan Olsson, 157-192 (Bloomington, IN: Indiana University Press, 2005).

26 Olsson, "Framing Silent Calls," in *Allegories of Communication*, eds. Fullerton and Olsson, 166.

Ernst Lubitsch) and vertical ones like those portraying the attorneys' pleas in *Les deux timides* (Films Albatros, 1928, dir. René Clair) come to mind. Its culmination reached the composition in separate panels in the monumental triptych finale of Abel Gance's *Napoléon* (Ciné France et al., 1927).

Far more common than rectangular ones were irregularly shaped mattes, however, presumably because they allowed individual image components to merge more organically. They illustrate the hero's recollections in *Der verlorene Schatten* (PAGU, 1921, dir. Rochus Gliese), the judgment of Paris in *Helena* (Bavaria Film, 1924, dir. Manfred Noa), the greedy eyes of the robot's audience in *Metropolis* (1927), or the world listening to the rocket launch in Fritz Lang's *Frau im Mond* (Fritz Lang Film/Ufa, 1929). Shots in which image components from different sources are placed side by side also appear in experimental films like Man Ray's *L'étoile de mer* (1928), Germaine Dulac's *La Coquille et le clergyman* (Délia Film, 1928), and Dziga Vertov's *Chelovek s kinoapparatom* (VUFKU, 1929) as well as advertising films like Julius Pinschewer's *Die chinesische Nachtigall* (1928) and Hans Richter's *Bauen und Wohnen* (1928). The pervasiveness of this device also attests to the inseparability of avant-garde and commercial practices in European silent film culture, which Sabine Hake, Malte Hagener, and Michael Cowan have emphasized.[27]

Although shots that perceptibly juxtapose different image components appear frequently in European silent films, scholars do not usually consider them a discrete phenomenon. Instead, they have been construed retroactively from a classical Hollywood perspective. Barry Salt for instance asserts:

> *Die Strasse* (Karl Grune, 1923) contains one of the very first fully realized montage sequences in the classical form, in which the dissolves take place absolutely continuously, so that there is always a changing sequence of superimposed images present on the screen. [...]. In some parts this sequence is further complicated by splitting the frame into multiple images side by side. A similarly used montage sequence in Murnau's *Sunrise* is also very similar in form and content, though more precise in execution, and far more elegant visually.[28]

27 Sabine Hake, "Das Kino, die Werbung und die Avantgarde," in *Die Spur durch den Spiegel: Der Film in der Kultur der* Moderne, eds. Malte Hagener, Johann N. Schmidt, and Michael Wedel, 193-206 (Berlin: Bertz und Fischer, 2004); Hagener, *Moving Forward, Looking*, 44–50; 138-140; Michael Cowan, *Walter Ruttmann and the Cinema of Multiplicity: Avant-Garde – Advertising –Modernity* (Amsterdam: Amsterdam University Press, 2014).

28 Barry Salt, *Film Style and Technology: History and Analysis* (London: Starwood, 1983), 194.

Salt mentions split screen composites merely as a "complication" within a montage sequence. Likewise, Kristin Thompson acknowledges them only within that context:

> Typically the German montage sequence involves a group of images, often canted or otherwise presented from an unusual angle, scattered about the screen against a black background, and often overlapping and changing within what is essentially a single shot. [...] Clearly this convenient way of telegraphing information quickly emerged as a convention.[29]

As Thompson points out herself, what she calls "montage sequence" is "essentially a single shot." Given that the name "montage sequence" suggests a succession of shots, the juxtaposition of image fragments from different sources in a single shot is more aptly described as a "montage shot."[30] Since the 1930s, when the Hollywood montage sequence became codified, montage shots were most frequently deployed in this context, but they constitute a distinct phenomenon. In European silent cinema, montage shots were not usually sequestered from narrative scenes nor used to summarize a string of events, as is the case with montage sequences. Bringing together a collection of picture fragments in the same shot, montage shots assert the primacy of the individual frame as carrier of meaning.

Seeber was an authority on montage shots and devoted much effort to this device, experimenting with production methods and deliberating on application areas. *Der Trickfilm* explains in detail how to split the image field with irregular masks and how to devise a "technical score" to time the image components.[31] For Seeber, montage shots constituted one of the principal "trick techniques of tomorrow," because they were capable of rendering abstract thought. By juxtaposing different picture components in the same shot, they can articulate ideas not contained in each source image separately.[32] What Seeber envisioned bears resemblance to ideas about

29 Kristin Thompson, *Herr Lubitsch Goes to Hollywood: German and American Film after World War I* (Amsterdam: Amsterdam University Press, 2005), 121.

30 For the function of montage sequences in "classical" cinema, see David Bordwell, *Narration in the Fiction Film* (Madison, WI: University of Wisconsin Press, 1985), 158-163; 186-188 and David Bordwell, "The Classical Hollywood Style, 1917-1960," in David Bordwell, Janet Staiger and Kristin Thompson, *The Classical Hollywood Cinema: Film Style and Mode of Production to 1960*, 1-84 (New York, NY: Columbia University Press, 1985), 28. For an early attempt to clarify the differences between montage theory, montage sequences and montage shots, see Karl Freund, "Just What Is 'Montage'?," *American Cinematographer* 5 (September 1934): 204; 210.

31 Seeber, *Der Trickfilm*, 244; 248; 251.

32 Seeber, *Der Trickfilm*, 240.

cinema as hieroglyphic writing as suggested for example by Vachel Lindsay.[33] More specifically, it is closely related to Eisenstein's notion of the ideogram as the starting point for the "intellectual cinema." In "The Cinematographic Principle and the Ideogram," Eisenstein writes:

> The point is that the copulation (perhaps we had better say, the combination) of two hieroglyphs of the simplest series to be regarded not as their sum, but as their product, i.e., as value of another dimension, another degree; each, separately, corresponds to an *object*, to a fact, but their combination corresponds to a *concept*. From separate hieroglyphs has been fused—the ideogram. By the combination of two "depictables" is achieved the representation of something that is graphically undepictable.[34]

Ideograms are a key feature in Seeber's promotional film for the 1925 Kipho cinema and photo exhibition in Berlin. The film, which is at once a commercial and a milestone of German avant-garde filmmaking, was supposed to actualize a "cross-section through the [medium of] film at a furious pace."[35] Seeber explained his choice of montage shots stemmed from the goal "to find a new form, which through its mode of representation and image sequencing could give objects a kind of tempo that fast, skilful sequential editing could not even remotely approximate."[36] The roughly 5.5-minute-long production is simply titled *Film*. By his own account, Seeber was inspired by contemporary French experimental films, in particular *Ballet mécanique* (Synchro-Ciné, 1924, dirs. Ferdinand Léger and Dudley Murphey). Technically, Seeber's and Léger's films work differently, however. *Ballet mécanique* employs lens attachments for its kaleidoscopic effects, while approximately half of the roughly fifty shots in Seeber's Kipho film were realized by means of split screen composites with three or more image components.[37] The combination of image fragments from the same source in

33 Vachel Lindsay, *The Art of the Moving Picture* [1915] (New York, NY: Liveright Publishing, 1970), 199-216. See also Miriam Hansen, "Mass Culture as Hieroglyphic Writing: Adorno, Derrida, Kracauer," *New German Critique* 56 (Spring/Summer 1992): 42-73.

34 Sergei Eisenstein, "The Cinematographic Principle and the Ideogram [1929]," in Sergei Eisenstein, *Film Form: Essays in Film Theory*, ed. and trans. Jay Leyda, 28-44 (New York, NY: Harcourt, Brace & World, 1949), 29-30.

35 Seeber, *Der Trickfilm*, 246.

36 Seeber, *Der Trickfilm*, 246.

37 For a discussion of the influence of *Ballet mécanique* on Seeber's Kipho film see Standish D. Lawder, *The Cubist Cinema* (New York, NY: New York University, 1975), 177-181.

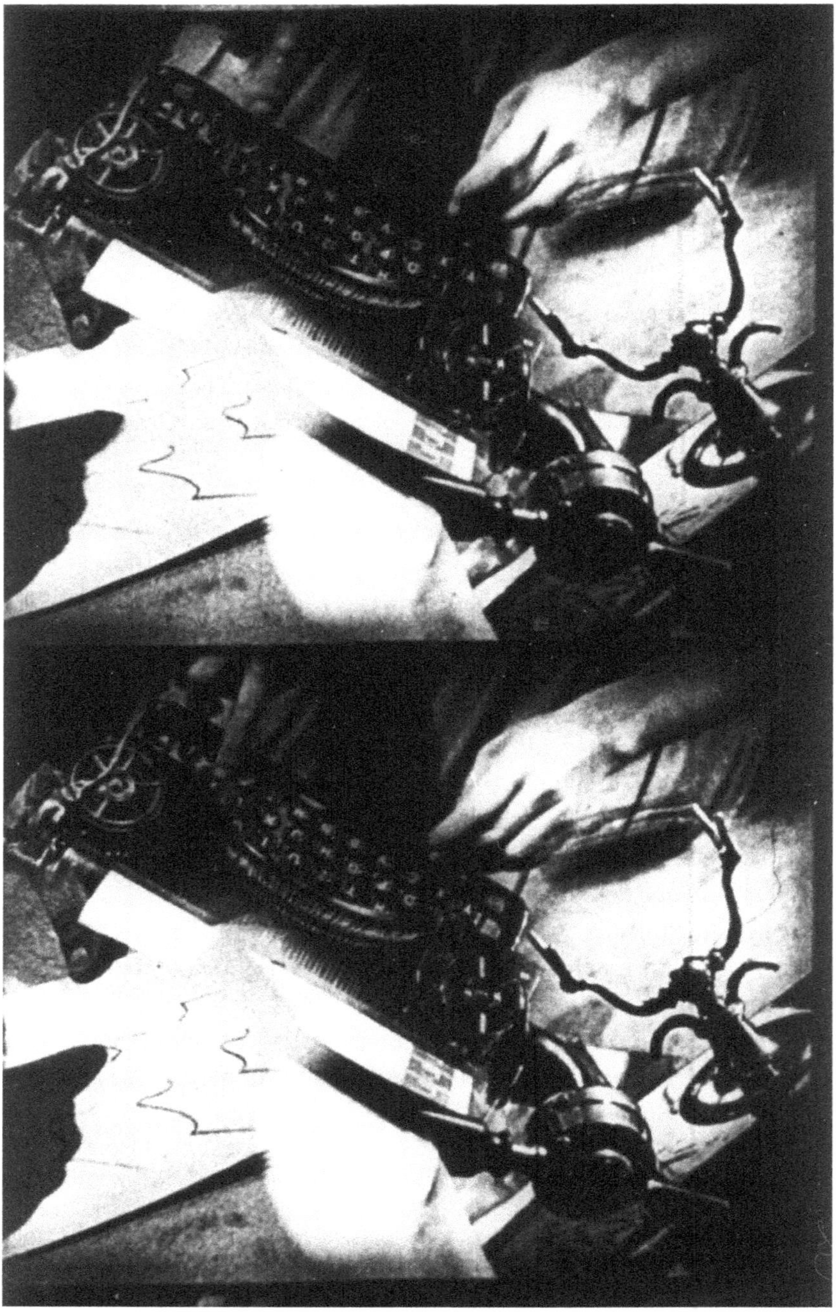

Fig. 2.2. Montage shot illustrating a step in the film production process in Guido Seeber's *Kipho* film (1925).
Source: Deutsche Kinemathek.

Ballet mécanique does not prompt the audience to create mental connections between them, while Seeber's Kipho film aims to achieve precisely that.

The individual ideograms of the Kipho film are relatively simple. They represent various steps in the film (and also photo) production process, including set construction, recording, and development, lighting, makeup, and costume, script writing, editing, title design, and projection. Each concept is captured in a single montage shot. The idea of projection, for instance, is represented by a fast, clockwise spinning shutter wheel in the bottom left corner, a slower, counter-clockwise spinning film reel in the top right corner and a projectionist closing the door of a lamp-house in the top left corner. In addition to the production process, the Kipho film shows filmic precursors like Anschütz's Tachyscope and makes references to recent German successes such as *Fridericus Rex* (1922/1923), *Die Nibelungen: Siegfried* (1924), *Der letzte Mann* (1924) and *Wege zu Kraft und Schönheit* (Ufa, 1925, dirs. Nicholas Kaufmann and Wilhelm Prager). Also included are text elements like "Liiight!," "Action!," a mock credit sequence, and the recurrent address "you" and "you must" which is eventually revealed as part of the phrase "You must go to Kipho!," an allusion to "You must become Caligari!" from *Das Cabinet des Dr. Caligari* (Decla, 1920, dir. Robert Wiene).[38]

The principal visual motif of *Film* is rotational motion. Everything is spinning: the film reels, shutters, camera, and tripod cranks, intermittent mechanisms, rewinds, processing drums, and praxinoscopes. The impression of omnipresent circular movement is intensified by the montage shots, which frequently juxtapose disks, cylinders, and cranks turning at different speeds and in different directions. Not only is much of cinematic hardware characterized by rotational motion, film technology also causes other things, like the takes on the editing table, Königsberg castle, carbon arc lights, or the film script, to revolve as well. The latter is additionally labelled "*Drehbuch*," which makes apparent that the film's ubiquitous circular motion also constitutes a verbal pun. After all, the German verb *drehen* means both "to rotate" and "to film." Beyond constituting a play on words, the incessant spinning of every part of the cinematic apparatus also brings to mind the Aristotelian view according to which uniform circular motion is perfect and eternal. Indeed, considering that nothing on earth revolves naturally around its own axis, the all-pervading rotations of its apparatus endow

38 For links between the *Caligari* reference, hypnosis, and advertising, see Hagener, *Moving Forward, Looking Back*, 139-140 and Michael Cowan, "Advertising, Rhythm and the Filmic Avant-Garde in Weimar Guido Seeber and Julius Pinschewer's Kipho Film," *October* 131 (Winter 2010): 23-50, 40-43.

the filmic medium with a transcendental quality. Observing them has a hypnotic effect. Simultaneously, rotational movement also lies at the core of the industrial revolution. Helmut Müller-Sievers has argued,

> It is the epochal achievement of nineteenth-century machines and their cylindrical components to have made rotation universally available and at the same time to have brought to light the limits of technics: it begins where the body ends. The machines born in the nineteenth century [...] disrupt the imaginary continuity of nature and human being and introduce with their motion a literally "inhuman" element into the world.[39]

The Kipho film makes tangible the links between cinema and kinematics. Notably its montage shots portray the medium as an intricate machine, yet one that is inseparable from its mesmerizing infinitude and the ever-present human eyes, hands, and arms that operate it. In the Kipho film, as Malte Hagener puts it, "cinema perfectly integrates the rational and industrial product with irrational and hypnotizing forces."[40] The film's montage shots convey a complex conception of cinema's nature, which exhibits close links to avant-garde explorations of the medium such as Vertov's *Chelovek s kinoapparatom* (1929). When rendering abstract thought by means of moving ideograms and making tangible the transcendent quality of cinematic motion, Seeber's Kipho film employs technology as a tool for capturing something in the ideal realm, thus actualizing techno-romantic tenets.

Seeber also used montage shots in narrative features like *Geheimnisse einer Seele* (Neumann-Film-Produktion/Ufa, 1926, dir. G.W. Pabst) and *Dirnentragödie* (Pantomim-Film, 1927, dir. Bruno Rahn), where the ideograms become substantially more complex. *Dirnentragödie*, one of Asta Nielsen's final films, tells the story of Auguste, an aging prostitute in love with a much younger man. Auguste plans a new life as a pastry shop owner with him, only to discover that during her absence he and her pretty young friend have locked themselves into her bedroom. The following forty-five-second-long climactic sequence condenses Auguste's anguish. Her downstairs neighbour is playing the hit waltz *Wenn die Liebe nicht wär'* ("If love didn't exist") by Walter Bromme, which Auguste requested to accompany her anticipated engagement celebration. To the sounds of the piano she breaks down in front of her locked bedroom door. The sequence consists of thirty-six shots

39 Helmut Müller-Sievers, *The Cylinder: Kinematics of the Nineteenth Century* (Berkeley, CA: University of California Press, 2012), 10-11.
40 Hagener, *Moving Forward, Looking Back*, 140.

that alternate between Auguste on the floor, the neighbour on the piano, animated titles of her pleas for him to stop, her fists drumming on the floor, and a series of nine, extremely brief shots of just a few frames each that summarize the events leading up to this moment. In addition, the sequence contains five montage shots that portray Auguste's subjective experience. They consist of keys, hammers, and pedals of the piano, a female ear, the score of *Wenn die Liebe nicht wär,'* the word "stop!" (*"Aufhören!"*), and eventually Auguste's face. The sequence integrates montage shots with sequential montage whose cutting rate challenges the viewers' perceptual capacity. Since viewers are not given time to fully grasp their meaning, they appear like subliminal memories of past events. The montage shots, in contrast, illustrate Auguste's sense perceptions. They function in two different registers: On the one hand, viewers are asked to evaluate the connections between the image components and grasp the situation accordingly. At the same time, the composites also have a strong visceral effect. This impression arises from the way the components are arranged and move in relation to each other. What is more, their emotional intensity increases over the course of the sequence. Initially, the audience is shown portions of the piano, then the ear, the writing, and the score. Eventually, Auguste's face, surrounded by piano parts, occupies the centre of the frame, and gradually increases in size. Covering her face with superimposed hands, she stares directly into the camera.

As the examples of the Kipho film and *Dirnentragödie* show, Seeber's montage shots employ trick technology to constitute a type of thinking in images that is at once apprehensible through ideation and sensory perception. Indeed, as Michael Cowan has noted, "Like Eisenstein's intellectual montage, Seeber's visual concepts occupied a middle ground between sensuous image and abstract thought."[41] Viewers are expected to generate meaning by establishing connections between image components. At the same time, the composites also facilitate visceral, intuitive comprehension. The synchronicity of impressions approximates our mental images and has an immersive effect. Montage shots share key characteristics with Eisenstein's conception of editing. They too seek "to create *a new quality of the whole from a juxtaposition of the separate parts*," by placing image fragments side by side and inviting viewers to synthesize them into abstract thought.[42] However, Eisensteinian montage

41 Cowan, "Moving Picture Puzzles," 204.

42 Sergei Eisenstein, "Dickens, Griffith and the Film Today [1944]," in *Film Form: Essays in Film Theory*, ed. and trans. Jay Leyda, 195-255 (New York, NY: Harcourt, Brace & World, 1949), 238.

is based on the succession of images and, as Yuri Tsivian has emphasized, "it is not so much shots per se that the montage director taps for meaning as cuts between shots."[43] In contrast to Eisensteinian montage, Seeber's montage shots generate meaning from direct juxtaposition and facilitate, as Seeber wrote in 1930, a "total impression."[44] John Belton's comparison of Eisenstein's and Murnau's cinemas thus also pertains more generally to the difference between montage shots and sequential, Soviet-style montage: "For Eisenstein, the shot is representational; it is a photo-fragment of reality whose meaning derives from its sequential combination with other shots. For Murnau, the shot becomes a unit of expression within which reality is transformed. Eisenstein's narration occurs between shots; Murnau's within them."[45]

Montage shots render possible the *concurrent* representation of the subjective and the objective, of action and reaction, of details and totality, thereby transcending the expressive limitations of "straight" photography. The placement of montage units side by side encourages viewers to contemplate—even if only for a brief moment—the relations between them. Because meaning is derived from concurrent juxtaposition, montage shots take on the character of self-contained images not unlike photomontages in motion. Like the photomontage, which Dada artist Raoul Hausmann called a "static film,"[46] but in contrast to sequential editing, montage shots rupture the illusion of the image as a view of reality, highlighting their own constructedness.

As a sophisticated trick technological tool for rendering abstract ideas and sensations, montage shots reflect Seeber's techno-romantic zeal for expanding the camera's ability to render the intangible. For him, the noblest task of the cameraman was the development of innovative cinematic devices and their imaginative application in the service of film art. Seeber's confidence in the camera as an artistic instrument, his commitment to technical excellence, perpetual quest for new discoveries, and belief in shared knowledge

Emphases in the original.

43 Yuri Tsivian, "Cyberspace and its Precursors: Lintsbach, Warburg, Eisenstein," in *Transmedia Frictions: The Digital, the Arts, and the Humanities*, eds. Marsha Kinder and Tara McPherson, 80-99 (Berkeley, CA: University of California Press, 2014), 90.

44 Guido Seeber, *Kamera-Kurzweil: Allerlei interessante Möglichkeiten beim Knipsen und Kurbeln* (Stuttgart: Union Deutsche Verlagsgesellschaft, 1930), 226.

45 John Belton, *Cinema Stylists* (Metuchen, NJ: Scarecrow Press, 1983), 14.

46 Raoul Hausmann, "Photomontage [1931]," in *The Weimar Republic Sourcebook*, eds. Anton Kaes, Martin Jay, and Edward Dimendberg, 651-653 (Berkeley, CA: University of California Press, 1994), 651.

set the tone for film technicians in the German film industry and shaped attitudes and objectives for generations to come.

Eugen Schüfftan: The art of the image

Like Seeber, Eugen Schüfftan was an avid innovator, problem solver, and camera virtuoso dedicated to techno-romantic tenets. In contrast to Seeber, however, who saw himself above all as a technician, Schüfftan's professional ethos was shaped by his background in painting, architecture, and set design. Nonetheless, he showed interest in film technological research early on and patented a motorized projector that could alternately project two reels in 1912.[47] After the war he collaborated with his former teacher, star architect Hans Poelzig on the acclaimed conversion of Max Reinhardt's Großes Schauspielhaus (1919) in Berlin. While working as an animator for the studio Deutsche Lichtbild-Gesellschaft, he started devising the trick technique that would make him famous. The first patent application specifying the Schüfftan process was filed in September 1922 and followed by three patents of addition.[48] They were held by Schüfftan's effects company and subsequently by the studio Tobis Film, which allowed them to expire in 1940.[49] Together, these patents detail the functional principles of Schüfftan's invention.

As illustrated in Figure 2.3., for a basic Schüfftan shot, a screen (9) is placed in front of the camera (2), while a live-action scene (4) is staged to the side of it. A front-silvered mirror (10) is positioned at a forty-five degree angle in front of the camera. Parts of the mirror's reflective surface are removed

47 Eugen Schüfftan, "Mein Verfahren," *Film-Kurier* 24 (19 November 1926).

48 Willy Köhler, *Verfahren und Einrichtung für kinematographische Aufnahmen*, Deutsches Reichspatent 428,589, filed 14 September 1922, and re-issued 10 May 1926; Willy Köhler, *Verfahren und Einrichtung für kinematographische Aufnahmen*, Deutsches Reichspatent 429,753 (addition to patent 428,589), filed 11 February 1923, and issued 2 June 1926; Willy Köhler, *Verfahren für kinematographische Aufnahmen*, Deutsches Reichspatent 429,754 (addition to patent 428,589), filed 10 May 1923, and issued 2 June 1926; Willy Köhler, Einrichtung für kinematographische Aufnahmen, Deutsches Reichspatent 429,755 (addition to patent 428,589), filed 4 March 1924, and issued 2 June 1926.

49 See entries in the patent register for DE 428,589, DE 429,753, DE 429,754 and DE 429,755 in the archive of the German Patent and Trade Mark Office, Berlin. See "Angebot der Universum-Film AG an die Deutsche Spiegeltechnik GmbH &Co, Berlin W9, Köthenerstrasee 1/4 vom 15. September 1926." *Files of the Deutsche Spiegeltechnik GmbH & Co. K.-G. Berlin* (R109/I2070). Bundesarchiv Berlin; "Auflösungsvertrag vom 28. Februar 1928." *Files of the Deutsche Spiegeltechnik GmbH & Co. K.-G. Berlin in the Bundesarchiv Berlin* (R109/I2070). Bundesarchiv Berlin.

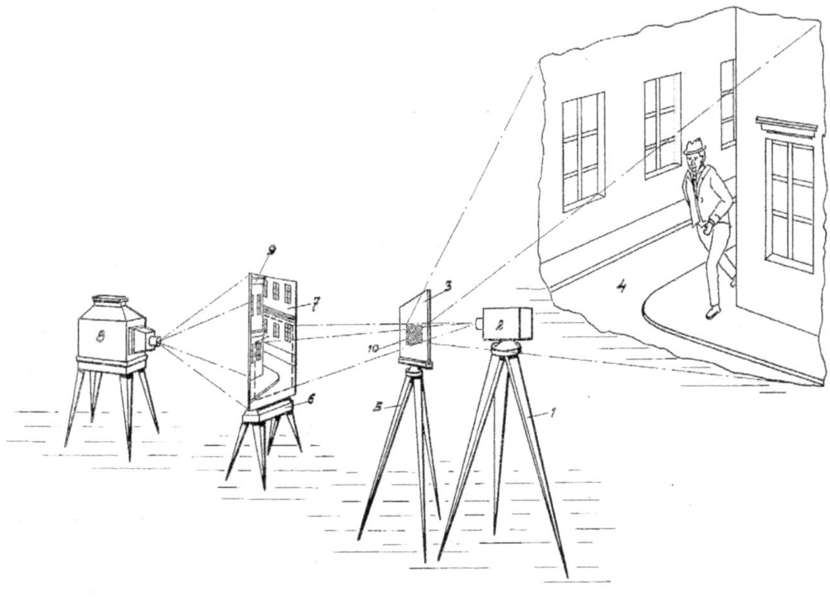

Fig. 2.3. Patent drawing of the Schüfftan process.
Source: Eugen Schüfftan, *Art of Making Motion Pictures*, U.S. patent 1,690,039, filed September 15, 1923, and issued October 30, 1928.

(3). A projector (8) projects an image (7) onto the screen, which the camera records through the transparent portions of the mirror. The live-action scene is reflected in the mirror's remaining silvered portions and replaces parts of the projected image in the camera's view. The mirror is positioned quite close to the camera's wide-angle lens, which is focused on infinity. As a result, the reflection of the live-action scene is in focus, whereas the mirror's surface itself is blurred. Optically, the two scenes merge and are recorded simultaneously. The positions of the live-action scene and the image can be reversed and instead of a rear-projected still image, a model, photograph, or rear-projected moving image can be used.

Schüfftan's goal was the creation of an all-purpose, indiscernible trick technical tool that offered new artistic means of expression as well as financial benefits.[50] He spent the 1920s refining, promoting, and implementing his invention, yet eventually abandoned his special-effects business venture. Instead, he launched a career as a cinematographer with *Menschen am Sonntag* (*People on Sunday*, Filmstudio 1929, 1930, dirs. Robert Siodmak et al.). Quickly establishing himself as a cameraman of distinction, Schüfftan

50 See Eugen Schüfftan, "Mein Verfahren," *Film-Kurier* 24 (19 November 1926).

photographed about a dozen films and directed one (*Das Ekel*, Ufa, 1931) before the National Socialists forced him into exile in 1933. He worked repeatedly with Max Ophüls, Robert Siodmak, and Georg Wilhelm Pabst, and photographed several masterpieces of poetic realist cinema, most notably *Drôle de drame* (Productions Corniglion-Molinier, 1937, dir. Marcel Carné) and *Le Quai des brumes* (Ciné-Alliance, 1938, dir. Marcel Carné).

Unlike Seeber, whose professional principles are easily discernible from his numerous articles and books, Schüfftan kept a low public profile. Most of what we know about his goals and convictions has been conveyed by his student, French camera legend Henri Alekan (1909-2001). Alekan, who assisted Schüfftan on nine productions during the 1930s, never tired of highlighting the "enormous influence" of "my master Eugen Shuftan" on his thinking and work as a cinematographer.[51] In 1984, Alekan published a magisterial work on lighting in cinematography, *Des lumières et des ombres,* realizing a book project that Schüfftan had envisioned but never completed.[52] Alekan emphasized that its "essential ideas are based on Schüfftan's lessons and it is therefore dedicated to him."[53] According to Alekan, it was Schüfftan "who revealed the art of the image to me."[54]

Schüfftan imparted two principles to Alekan, namely "the significance of painting for our profession,"[55] and "how the director of photography lights according to a theme, and every time differently."[56] As a former painter, Schüfftan held in highest regard the concurrence of artistic vision and technical skill he found in the work of the distinguished masters of the early modern period like da Vinci or Rembrandt. He taught Alekan the "observation of the composition and the lighting of the Old Masters of painting; transposition and application to the cinema of the rules revealed

51 Heidi Wiese and Henri Alekan, "Man muß die Technik überwinden, um zur Kunst zu gelangen: Ein Gespräch mit dem Kameramann Henri Alekan," in *Die Metaphysik des Lichts: Der Kameramann Henri Alekan,* ed. Heidi Wiese, 22-57 (Marburg: Schüren-Verlag, 1996), 52; Henri Alekan, *Le vécu et l'imaginaire: Chroniques d'un homme d'images* (Geneva: Publishing La Sirène, 1999), 15.

52 Henri Alekan, *Des lumières et des ombres* (Paris: Librairie du Collectionneur, 1991). For Schüfftan's book project, see Werner Zurbuch, "Eugen Schüfftan – Meister der Filmtechnik," *Filmtechnikum* (June 1962): 179-180; and Wiese and Alekan, "Man muß die Technik überwinden," in *Die Metaphysik des Lichts,* ed. Wiese, 52.

53 Wiese and Alekan, "Man muß die Technik überwinden," in *Die Metaphysik des Lichts,* ed. Wiese, 24.

54 Alekan, *Des lumières et des ombres,* 3.

55 Wiese and Alekan, "Man muß die Technik überwinden," in *Die Metaphysik des Lichts,* ed. Wiese, 40.

56 "Interview avec Henri Alekan," *Cinématographe* 68 (June 1981): 23-28.

by paintings."[57] Alekan perceived the resultant non-realist lighting designs as game changing: "Schüfftan invented and applied what I later called the *effect of aestheticizing light*," which "does not emanate from a specific natural source, but rather from the ideas of its creator, the ideas of the artist, who thinks there should be a light spot, even if it is not logical."[58]

While *Menschen am Sonntag* (1930) demonstrates Schüfftan's mastery in capturing the poetry of the everyday, for him cinematography was not about reproducing the physical world, but instead, in accordance with techno-romantic tenets, externalizing an artistic vision, rendered possible by the cinematographer's command of their technical tools. As Alekan put it, the filmic image is "not what you can also see in reality, but that which the cinematographer has thought up."[59] In France, Schüfftan reached the zenith of his career. Max Ophüls called him "the master and patriarch of German cinematographers at the time," which, given that the likes of Karl Freund, Carl Hoffmann, and Fritz Arno Wagner might have also been considered for the epithet, speaks to Schüfftan's stature.[60] However, the German invasion of France in 1940 forced him to flee once again, this time to the United States, where his career came to a virtual standstill. Because of Hollywood studios' closed shop labour environment and the rigid protectionism of the International Photographers Union I.A.T.S.E. Local 659, Schüfftan was only able to find sporadic and mostly uncredited employment as a cameraman or special-effects supervisor. After the war, he worked primarily in Europe, but could never reclaim his former position.

In contrast to Seeber, who was above all concerned with the camera as an image-forming device, Schüfftan aimed his attention at shaping the onscreen space. As Alekan's recollections indicate, in his capacity as cinematographer, Schüfftan was preoccupied with questions of composition and light. As a film technician, he utilized the laws of optics to composite spatially disjoined, virtual, and real image components. As opposed to Seeber's montage shots, which generate meaning from the conspicuous juxtaposition of pictorial elements, Schüfftan shots create an illusion of coherent diegetic space. They

57 Henri Alekan quoted in René Prédal, "Les Grands Opérateurs (IV): Henri Alekan," *Cinéma* 173 (February 1973): 88-95, 93.

58 Wiese and Alekan, "Man muß die Technik überwinden," in *Die Metaphysik des Lichts*, ed. Wiese, 37. Emphasis in the original.

59 Wiese and Alekan, "Man muß die Technik überwinden," in *Die Metaphysik des Lichts*, ed. Wiese, 38.

60 Max Ophüls, *Spiel im Dasein: Eine Rückblende* [1959], ed. Helmut G. Asper (Berlin: Alexander Verlag, 2015), 112.

capitalize on the tension between the actual flatness of the filmic image and the imaginary three-dimensionality of onscreen space.[61] As Schüfftan explained in one of his rare interviews:

> My invention relied on bringing together objects in space. Space is the basis of my invention. [...] In painting one always strove for three-dimensionality. Film was actually an expression of the depth of the image in space. I was always interested in the image in space, which became possible through cinematography. The life of objects in space.[62]

Compositing with mirrors

Schüfftan's compositing tool, the mirror, is distinguished by its ostensibly paradoxical nature: The device appears simultaneously rational and metaphorical, scientific and mystical. Rainer Maria Rilke famously coined the following aphorism: "Mirrors: no one who's tried to solve you has yet unlocked your true being."[63] In philosophy and art, mirrors have served as an important epistemological symbol that can equally stand for deception or (divine) truth. They stand for the soul or self as well as the other or a world beyond, thus linking the outside and the inside. At the same time, mirrors have played a crucial role in science and engineering. Archimedes is said to have burned down the Roman fleet during the Second Punic War using mirrors as a parabolic reflector. Mirrors are key components of Newtonian telescopes, periscopes, heliographs, searchlights, arc lamps and more recently of photovoltaic systems, interferometers, solar sails, lasers, high definition television sets, and numerous moving image projection technologies.

The mirror embodies what Tom Gunning has described as the "central paradox of modern optical magic: It can be simultaneously rational in its

61 For an in-depth discussion about filmic space and techniques of depth, see Jacques Aumont, Alain Bergala, Michel Marie, and Marc Vernet, *Aesthetics of Film* [1983], transl. Richard Neupert (Austin, TX: University of Texas Press, 2004), 9-25.

62 Eugen Schüfftan quoted in Helmut G. Asper, "'Man muss eben struggeln, um oben zu bleiben.' Eugen Schüfftan und Siegfried Kracauer im amerikanischen Exil," in *Nachrichten aus Hollywood, New York und anderswo: Der Briefwechsel Eugen und Marlise Schüfftans mit Siegfried und Lili Kracauer*, ed. Helmut G. Asper, 3-30 (Trier: Wissenschaflicher Verlag, 2003), 5.

63 Rainer Maria Rilke, *Sonnets to Orpheus* [1922], trans. Edward Snow (New York, NY: North Point Press, 2004), 65.

method and seemingly supernatural in its effect."[64] Maybe it is precisely for this reason that mirrors are integral parts of optical toys like Kaleidoscopes and Praxinoscopes and fairground attractions like halls of mirrors and mirror mazes. Nineteenth-century magicians relied on mirrors for various levitation, vanishing, decapitation, and multiplication acts. The best-known trick for combining visual components live on stage makes use of a partially reflective mirror (i.e., a pane of glass) and is commonly referred to as Pepper's Ghost.[65]

Mirrors were not only omnipresent in optical illusions and stage magic, they were also used long before Schüfftan for image compositing in photography and film. Widely publicized, for instance, was the mirror trick behind the confrontation between a human-sized smoker and live-action miniature fairies in the most celebrated trick film of its day, *Princess Nicotine; or, The Smoke Fairy* (Vitagraph, 1909, dir. J. Stuart Blackton).[66] Here, a mirror that reflected actors performing next to the camera was imperceptibly integrated into the set. As the example of *Princess Nicotine* suggests, the most important innovation of the Schüfftan process was the placement of the mirror in the foreground rather than in the background of the scene while avoiding superimposition effects à la Pepper's Ghost. Because mirror effects were common in filmmaking early on, the novelty of the Schüfftan process was periodically called into question. Special-effects pioneer Norman O. Dawn and cinematographer Carl Louis Gregory, for instance, claimed to have employed the technique previously.[67] American director J. Searle Dawley held a patent that was sometimes mentioned as evidence for Schüfftan's lack of originality.[68] However, as his patent shows, Dawley failed to solve the central problem of all mirror composites, namely how to integrate reflection and live-action scenes indiscernibly. In fact, there is no evidence

64 Tom Gunning, "Uncanny Reflections, Modern Illusions: Sighting the Modern Optical Uncanny," in *Uncanny Modernity: Cultural Theories, Modern Anxieties*, eds. Jo Collins and John Jervis, 68-90 (New York, NY: Palgrave Macmillan, 2008), 82.

65 For the origins of the Pepper's Ghost illusion, see Jeremy Brooker, "The Polytechnic Ghost: Pepper's Ghost, Metempsychosis and the Magic Lantern at the Royal Polytechnic Institution," *Early Popular Visual Culture* 2 (July 2007): 189-206.

66 J. Stuart Blackton and Albert E. Smith, "Some Tricks of the Moving Picture Maker," *Scientific American*, (26 June 1909): 476. See also Frederick A. Talbot, *Moving Pictures: How They Are Made and Worked* (London: William Heinemann, 1912), 242-53.

67 "Inventory List." *Norman O. Dawn Collection of Special Effects Cinematography.* Harry Ransom Center at the University of Texas at Austin; Carl Louis Gregory, "Trick Photography Methods Summarized," *American Cinematographer* (June 1926): 21–22.

68 James Searle Dawley, *Art of Making Motion Pictures*, U.S. patent 1,278,117, filed 17 August 1914, and issued 10 September 1918.

that others anticipated Schüfftan's approach to mirror compositing and he consequently won all patent litigations.[69]

Between 1924 and 1930, Schüfftan was granted at least forty patents in seven leading film nations for variations of this process. Many are quite complex, involving multiple mirrors, including tessellated reflecting fragments, sliding, and two-way mirrors, lenses, masks, prisms, and projectors. They facilitate the simultaneous or successive combination of a variety of visual components, such as live-action scenes, photographs, models, and projected images, both still and in motion. During the 1920s, Schüfftan also obtained multiple patents for other compositing devices and techniques, including optical and contact-printing methods, a dual-lens photographic system, and two travelling matte processes. Most film technicians only patented a small number of their innovations: Although an industrious inventor, Seeber, for instance, took out only a total of six patents in two countries. Schüfftan's patenting efforts, in contrast, rival those of Germany's film-technological pioneer Oskar Messter and testify not just to his ardour as an experimenter but also his commercial ambitions on an international scale.

The Schüfftan is versatile and the process provided filmmakers with numerous new opportunities. In particular, supernatural events that were traditionally represented by dissolves could be depicted more convincingly. For example, the Schüfftan process made it possible for characters to walk through walls, as can be seen in F. W. Murnau's *Faust* (Ufa, 1926) or in the much later East German television series *Spuk im Hochhaus* (DEFA, 1982, dir. Günter Meyer). It also portrayed metamorphoses more persuasively than any other existing technique of the 1920s. This is particularly apparent from its first application in Fritz Lang's *Siegfried* (Ufa, 1924).

Here, the eponymous hero slays Alberich, king of the Nibelungs, whereupon the dwarfs who are chained to the king's treasure basin turn into stone (Figures 2.4.-2.6.). To achieve the dwarfs' petrifaction, according to art director Erich Kettelhut, two treasure basins were constructed, one in front of the camera for the live actors and a replica, complete with petrified dwarfs, was situated to the side.[70] Two mirrors were used: one fixed and fully reflective and one that was partially transparent and could slide up and down. Initially, the live-action scene was shot through the transparent portions of the sliding mirror. At the moment of

69 See Rolf Giesen, "Eugen Schüfftan," in *Cinegraph: Lexikon zum deutschsprachigen Film*, ed. Hans-Michael Bock, vol. 7 (Munich: edition text + kritik, 1984-), E2.

70 Erich Kettelhut, *Der Schatten des Architekten*, ed. Werner Sudendorf (Munich: Belleville, 2009), 93.

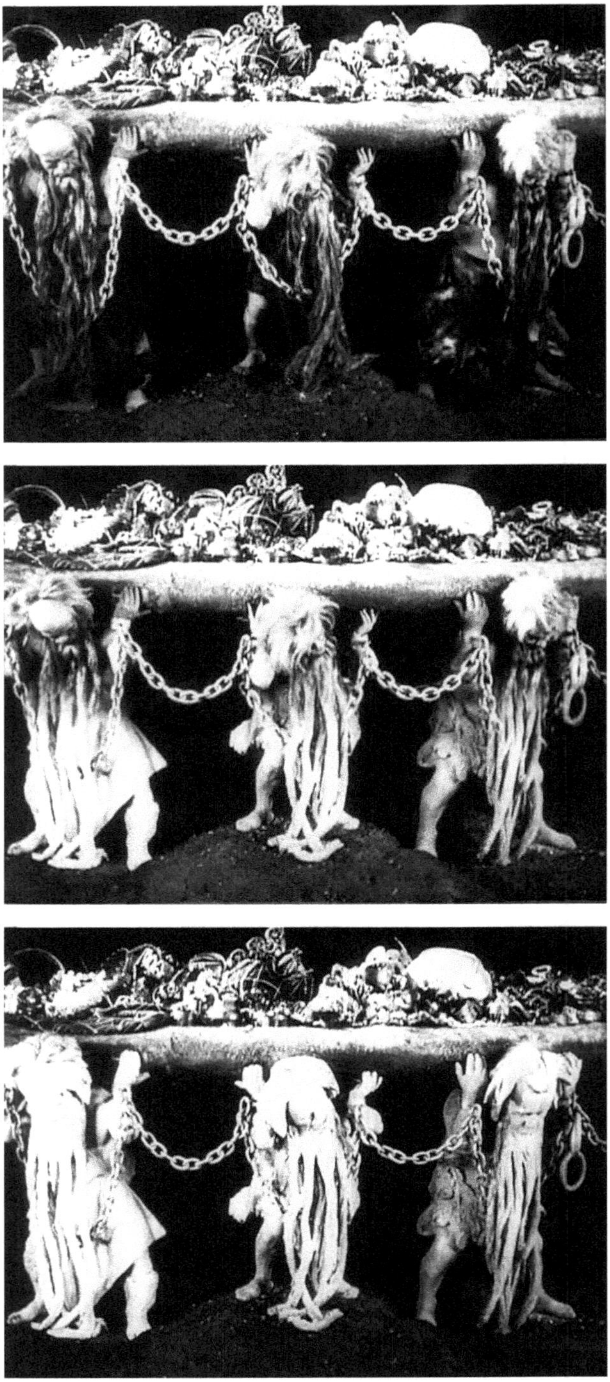

Figs. 2.4-2.6. Petrification of the dwarfs in *Siegfried* (1924), facilitated by the Schüfftan process. Courtesy of Friedrich Wilhelm Murnau Foundation.

transformation, the mirror slowly moved up, allowing the reflective part to replace the live actors with their plaster copies. Conventionally, this effect could have been realized by means of a dissolve from the live actors to their stone replicas. The Schüfftan process, however, facilitated a previously unattainable effect: The viewers can follow the metamorphosis in "real time." We see the dwarfs' feet turn to stone, then the tips of their long beards, then their midsections and their chains. We observe the horror on their faces as they track—like us—step by step what is happening to them. This bottom-up transformation seems to approximate how such an event might look in real life, which gives the shot an eerily authentic quality and heightens the viewers' emotional engagement.[71]

The Schüfftan process not only lends itself to fantastic transformations, but also to the portrayal of extreme size differences. Although *Gulliver's Travels*, one of Schüfftan's pet projects and originally a key selling point of his invention, was never realized with its help, the Schüfftan process was used to this effect repeatedly after World War II. Colour films pose particular challenges to colour compositing techniques like blue screen and filmmakers therefore resorted to Schüfftan shots to portray meetings with persons of unusually large or small stature. For instance, the Schüfftan process was used to depict the hero's encounters with the little glass imp in *Das kalte Herz* (DEFA, 1950, dir. Paul Verhoeven), the Greeks' perilous confrontation with the Cyclops Polyphemus in *Ulisse* (Lux Film/Paramount Pictures et al., 1954, dir. Mario Camerini), Darby's visit to the leprechaun cave in *Darby O'Gill and the Little People* (Walt Disney Productions, 1959, dir. Robert Stevenson), or the giant magic dogs in *Das Feuerzeug* (DEFA 1959, dir. Siegfried Hartmann). As Schüfftan later suggested, the invention was originally prompted by techno-romantic impulses. He had intended to employ film technology to facilitate representations of the unreal and impossible, to "visualize the imagination."[72] At the same time, however, Schüfftan was fully aware that the principal selling point of his process was cost reduction and increased production value.[73]

Although the Schüfftan process had been used to great effect in *Siegfried*, this application was not publicized. Schüfftan began promoting the technique in the context of Karl Grune's *Eifersucht* (Stern Film, 1925), where it was

71 The effect was recreated in the East German fairy-tale production *Der Teufel vom Mühlenberg* (DEFA, 1955, dir. Herbert Ballmann), where a wicked miller is turned into a pillar of stone.

72 Gertrud Isolani, "Gespräch mit Eugen Shuftan," *Basler Nachrichten* (19 October 1965): 9.

73 Eugen Schüfftan, "Verbesserungen in der Miniatur-Aufnahmetechnik," in Friedrich Porges, *Mein Film-Buch*, 35-37 (Vienna: Mein Film Verlag, 1929), 36.

employed to extend the set of a theatre building.[74] Critics were impressed with the overall execution of the shot, yet also criticized its lack of aerial perspective.[75] Set extensions subsequently became the technique's primary application area. The monumental cityscapes of Fritz Lang's *Metropolis* (1927), as we will see in Chapter 5, would have been inconceivable without reflections of three-dimensional models completing sets like the stadium in the Club of Sons, the workers' apartment buildings, and the interior of the cathedral.

For optical reasons, the Schüfftan process could initially only be used with large three-dimensional models. For instance, production photos show that the miniature engines in the upper part of the machine halls of Metropolis were over three metres high. Schüfftan was eventually able to reduce the model scale necessary to keep all image components in focus by a factor of four by placing a lens between the camera and the mirror.[76] The lens also helped when utilizing photographs and films instead of miniatures. For E. A. Dupont's first American feature, *Love Me And The World Is Mine* (Universal Pictures, 1927), which is set in Vienna, for example, all location scenes were realized with reflected photographs.[77] Similarly, because the lighting conditions inside the British Museum prevented shooting on location, the climactic final sequence of Alfred Hitchcock's *Blackmail* (BIP, 1929) relied on photographs for nine shots. As Hitchcock explained to François Truffaut and Peter Bogdanovich, the photographs, which required exposure times of thirty minutes, were turned into backlit transparencies of about twelve by fourteen inches in size and combined with the live action by means of Schüfftan mirrors.[78] Given that most of the frame was replaced, there was, as Hitchcock recalled, "barely any set that could be seen on the stage."[79]

Indeed, observers were often struck by the "alarmingly empty"[80] stages they encountered when the Schüfftan process was being employed. This contributed to a general sense that this new technique implicated a radical

74 See illustration in A. Lion, "Das Schüfftan-Verfahren: Umwälzung in der Filmaufnahme," *Die Umschau* (1927): 248-251.

75 See for instance W...f, "Zum ersten Male Schüfftansches Patent," *Die Kinotechnik* 18 (25 September 1925): 456.

76 See Schüfftan, "Verbesserungen," in Porges, *Mein Film-Buch*, 36.

77 See Leo Wiltin, "Die Technik," *Film-Kurier* 24 (19 November 1926).

78 François Truffaut, *Hitchcock* (New York, NY: Simon & Schuster, 1967), 64; Peter Bogdanovich, *Who The Devil Made It* (New York, NY: Alfred A. Knopf, 1997), 495. See also Tom Ryall, "Blackmail," in *British Film Institute Film Classics, vol. 1*, eds. Rob White and Edward Buscombe (London: British Film Institute, 2003), 99.

79 Hitchcock in Bogdanovich, *Who The Devil Made It*, 495.

80 "Das Ende der Dekoration," *Kinematograph* 950 (3 May 1925).

change in conventional film production. "[Karl] Grune rightly said the future would bring the 'empty studio.' [...] Today it is already a cinch to predict that the future belongs to this invention [the Schüfftan process] and that the set designer and the huge set will be ousted with its help.'[81] In the imagination of many in the film industry, the Schüfftan process heralded a future shaped by virtual images.

A new era of film production

Precisely because the Schüfftan process was taken as a harbinger of radical change, reactions differed widely. On the one hand, the technique was perceived as a hallmark of German technological leadership and as such inspired considerable national pride. Furthermore, for observers like Austrian journalist Friedrich Porges the process manifested as techno-romantic imagination incarnate:

> The man, the magician of the castles in the air, is called Schüfftan. His process is known as 'Schüfftan film model and mirror reflex process.' The effect, however, is poetic. Schüfftan conjures fairy palaces onto the screen, buildings of fantastic proportions, which never give the impression of hocus-pocus. They look like veritable, realistic monumental buildings. [...] Romanticism has united with technology to realize the extraordinary.[82]

Also prominent filmmakers expressed their delight with the Schüfftan process, particularly about the prospect of eliminating the prevailing expensive and cumbersome monumental sets. Karl Freund hoped that virtual sets would allow cinematographers to return their attention to their core province, to image composition and to actors.[83] Similarly, art director Walter Reimann seemed weary of the present material-intensive mode of production:

> The film of the future will no longer need huge studios and warehouses full of props. The moving camera, the mirror image, the painted backdrop, veils, smoke, pyrotechnics, curtains and a few rags – and in between

81 "Das Ende der Dekoration," *Kinematograph* 950 (3 May 1925).

82 Friedrich Porges, "Luftschlösser im Film," *Die Bühne* (Vienna) 120 (25 February 1927).

83 Alexander Kossowsky, "Die Männer der Kurbel IV: Karl Freund," *Film-Kurier* 126 (30 May 1925).

> human beings, real human beings, with incredible expressivity and light in all nuances.[84]

Freund and Reimann saw the Schüfftan process as a pathway to a more authentic, intimate, and poetic style of filmmaking. Seeber, in contrast, regarded it with considerable contempt. His aversion highlights an important difference between two kinds of effects that, given that they are all achieved on the original negative, are often classified together under the label "in-camera effects." As we have seen, Seeber was devoted to techniques that manipulate the image photo- or cinematographically such as multiple exposure composites, stop-action, slow motion, or time-lapse photography. These effects tend to acknowledge alterations to the image and may give cinematographers the opportunity to put their virtuosity on display. Seeber was less appreciative of techniques like forced perspective, glass shots, hanging miniatures, front and rear projection, and Schüfftan shots, which are based on monocular cues of depth perception and function like optical illusions. They are contingent upon the camera's vision, but, in contrast to the first group, do not rely on the camera as an image capturing and rendering device. The illusive scene, which is typically not supposed to be discernible as such, is set up *for* the camera, but it is recorded regularly. As a result, as Seeber remarked, "the cameraman has to content himself with turning his handle in the usual way. His personal skills come less into their own, because the process [...] already prefabricates everything before the take."[85] Because of the central role of the scenery in the creation of such perceptual effects, during the silent era they were typically created under the aegis of the art director in collaboration with the cinematographer or, particularly later on, entrusted to specialized personnel.

While Seeber found the Schüfftan process for the most part simply uninspiring, other members of the industry viewed it as a veritable threat to their livelihood. Trade press articles sought to assure fearful staff that the new technique was no job killer:

> Reflections of miniatures or actors can save a lot of time for multiple exposure composites and construction material, but they will not put anyone out of a job. After all, the foundation structures have to be built anyways and the combination of mirror views from three or four different

84 Walter Reimann, "Filmarchitektur – heute und morgen," *Filmtechnik* (20 February 1926).
85 Seeber, *Der Trickfilm*, 141-142.

> settings requires many operating crews to supervise the orientation of the models and mirrors and the precise illumination of the total image [...].[86]

Not only workers but also studio managers anticipated a major impact on film production, which is apparent from Ufa's major investments in the Schüfftan process.

In April 1925, the studio acquired the license to commercialize the Schüfftan patents with a view to the anticipated set expenditures for *Metropolis*, which started principal photography only one month later.[87] Ufa also employed the process in films shot alongside *Metropolis* like *Varieté* (1925 , dir. E. A. Dupont), *Manon Lescaut* (1925, dir. Artur Robison), and *Die Brüder Schellenberg* (1926, dir. Karl Grune). Shortly before *Metropolis* concluded principal photography in October 1926, a joint venture between Ufa and a Schüfftan-founded share corporation was established. The new limited partnership was named Spiegeltechnik GmbH & Co. and was supposed to produce made-to-order Schüfftan shots for the entire German film industry. As one of Spiegeltechnik's shareholders, Ufa had an exclusive contract that entitled the studio to 1000 metres of Schüfftan footage (approximately 45 minutes of running time) per year for an annual guarantee sum of 100,000 Reichsmark (equivalent to $25,000 in 1927). All other licensees were charged around 125 Reichsmark ($31.25 in 1926) per metre of Schüfftan footage produced, which covered the license, rental fees for mirrors and cameras, salaries for the technicians, and expenses.[88]

German studios took advantage of Spiegeltechnik's services, but not at the anticipated rate. In 1927, 242 features were produced in Germany and around twenty of them contained Schüfftan shots.[89] Considering the

86 Waldemar Lydor, "Kristallene Filmbauten," *Das Kino-Journal* (Vienna) 848 (30 October 1926): 14-15.

87 See "Bericht über Zwischenrevision der Geschäftsbücher und der Bilanz per 28. Februar 1927." *Files of the Deutsche Spiegeltechnik GmbH & Co. K.-G. Berlin* (R109/I2455). Bundesarchiv Berlin.

88 See "Angebot der Universum-Film AG an die Deutsche Spiegeltechnik GmbH &Co, Berlin W9, Köthenerstrasee 1/4 vom 15. September 1926." *Files of the Deutsche Spiegeltechnik GmbH & Co. K.-G. Berlin* (R109/I2070). Bundesarchiv Berlin; "Auflösungsvertrag vom 28. Februar 1928." *Files of the Deutsche Spiegeltechnik GmbH & Co. K.-G. Berlin in the Bundesarchiv Berlin* (R109/I2070). Bundesarchiv Berlin.

89 Exact numbers cannot be ascertained because Ufa's reports list studios as clients. See "Bericht ... 28. Februar 1927"; "Bericht über Revision der Geschäftsbücher und der Bilanz für das am 30. September 1927 abgelaufene Geschäftsjahr." *Files of the Deutsche Spiegeltechnik GmbH & Co. K.-G. Berlin* (R109/I2456). Bundesarchiv Berlin. For the German feature film production see Alexander Jason, *Handbuch der Filmwirtschaft: Filmstatistiken und Verzeichnisse der*

expectations on the part of Ufa, this outcome was disappointing. What is more, the studio's own production units actively avoided the process. A 1927 internal report pointed out:

> Certain circles within [Ufa's] production departments are sabotaging the Schüfftan process. Despite the sympathetic support from the head of production, Dr. Grau, it has not been possible to obtain orders from directors or set designers, even though there had been plenty of opportunities to shoot scenes using the Schüfftan process.[90]

Senior management was evidently caught off guard by staff opposition to the Schüfftan process, which likely resulted from workers' fear of losing their livelihood. What is more, Ufa had been ambitious regarding its demand estimation. In order for the arrangement with Spiegeltechnik GmbH & Co. to pay off, approximately half of Ufa's features (twenty-nine titles in 1925) would have had to include around three minutes of Schüfftan footage each.[91] In actual fact, however, only about one minute of Schüfftan footage was used in one quarter of Ufa's features in 1925. The investment in Spiegeltechnik was deemed unprofitable and Ufa withdrew its participation in February 1928.[92] Subsequent Ufa productions like *F.P.1 antwortet nicht* (1932, dir. Karl Hartl), *Amphitryon* (1935, dir. Reinhold Schünzel) or *Münchhausen* (1943, dir. Josef von Báky) continued to employ the process, but the studio no longer backed Schüfftan's business venture.

Schüfftan's audacious yet ultimately unsatisfactory partnership with Ufa bears some resemblance with his concurrent cooperation with Universal Pictures. Carl Laemmle became aware of the Schüfftan process in August 1925 and obtained a two-year license for commercializing the patents in North America, "primarily to film *Gulliver's Travels*," as the New York Times reported.[93] The scenes from *Gulliver's Travels* featured in Schüfftan's demo reel had apparently convinced Laemmle that a film adaptation had become finally feasible.[94] He announced that *Gulliver's Travels* was to

Filmschaffenden, Filmfirmen, der Filme und der Tonfilmkinos, vol. 2, Film-Europa (Berlin: Verlag für Presse, Wirtschaft und Politik, 1931), 17.

90 See "Bericht ... 28. Februar 1927."

91 For Ufa's 1925 annual production, see Kinematheksverbund, Arbeitsgruppe Deutsche Filmografie, "Deutsche Spielfilme 1925: Jahresproduktion und Filmbestand Bundesarchiv," Bundesarchiv-Filmarchiv, October 2007.

92 See "Auflösungsvertrag vom 28. Februar 1928" and "Bericht ... 30. September 1927."

93 See Frederic Wynne-Jones, "When the Camera's Eye Lies for Entertainment," *The New York Times* (2 May 1926): X5.

94 See Herbert Moulton, "Magical Effects Brought to Screen by Unique Process," *Los Angeles Times* (18 April 1926): C25–6.

become one of three Universal "Super-Jewels" in celebration of the Swift bicentenary in 1926.[95] Starting in late 1925, Schüfftan and his collaborator Ernst Kunstmann spent about ten months in Hollywood where they worked on eight Universal productions, including *Rolling Home* (dir. William A. Seiter), *The Ice Flood* (dir. George B. Seitz), and *The Love Thief* (dir. John McDermott). For the independently produced Corinne Griffith vehicle *Into Her Kingdom* (dir. Svend Gade) they also created effects in Technicolor. Still, the *Gulliver*-project was eventually abandoned and Universal did not continue its collaboration with Schüfftan. On the one hand, as Schüfftan recalled, American producers were quite open to his invention.[96] On the other hand, however, as I will discuss in more detail in Chapter 6, American trade press articles about foreign film technology reveal much contempt with unmistakable xenophobic undertones and it is hardly surprising that the technique did not prevail in this environment.

As opposed to the resistance encountered in Hollywood and Berlin, the Schüfftan process prospered in Britain. In January 1927, British National, which was promptly taken over by John Maxwell's British International Pictures (BIP), acquired the world rights to the Schüfftan patents (outside of Canada and the States).[97] Replicating Schüfftan's German business strategy, a subsidiary company, the British Schüfftan Process Co, Ltd, was established, which offered Schüfftan shots to all interested studios.[98] The process was first employed in Herbert Wilcock's *Madame Pompadour* (British National, 1927) and subsequently in films like E. A. Dupont's *Atlantic* (BIP, 1929) and *Things to Come* (London Film, 1936, dir. William Cameron Menzies). Most of Alfred Hitchcock's British films, including *The Ring* (BIP, 1927), *Rich and Strange* (BIP, 1932), *The Man Who Knew Too Much* (Gaumont British, 1934) and *The 39 Steps* (Gaumont British, 1935), contain multiple Schüfftan shots each. Indeed, the Schüfftan process served as the special-effects staple in British cinema into the 1940s. Art director Edward Carrick for instance noted: "I have employed this process for innumerable subjects and am convinced that when care is taken in its use it is the most adaptable and controllable process in use to-day."[99]

95 See Grace Kingsley, "Flashes," *Los Angeles Times* (17 September 1925): A11.

96 Eugen Schüfftan, "Mein Verfahren," *Film-Kurier* 24 (19 November 1926).

97 See "British National Strengthened," *Kinematograph Weekly* (13 January 1927): 56; and Rachel Low, *The History of British Film: 1918-1929* (London: George Allen & Unwin, 1971), 186.

98 See Hans Nieter, "The Schüfftan Process of Model Photography," *Photographic Journal*, London (January 1930): 16; and "Schufftan Process Acquired," *Kinematograph Weekly* (13 January 1927): 56.

99 Edward Carrick, *Designing for Films* [1941] (London: The Studio Publications, 1949), 99.

In mainstream filmmaking, the Schüfftan process was eventually superseded by more automatized techniques like chroma keying and rear projection. At the same time, the method was a crucial factor in the development of front projection compositing systems, the first of which were patented in the 1930s, but not widely used until decades later.[100] In the 1950s, Will Jenkins and Henri Alekan pioneered the Scotchlight system of front projection, which offered several advantages over chroma keying and rear projection. Both referenced the Schüfftan process as a precursor technique in their patent applications.[101] The Schüfftan process remained in use especially where resources were scarce like low-budget film production, amateur filmmaking, and analogue television. For instance, filmmakers working with small budgets like Mario Bava in *Terrore nello spazio* (AIP (Italy) et al., 1965) or Roberto Rossellini in *La Prise de pouvoir par Louis XIV* (ORTF (France), 1966) and *Socrate* (RAI (Italy), 1971) used Schüfftan shots for set extensions. Even in the digital age, filmmakers have turned to the Schüfftan process. Visual effects supervisors Robert and Dennis Skotak utilized it in James Cameron's *Aliens* (Twentieth Century Fox Film Corporation et al., 1986) and, reportedly, it was also employed in Peter Jackson's *The Lord of the Rings: The Return of the King* (New Line Cinema et al., 2003).[102] The Skotaks explain their predilection for methods like the Schüfftan process in the following way: "Even in the digital age, these techniques still work perfectly, and you can get your shot done in one take without any additional processes."[103] This assessment points to some of the causes behind the cult following that the Schüfftan process has gained from cinephile practitioners and audiences in recent years: The principles of the method are simple, yet its effects can be startling. Viewers can easily appreciate its mechanisms and impressions, while practitioners can realize basic applications during

100 See Walther Thorner, *Verfahren und Vorrichtung zur Herstellung kinematographischer Kombinationsaufnahmen*. Deutsches Reichspatent 598,712, filed 19 May 1932, and issued 18 June 1934.

101 Will F. Jenkins, *Apparatus for Production of Light Effects in Composite Photography*. United States Patent 2,727,427, filed 3 March 1952, and issued 20 December 1955; Henri Alekan, *Procédé et dispositif de prise de vues combinées*. République Française Brevet d'invention 1,098,128, filed 7 January 1954, and issued 2 March 1955.

102 Sheldon Teitelbaum, "Special Effects: Aliens," *Cinefantastique* 16 (October 1986): 122; For the use of the Schüfftan process in *The Lord of the Rings: The Return of the King*, see Carl Casinghino and Geoffrey Nowell-Smith, who do not reveal their sources nor the specific shots, however. Carl Casinghino, *Moving Images: Making Movies, Understanding Media* (Clifton Park, NY: Delmar, 2010), 309; Geoffrey Nowell-Smith, *The History of Cinema: A Very Short Introduction* (Oxford: Oxford University Press, 2017), 23.

103 Quoted in Richard Rickitt, *Special Effects: The History and Technique* (New York, NY: Billboard Books, 2007), 91.

principal photography without a complex technical apparatus or specialized labour. Arguably a factor in the interest in the Schüfftan process is also its use of mirrors, which are simultaneously commonplace and a timelessly fascinating tool of optical magic. These features endow the Schüfftan process with an air of tangibility and quaintness and dovetail with a more general sense of nostalgia for bygone eras of (film) history.

Guido Seeber and Eugen Schüfftan each had a transformative impact on the production, perception, and evolution of special effects in Germany and beyond. Seeber was a technician at heart, but also saw the cinematographer as an artist, a visionary master of camera technique. Accordingly, he favoured methods like multiple exposure composites that allow the cinematographer to excel both technically and creatively. As we will see in the next chapter, his doppelganger shots in *Der Student von Prag* not only surpassed all special effects feats hitherto seen on German screens, they also corroborated the techno-romantic paradigm by highlighting technology's capacity to reveal ideas and emotions, which in turn allowed the film to meet key stipulations of art. As a mentor and a publicist, Seeber was instrumental in establishing standards of effects production, elevating the status of technicians, and defining the key position of technology in German art cinema. Compared to Seeber's general and pervasive influence, Schüfftan's contribution was more distinct. He invented the only widely used commercial special-effects technique originating in Europe. Guided by the sensibility of a visual artist, Schüfftan concerned himself with forging convincing composite spaces on screen, which made his method particularly suitable for inconspicuous uses. The Schüfftan process established set extensions as standard practice in Europe and thus launched a new era of effects production. Seeber and Schüfftan can be regarded as representatives of two consecutive generations of film technicians, Seeber typifying the camera pioneer and Schüfftan the later effects specialist. Nonetheless, in similar ways Seeber's photographic and Schüfftan's perceptual effects construe technology as cinema's core creative tool and the cinematic image as fundamentally malleable, a stance that can be characterized as the conceptual bedrock of German silent film culture. Both shared techno-romantic views, which is apparent from their devotion to the goal of film art and commitment to devising medium-specific means for transcending material reality and expressing emotions and ideas. Like all great German film technicians, Seeber and Schüfftan understood themselves as master-craftsmen who, on account of their experience, eminent skill, and ingenuity, were capable of transcending physical reality. This self-conception may be best encapsulated in the words of Günther Rittau who asserted: "The cameraman is the modern magician. There is no such thing as 'impossible' for him."[104]

Bibliography

Henri Alekan, *Des lumières et des ombres* (Paris: Librairie du Collectionneur, 1991).

Henri Alekan, *Le vécu et l'imaginaire: Chroniques d'un homme d'images* (Geneva: Publishing La Sirène, 1999).

Henri Alekan, *Procédé et dispositif de prise de vues combinées*. République Française Brevet d'invention 1,098,128, filed 7 January 1954, and issued 2 March 1955.

"Angebot der Universum-Film AG an die Deutsche Spiegeltechnik GmbH &Co, Berlin W9, Köthenerstrasee 1/4 vom 15. September 1926." *Files of the Deutsche Spiegeltechnik GmbH & Co. K.-G. Berlin* (R109/I2070). Bundesarchiv Berlin.

Helmut G. Asper, "'Man muss eben struggeln, um oben zu bleiben.' Eugen Schüfftan und Siegfried Kracauer im amerikanischen Exil," in *Nachrichten aus Hollywood, New York und anderswo: Der Briefwechsel Eugen und Marlise Schüfftans mit Siegfried und Lili Kracauer*, ed. Helmut G. Asper, 3-30 (Trier: Wissenschaflicher Verlag, 2003).

"Auflösungsvertrag vom 28. Februar 1928." *Files of the Deutsche Spiegeltechnik GmbH & Co. K.-G. Berlin in the Bundesarchiv Berlin* (R109/I2070). Bundesarchiv Berlin.

Jacques Aumont, Alain Bergala, Michel Marie, and Marc Vernet, *Aesthetics of Film* [1983], trans. Richard Neupert (Austin, TX: University of Texas Press, 2004).

John Belton, *Cinema Stylists* (Metuchen, NJ: Scarecrow Press, 1983).

A. Berein, "Kinematographisches," *Der Komet* 1109 (23 June 1906): 6.

"Bericht über Revision der Geschäftsbücher und der Bilanz für das am 30. September 1927 abgelaufene Geschäftsjahr." *Files of the Deutsche Spiegeltechnik GmbH & Co. K.-G. Berlin* (R109/I2456). Bundesarchiv Berlin.

"Bericht über Zwischenrevision der Geschäftsbücher und der Bilanz per 28. Februar 1927." *Files of the Deutsche Spiegeltechnik GmbH & Co. K.-G. Berlin* (R109/I2455). Bundesarchiv Berlin.

J. Stuart Blackton and Albert E. Smith, "Some Tricks of the Moving Picture Maker," *Scientific American* (26 June 1909): 476.

Peter Bogdanovich, *Who the Devil Made It* (New York, NY: Alfred A. Knopf, 1997).

David Bordwell, *Narration in the Fiction Film* (Madison, WI: University of Wisconsin Press, 1985).

104 Günther Rittau, "Special Effects in *Metropolis* [1927]," in *Fritz Lang's Metropolis: Cinematic Visions of Technology and Fear*, eds. Michael Minden and Holger Bachmann, 78-80 (Rochester, NY: Camden House, 2002), 78.

David Bordwell, "The Classical Hollywood Style, 1917-1960," in David Bordwell, Janet Staiger, and Kristin Thompson, *The Classical Hollywood Cinema: Film Style and Mode of Production to 1960*, 1-84 (New York, NY: Columbia University Press, 1985).

Thomas Brandlmeier, *Kameraautoren: Technik und Ästhetik* (Marburg: Schüren Verlag, 2008).

"British National Strengthened," *Kinematograph Weekly* (13 January 1927): 56.

W. Stephen Bush, "'The Life of Richard Wagner' Made by Messter & Co. in Four Reels and Imported by Klaw and Erlanger," *Moving Picture World* 8-13 (1913): 995.

Edward Carrick, *Designing for Films* [1941] (London: The Studio Publications, 1949).

Carl Casinghino, *Moving Images: Making Movies, Understanding Media* (Clifton Park, NY: Delmar, 2010).

Michael Cowan, "Advertising, Rhythm and the Filmic Avant-Garde in Weimar Guido Seeber and Julius Pinschewer's Kipho Film," *October* 131 (Winter 2010): 23-50.

Michael Cowan, "Moving Picture Puzzles: Training Urban Perception in the Weimar 'Rebus Films,'" *Screen* 51:3 (Autumn 2010): 197-218.

Michael Cowan, *Walter Ruttmann and the Cinema of Multiplicity: Avant-Garde - Advertising - Modernity* (Amsterdam: Amsterdam University Press, 2014).

"Das Ende der Dekoration," *Kinematograph* 950 (3 May 1925).

James Searle Dawley, *Art of Making Motion Pictures*, U.S. patent 1,278,117, filed 17 August 1914, and issued 10 September 1918.

Sergei Eisenstein, "Dickens, Griffith and the Film Today [1944]," in *Film Form: Essays in Film Theory*, ed. and trans. Jay Leyda, 195-255 (New York, NY: Harcourt, Brace & World, 1949).

Sergei Eisenstein, "The Cinematographic Principle and the Ideogram [1929]," in Sergei Eisenstein, *Film Form: Essays in Film Theory*, ed. and trans. Jay Leyda, 28-44 (New York, NY: Harcourt, Brace & World, 1949).

Lotte Eisner, "Der produktive Kameramann wird zum Mitschöpfer," *Film-Kurier* 272 (26 November 1931).

Karl Freund, "Just What Is 'Montage'?," *American Cinematographer* 5 (September 1934): 204; 210.

Rolf Giesen, "Eugen Schüfftan," in *Cinegraph: Lexikon zum deutschsprachigen Film*, ed. Hans-Michael Bock, vol. 7 (Munich: edition text + kritik, 1984-).

Carl Louis Gregory, "Trick Photography Methods Summarized," *American Cinematographer* (June 1926): 21-22.

Tom Gunning, "Uncanny Reflections, Modern Illusions: Sighting the Modern Optical Uncanny," in *Uncanny Modernity: Cultural Theories, Modern Anxieties*, eds. Jo Collins and John Jervis, 68-90 (New York, NY: Palgrave Macmillan, 2008).

Malte Hagener, *Moving Forward, Looking Back: The European Avant-Garde and the Invention of Film Culture 1919-1939* (Amsterdam: Amsterdam University Press, 2007).

Sabine Hake, "Das Kino, die Werbung und die Avantgarde," in *Die Spur durch den Spiegel: Der Film in der Kultur der* Moderne, eds. Malte Hagener, Johann N. Schmidt, and Michael Wedel, 193-206 (Berlin: Bertz und Fischer, 2004).

Miriam Hansen, "Mass Culture as Hieroglyphic Writing: Adorno, Derrida, Kracauer," *New German Critique* 56 (Spring/Summer 1992): 42-73.

Raoul Hausmann, "Photomontage [1931]," in *The Weimar Republic Sourcebook*, eds. Anton Kaes, Martin Jay, and Edward Dimendberg, 651-653 (Berkeley, CA: University of California Press, 1994).

"Interview avec Henri Alekan," *Cinématographe* 68 (June 1981): 23-28.

"Inventory List." *Norman O. Dawn Collection of Special Effects Cinematography.* Harry Ransom Center at the University of Texas at Austin.

Gertrud Isolani, "Gespräch mit Eugen Shuftan," *Basler Nachrichten* (19 October 1965): 9.

Alexander Jason, *Handbuch der Filmwirtschaft: Filmstatistiken und Verzeichnisse der Filmschaffenden, Filmfirmen, der Filme und der Tonfilmkinos*, vol. 2, Film-Europa (Berlin: Verlag für Presse, Wirtschaft und Politik, 1931).

Will F. Jenkins, *Apparatus for Production of Light Effects in Composite Photography.* United States Patent 2,727,427, filed 3 March 1952, and issued 20 December 1955.

Erich Kettelhut, *Der Schatten des Architekten*, ed. Werner Sudendorf (Munich: Belleville, 2009).

Kinematheksverbund, Arbeitsgruppe Deutsche Filmografie, "Deutsche Spielfilme 1925: Jahresproduktion und Filmbestand Bundesarchiv," Bundesarchiv-Filmarchiv, October 2007.

Grace Kingsley, "Flashes," *Los Angeles Times* (17 September 1925): A11.

Willy Köhler, *Verfahren und Einrichtung für kinematographische Aufnahmen*, Deutsches Reichspatent 428,589, filed 14 September 1922, and re-issued 10 May 1926.

Willy Köhler, *Verfahren und Einrichtung für kinematographische Aufnahmen*, Deutsches Reichspatent 429,753 (addition to patent 428,589), filed 11 February 1923, and issued 2 June 1926.

Willy Köhler, *Verfahren für kinematographische Aufnahmen*, Deutsches Reichspatent 429,754 (addition to patent 428,589), filed 10 May 1923, and issued 2 June 1926.

Willy Köhler, *Einrichtung für kinematographische Aufnahmen*, Deutsches Reichspatent 429,755 (addition to patent 428,589), filed 4 March 1924, and issued 2 June 1926.

Alexander Kossowsky, "Die Männer der Kurbel II: Reimar Kuntze," *Film-Kurier* (14 May 1925).

Alexander Kossowsky, "Die Männer der Kurbel III: Werner Brandes," *Film-Kurier* (19 May 1925).

Alexander Kossowsky, "Die Männer der Kurbel IV: Karl Freund," *Film-Kurier* (30 May 1925).

Alexander Kossowsky, "Die Männer der Kurbel VI: Guido Seeber," *Film-Kurier* (22 August 1925).

Émile Kress, "Trucs et illusions. Applications de l'optique et de la mécanique au cinématographe (c. 1912)," *1895. Revue d'histoire du cinéma* 27 (September 1999), 7-20.

Standish D. Lawder, *The Cubist Cinema* (New York, NY: New York University, 1975).

Vachel Lindsay, *The Art of the Moving Picture* [1915] (New York, NY: Liveright Publishing, 1970).

A. Lion, "Das Schüfftan-Verfahren: Umwälzung in der Filmaufnahme," *Die Umschau* (1927): 248-251.

Rachel Low, *The History of British Film: 1918-1929* (London: George Allen & Unwin, 1971).

Herbert G. Luft, "Der Mann hinter der Kamera: Karl Freund – schon Filmgeschichte," *Der Filmkreis* (July 1963): 16-19.

Waldemar Lydor, "Kristallene Filmbauten," *Das Kino-Journal* (Vienna) 848 (30 October 1926): 14-15.

Herbert Moulton, "Magical Effects Brought to Screen by Unique Process," *Los Angeles Times* (18 April 1926): C25-26.

Corinna Müller, *Frühe deutsche Kinematographie: formale, wirtschaftliche und kulturelle Entwicklungen 1907-1912* (Stuttgart: Metzler-Verlag, 1994).

Helmut Müller-Sievers, *The Cylinder: Kinematics of the Nineteenth Century* (Berkeley, CA: University of California Press, 2012).

Hans Nieter, "The Schüfftan Process of Model Photography," *Photographic Journal*, London (January 1930): 16.

Geoffrey Nowell-Smith, *The History of Cinema: A Very Short Introduction* (Oxford: Oxford University Press, 2017).

Jan Olsson, "Framing Silent Calls: Coming to Cinematographic Terms with Telephony," in *Allegories of Communication: Intermedial Concerns from Cinema to the Digital*, eds. John Fullerton and Jan Olsson, 157-192 (Bloomington, IN: Indiana University Press, 2005).

Max Ophüls, *Spiel im Dasein: Eine Rückblende* [1959], ed. Helmut G. Asper (Berlin: Alexander Verlag, 2015).

Friedrich Porges, "Luftschlösser im Film," *Die Bühne* (Vienna) 120 (25 February 1927).

René Prédal, "Les Grands opérateurs (IV): Henri Alekan," *Cinéma* 173 (February 1973): 88-95.

Stephen Prince, *Digital Visual Effects in Cinema: The Seduction of Reality* (New Brunswick, NJ: Rutgers University Press, 2011).

Walter Reimann, "Filmarchitektur – heute und morgen," *Filmtechnik* (20 February 1926).

Richard Rickitt, *Special Effects: The History and Technique* (New York, NY: Billboard Books, 2007).

Rainer Maria Rilke, *Sonnets to Orpheus* [1922], trans. Edward Snow (New York, NY: North Point Press, 2004), 65.

Günther Rittau, "Günther Rittau über Filmkunst und Filmtechnik," *Filmtechnik* 3 (5 February 1926): 50.

Günther Rittau, "Special Effects in *Metropolis* [1927]," in *Fritz Lang's Metropolis: Cinematic Visions of Technology and Fear,* eds. Michael Minden and Holger Bachmann, 78-80 (Rochester, NY: Camden House, 2002).

Tom Ryall, "Blackmail," in *British Film Institute Film Classics, vol. 1,* eds. Rob White and Edward Buscombe (London: British Film Institute, 2003).

–s, "Kritik der Leinwand," *Die Filmwoche* 22 (1925): 519.

Barry Salt, *Film Style and Technology: History and Analysis* (London: Starwood, 1983).

Eugen Schüfftan, "Mein Verfahren," *Film-Kurier* 24 (19 November 1926).

Eugen Schüfftan, "Verbesserungen in der Miniatur-Aufnahmetechnik," in Friedrich Porges, *Mein Film-Buch,* 35-37 (Vienna: Mein Film Verlag, 1929).

"Schufftan Process Acquired," *Kinematograph Weekly* (13 January 1927): 56.

Guido Seeber, *Der praktische Kameramann* [1927] (Frankfurt am Main: Deutsches Filmmuseum, 1980).

Guido Seeber, *Der praktische Kameramann*, vol. 2, *Der Trickfilm in seinen grundsätzlichen Möglichkeiten* [1927] (Frankfurt am Main: Deutsches Filmmuseum, 1979).

Guido Seeber, *Kamera-Kurzweil: Allerlei interessante Möglichkeiten beim Knipsen und Kurbeln* (Stuttgart: Union Deutsche Verlagsgesellschaft, 1930).

Stiftung Deutsche Kinemathek, ed. *Das wandernde Bild: Der Filmpionier Guido Seeber* (Berlin: Stiftung Deutsche Kinemathek, 1979).

Frederick A. Talbot, *Moving Pictures: How They Are Made and Worked* (London: William Heinemann, 1912).

Sheldon Teitelbaum, "Special Effects: Aliens," *Cinefantastique* 16 (October 1986).

"The German Kinematograph and Foreign Countries," *Der Kinematograph* 38 (18 September 1907).

Kristin Thompson, *Herr Lubitsch Goes to Hollywood: German and American Film after World War I* (Amsterdam: Amsterdam University Press, 2005).

Walther Thorner, *Verfahren und Vorrichtung zur Herstellung kinematographischer Kombinationsaufnahmen*. Deutsches Reichspatent 598,712, filed 19 May 1932, and issued 18 June 1934.

François Truffaut, *Hitchcock* (New York, NY: Simon & Schuster, 1985).

Yuri Tsivian, "Cyberspace and its Precursors: Lintsbach, Warburg, Eisenstein," in *Transmedia Frictions: The Digital, the Arts, and the Humanities*, eds. Marsha Kinder and Tara McPherson, 80-99 (Berkeley, CA: University of California Press, 2014).

Julie Turnock, "Patient Research on the Slapstick Lots: From Trick Men to Special Effects Artists in Silent Hollywood," *Early Popular Visual Culture* 2 (May 2015): 152-173.

Heidi Wiese and Henri Alekan, "Man muß die Technik überwinden, um zur Kunst zu gelangen: Ein Gespräch mit dem Kameramann Henri Alekan," in *Die Metaphysik des Lichts: Der Kameramann Henri Alekan*, ed. Heidi Wiese, 22-57 (Marburg: Schüren-Verlag, 1996).

W...f, "Zum ersten Male Schüfftansches Patent," *Die Kinotechnik* 18 (25 September 1925): 456.

Leo Wiltin, "Die Technik," *Film-Kurier* 24 (19 November 1926).

K. W. Wolf-Czapek, "Deutsche Films – technisch bewertet," *Lichtbild-Bühne* 21 (25 May 1912): 22-23.

Frederic Wynne-Jones, "When the Camera's Eye Lies for Entertainment," *The New York Times* (2 May 1926): X5.

Werner Zurbuch, "Eugen Schüfftan – Meister der Filmtechnik," *Filmtechnikum* (June 1962): 179-180.

3. The Uncanny Mirror: *Der Student von Prag* (1913)

Abstract
The first German film to excite art critics was simultaneously a milestone in the history of special effects. *Der Student von Prag* was co-created by some of Germany's most ardent early cinephiles with the goal to demonstrate the feasibility of film art. Proceeding from techno-romantic assumptions, they construed artistic filmmaking as the articulation of ideas and feelings through the imaginative application of the medium's technological assets, specifically location shots and trick effects. Consequently, *Der Student von Prag* depicts the intrusion of an uncanny doppelganger into a real-life setting, the mystical city of Prague. As a vehicle for abstract notions, the horrific double thus bore witness to cinema's ability to convey figurative meaning and participate in the life of the mind.

Keywords: doppelganger, double, film art, split-screen composites, location cinematography

The first German film to excite art critics across the board was simultaneously a milestone in the history of special effects. *Der Student von Prag* (Deutsche Bioscop, 1913) was co-created by some of Germany's most ardent early cinephiles, actor Paul Wegener, author Hanns Heinz Ewers, director Stellan Rye, and cinematographer Guido Seeber. Their goal was to demonstrate the feasibility of film art, and special effects played a central role in this endeavour. *Der Student von Prag* emerged from an art film movement that dominated German production between the fall of 1912 and the summer of 1914. Encouraged by new commercial opportunities arising in the wake of the feature-length film, ambitious producers engaged in the creation of "artistic" premium films. Specifically, they aspired to mitigating concerns about cinema's negative social and aesthetic implications and improving the medium's reputation with elite opinion leaders.

Loew, Katharina, *Special Effects and German Silent Film: Techno-Romantic Cinema.*
Taylor & Francis Group, 2021
DOI 10.5117/9789463725231_CH03

As I have shown in Chapter 1, aesthetic theory considered medium specificity as a prerequisite for art. Cinema's first critics identified two types of subject matter as cinema's proper areas of competence: nature cinematography and fantastic or mental imagery, realized by means of special effects. In line with this and proceeding from fundamentally techno-romantic premises, the creators of *Der Student von Prag* sought to devise an art film by capturing the intangible through cinema's unique technological assets. Their tale about a man haunted by his mirror image featured scenic shots of romantic Prague and the eerie figure of the doppelganger, created by means of split screen composites. Combining the two aspects that were considered genuinely cinematic, *Der Student von Prag* renders possible supernatural intrusions into an actual environment, which are both emotionally gripping and impossible in any other medium.

As one of the first feature-length films, *Der Student von Prag* portrayed the supernatural as menacing and induced fear in its audiences. In so doing, it spearheaded a new cinematic genre: horror. At the same time, the figure of the detached mirror image oscillates between monster and metaphor. The doppelganger is easily construable as a trope for personal identity and the depths of the human psyche, thus evincing cinema's capacity to convey figurative meaning.[1] Finally, the uncanny double can also be read as commentary on the nature of the cinematic medium: Because the supernatural appears as truthful as the location shots from which it emerges, the film calls into question the alleged objectivity of photographic images. By affording an affective and cognitive engagement with abstract concepts, *Der Student von Prag* demonstrated that film technology could reach beyond the concrete and tangible.

Precisely because *Der Student von Prag* tackled philosophical issues by cinematic means, it quickly established itself as a lodestar for artistic film production.[2] The film made an enormous impression on critics, changing virtually overnight widely held preconceptions about the nature and capabilities of the new medium. *Der Student von Prag* prefigured what would become the guiding principle for German artistic filmmaking, namely, as art director Walter Reimann formulated it in 1926, "to translate the mental into

1 For an in-depth discussion of the relationship between fantastic fiction and figurative discourse, see Tom Gunning, "Like Unto A Leopard: Figurative Discourse in *Cat People* (1942) and Todorov's *The Fantastic*," *Wide Angle* 3 (1988): 30-39.

2 See for instance Kristin Thompson, "Im Anfang war...: Some Links between the German Fantasy Films of the Teens and the Twenties," in *Before Caligari: German Cinema 1895-1920*, eds. Paolo Cherchi Usai and Lorenzo Codelli, 138-161 (Pordenone: Edizioni Biblioteca dell'Immagine, 1990).

the purely optical."[3] The conception of cinema as a means to give material form to the emotions and ideas reflects the techno-romantic outlook that shaped German film culture. In *Der Student von Prag*, quite likely for the first time in German cinema, film technology became an agent for the ideal.

The emergence of German art cinema

"Artistic" film production came to play a defining role in the German film industry for both ideological and economic reasons. As cinema emerged as a prominent form of mass entertainment in the mid-1900s, elite opinion leaders felt compelled to assert their position as arbiters of cultural value. As I have shown in Chapter 1, they vocally called for "high-quality" cultural production and expected cinema to conform with established aesthetic tenets. Cinema-friendly critics consequently resorted to techno-romantic lines of argument to reconcile the concomitant conceptual conflict between art and technology, to establish cinema's ties to the immaterial and thus its artistic potential. Film producers, for their part, saw quality, "artistic" filmmaking as the principal path to cultural legitimization and the necessary precondition for strengthening the German production sector and its export prospects.

By the early 1910s, German production was still severely underdeveloped. German theatres screened mostly foreign films and the few German production companies focused on non-fiction and quick-and-cheap amusement.[4] This began to change with Deutsche Bioscop's first Asta Nielsen series in 1911/1912. By 1913, Germany released a sizable number of ambitious feature films. Besides *Der Student von Prag*, films like the biopic *Richard Wagner* (Messter-Film, dirs. William Wauer and Carl Froelich), *Eva* (Messter-Film, dir. Curt A. Stark) starring Henny Porten, and the Hermann Sudermann adaptation *Die Sünden der Väter* (Deutsche Bioscop, dir. Urban Gad) with Asta Nielsen not only met with widespread approval from critics, they also attracted attention abroad.

3 Alexander Kossowsky, "Die Film-Architekten II: Walter Reimann," *Film-Kurier* (28 January 1926).

4 Corinna Müller, *Frühe deutsche Kinematographie: formale, wirtschaftliche und kulturelle Entwicklungen 1907-1912* (Stuttgart: Metzler-Verlag, 1994), 43-47. In 1912, 30% of footage shown in Germany was of French origin, 28% American, 21% Italian, 13% German, 5% Danish, and 3% British. See *Lichtbild-Bühne* quoted in Emilie Altenloh, *Zur Soziologie des Kino: Die Kino-Unternehmung und ihre Besucher* (Leipzig: Spamerische Buchdruckerei, 1913), 10.

These initial German advances into quality film production were inspired by the enormous popular success of early Danish multi-reel films, most notably Urban Gad's and Asta Nielsen's risqué melodrama *Afgrunden* (Kosmorama, 1910), Nordisk's sex-slave-themed box office sensation *Den hvide Slavehandel* (1910, dir. August Blom), and its sequel *Den hvide Slavehandels sidste Offer* (1911, dir. August Blom) as well as art films like Kinografen's *De fire djævle* (1911, dir. Alfred Lind et al). They forcefully demonstrated to German producers that film length, star power, and artistic aspirations could in fact serve as selling points. Although the share of Danish films in the German market was relatively small, Danish studios, artists, and methods had a decisive impact on the development of German cinema. The Danish Nordisk Films Kompagni became one of Germany's largest distributors and exhibitors and a large contingent of Danish filmmakers—actors like Asta Nielsen, Olaf Fønss, Valdemar Psilander, and Viggo Larsen, cinematographers like Axel Graatkjær and directors like Urban Gad, Holger-Madsen, Alfred Lind, and Stellan Rye—had a formative influence on German pre-war filmmaking.[5]

Game changing for the feeble German production sector was ultimately the establishment of a novel distribution system known as monopoly film (*Monopolfilm*). Monopoly distribution guaranteed a film's distributors and exhibitors exclusivity for a certain time in one geographical area, giving them an advantage in a highly competitive market. Producers profited from higher prices and a dependable number of prints in circulation. As a result, the domestic production sector began to expand substantially. Between 1912 and 1914 the number of German production companies more than doubled from eleven to twenty-five.[6] The conventional subscription-based distribution system of pre-compiled short film programmes facilitated cheap, diverse entertainment. The monopoly distribution system, on the other hand, construed cinema in a radically different way: as a big capital and bourgeois institution.

Although the monopoly distribution system strengthened the German film industry considerably, the need for further product differentiation quickly became evident. In response, a consortium of monopoly distributors led by Paul Davidson's Projektions-AG 'Union' (PAGU) began building up Germany's first film star.[7] Rather than distributing one multi-reel film at a

5 In 1917, when it sold its German operations to Ufa, Nordisk distributed one quarter of all German releases and owned 35% of German theatres.

6 Paul Rohnstein, "Beiträge zur wirtschaftlichen Entwicklung der Deutschen Film-Industrie (unter besonderer Berücksichtigung des Kinomatographentheatergewerbes [sic])," Ph.D. Dissertation, Julius-Maximilians-Universität Würzburg, 1922, 36.

7 Müller, Frühe deutsche Kinematographie, 144-157.

time, they put together a season package that could be booked in advance. All films in the series featured the same actress, Asta Nielsen, whose name consequently became the principal selling point. Deutsche Bioscop, the studio that would go on to make *Der Student von Prag*, was commissioned with the shooting of the first two Asta Nielsen series, which proved exceedingly lucrative. Deutsche Bioscop was put in a position to establish state of the art production facilities in Babelsberg near Berlin, a project that was supervised by the studio's technical director Guido Seeber in 1911/1912. Acquired by Ufa in 1920, the Studio Babelsberg is today considered the oldest large-scale film studio in the world.

The exceptional success of the Asta Nielsen vehicles made evident the value of name recognition. Almost immediately, industry rivals started grooming stars of their own, like Lissy Nebuschka (Dekage, 1912/1913) or Henny Porten (Messter, 1913/1914). Most monopoly series drew on established stage artists in much the same way as synchronized sound pictures (*Tonbilder*, 1903-1913) and French and Italian *films d'art* (since 1908). What was new, however, was the marketing: the artist had become the main attraction.

In January 1913, Deutsche Bioscop signed authors Victor Blüthgen and Hanns Heinz Ewers as well as noted stage actors Carl Clewing, Tilla Durieux, and Paul Wegener.[8] In terms of prestige, this line-up could not compete with theatre legend Albert Bassermann, whom Vitascope was about to showcase in *Der Andere* (1913, dir. Max Mack), or illustrious German-language poets like Arthur Schnitzler, Hugo von Hofmannsthal, and Felix Salten (of future *Bambi* fame) with whom Nordisk intended to collaborate.[9] But even at Bioscop's more modest level, series offered significant advantages since distribution could be secured for multiple films in advance, which in turn allowed for larger budgets and higher quality films. As a result, "artistic" actor-, director-, and author-series featuring more or less familiar names began to dominate domestic production. Although "artistic" aspirations often served primarily as an advertising point, a growing number of filmmakers were genuinely committed to the project of advancing film art. In 1913/1914, several of them joined forces at Deutsche Bioscop and briefly made the studio the most aesthetically innovative in the German industry.

8 See advertisements in *Der Kinematograph* 314 (1 January 1913) and *Der Kinematograph* 315 (8 January 1913).

9 For a reprint of Nordisk's November 1912 announcement, see Ludwig Greve, Margot Pehle, and Heidi Westhoff, eds. *Hätte ich das Kino! Die Schriftsteller und der Stummfilm* (Stuttgart: Klett-Verlag, 1976), 129.

Enacting film art: *Der Student von Prag*

Der Student von Prag was one of the first "art films" to be produced at Deutsche Bioscop. The studio's prioritization of "artistic" filmmaking likely traces back to the initiative of bestselling author Hanns Heinz Ewers, an unconventional and multifaceted character. Bourgeois literary circles disdained Ewers's work as trivial, gory, and obscene, yet the worldwide success of his scandalous novel *Alraune* (1911) had made him a celebrity. Strongly influenced by the radical individualist philosophy of Max Stirner (1806-1856) and Nietzschean moral philosophy, the well-travelled bohemian had strong interests in the occult, was an opium aficionado, nudist, gay rights activist with homosexual proclivities, and advocate of sexology. At the same time, Ewers was a devoted German nationalist and early Nazi follower. In 1932, he authored a novelistic biography of Nazi martyr Horst Wessel, which he subsequently adapted for the Nazi propaganda film *Hans Westmar. Einer von vielen. Ein deutsches Schicksal aus dem Jahre 1929* (Volksdeutsche Film 1933, dir. Franz Wenzler). Despite his early ties to the Nazi movement, in 1934 authorities deemed Ewers decadent and untrustworthy, and most of his works were banned. Ewers died in poverty in 1943.

As we have seen in Chapter 1, Ewers was also a pioneering cinephile and ardent advocate for film art, who, as early as 1907, had called on fellow artists to join the cause: "Where are the poets and painters who work for the cinema? [...] Here is a new sphere for art, a new field, who will help to plough it?"[10] Ewers's vision for film art was rooted in techno-romantic convictions that posited the technical possibilities of cinema as a path to a fantastic realm. His penchant for the supernatural left strong marks on Deutsche Bioscop. In 1913/1914 the studio released a striking number of fantastic feature-length films including *Ein Sommernachtstraum in unserer Zeit, Kadra Sâfa, Die Eisbraut, Die Augen des Ole Brandis* and *Erlkönigs Töchter*, most of which credited Ewers for the script and Stellan Rye for the direction. Ewers's and Rye's first collaboration, and the only one to survive, was *Der Student von Prag*.

The extent of Rye's contribution to *Der Student von Prag* has variously been called into question, most notably by Ewers himself. Rye, who died in November 1914 at the age of thirty-four as a prisoner of war in Flanders, never commented on the matter. Marguerite Engberg's and Casper Tybjerg's ground-breaking research notwithstanding, little is known about Rye's final

10 Hanns Heinz Ewers, "Der Kientopp," *Der Morgen* (Berlin) 18 (11 October 1907): 578.

years in Germany.[11] Persecuted as a homosexual in his native Denmark, the theatre director and former army officer arrived in Berlin presumably in late 1912. As one of three regular directors at Deutsche Bioscop, he supervised at least fifteen productions before joining the German army at the beginning of the war. Even though Rye's name was not mentioned in the original credits or publicity of *Der Student von Prag*, contemporaries were nonetheless aware of his involvement. Critic Horst Emscher for instance wrote: "For the implementation the courageous campaigner for art [Ewers] had the most excellent forces at his disposal, first and foremost the sophisticated Nordic director Stellan Rye, who worked with an unequalled devotion on this work, which presented difficulties like no other film before."[12] Guido Seeber likewise emphasized Rye's role in the creation of *Der Student von Prag*. In 1926, he publicly contradicted Ewers's claim to the film's directorship: "Even though the author received director's credit back in the day, in actual fact the extraordinarily gifted Danish director Stellan Rye, who was killed in the World War, directed the film."[13] At least in part due to the efforts of Ewers's biographer and advocate Wilfried Kugel, Rye's merits remain contested, which is also reflected in the Munich Film Museum's questionable decision to credit Stellan Rye anachronistically as "assistant director" in its most recent restoration and video edition of *Der Student von Prag*.[14]

Compared to the question of directorial credit, the genesis of the film itself is less controversial. Ewers had joined Deutsche Bioscop with the objective of demonstrating the feasibility of film art and creating what he deemed, in his words, "*must* become the world's *best film*."[15] The idea for a doppelganger film, on the other hand, was evidently Paul Wegener's, one of the leading character actors at Max Reinhardt's Deutsches Theater in Berlin. Wegener, who at the time was evidently not yet convinced of cinema's artistic merit, recalled in 1914: "I joined the film industry with a new idea that could not be realized on stage: playing scenes with myself. This

11 Marguerite Engberg, "*Studenten fra Prag* og den gådefulde Stellan Rye," in *Sekvens. Filmvidenskabelig årbog*, 161-185 (Copenhagen: Reitzel, 1982); Casper Tybjerg, "The Faces of Stellan Rye," in *A Second Life: German Cinema's First Decades*, eds. Thomas Elsaesser and Michael Wedel, 151-159 (Amsterdam: Amsterdam University Press, 1996).

12 Horst Emscher (= Josef Coböken),"Große Kunst im Film," *Zeit im Bild*, vol. 3 (1913): 2524-2525.

13 Guido Seeber, "Der Student von Prag," *Lichtbild-Bühne* (15 November 1926).

14 *Der Student von Prag*, dir. Hanns-Heinz Ewers (1913, Edition Filmmuseum 80, Munich: film&kunst, 2016), DVD.

15 Letter by Hanns Heinz Ewers to his mother Maria Ewers (1913) in the Hanns Heinz Ewers estate at the Heinrich Heine-Institut Düsseldorf, quoted in Wilfried Kugel, *Der Unverantwortliche: Das Leben des Hanns Heinz Ewers* (Düsseldorf: Grupello Verlag, 1992), 189. Emphases in the original.

attracted me as an actor and at the same time I saw new possibilities for the cinema to achieve certain artistic effects (insofar as it is at all possible)."[16] Two years later, however, as we have seen in Chapter 1, Wegener became one of the first German cultural figures to publicly champion the camera as cinema's central means of artistic expression.

As it happened, Deutsche Bioscop had a uniquely qualified trick specialist on staff, one of the few German cinematographers at the time who would have been capable of realizing Wegener's vision. Guido Seeber had previously realized several trick shorts for Deutsche Bioscop and just distinguished himself as the cameraman of eighteen Asta Nielsen films. Guido Seeber's stature at Deutsche Bioscop can be deduced from the fact that studio promotional material described the cinematographer as an artist on par with the director: "In contrast to the theatre, where three factors—poet, director, and actor—work together to create a whole, in the cinema a fourth is added, the cameraman, whose conception of art and artistic sensibility have to meet the same standards as the director's. This faculty, combined with a perfect mastery of everything technical, is represented sumptuously by Guido Seeber."[17] As I have indicated in Chapter 2, Seeber was particularly fond of the doppelganger illusion and engaged in it throughout his life. Surviving doppelganger self-portraits show him as a young film pioneer in 1898 and in 1912 he staged a photograph of himself in different representative functions in the new studios of Deutsche Bioscop (Figure 3.1.).[18] Seeber must have been delighted to create a feature-length doppelganger film.

The doppelganger theme constituted, as a contemporary observer remarked, an ingenious effort "to bring refined audiences to the movie theatres without displacing the mass audiences."[19] Given the contemporary fad of the mysterious and occult, supernatural subject matter could be expected to have mass appeal. Since the 1880s, fantastic literature had been experiencing a revival, producing successes such as Alfred Kubin's *Die andere Seite* (1909), Karl Hanns Strobl's *Eleagabal Kuperus* (1910), Ewers's

16 Paul Wegener, "Warum ich für den Film spiele," *Berliner Börsen-Courier* 321 (12 July 1914). Ewers later corroborated this claim. See Hanns Heinz Ewers, "Der Student von Prag," (1935), in the Hanns Heinz Ewers estate at the Heinrich Heine-Institut Düsseldorf, quoted in Kugel, *Der Unverantwortliche*, 188.

17 Deutsche Bioscop Gesellschaft, ed., *Unsere Künstler* (Promotion Booklet, 1913), 14-15.

18 For his early experiments with doppelganger self-portraits, see Guido Seeber, *Kamera-Kurzweil: Allerlei interessante Möglichkeiten beim Knipsen und Kubeln* (Stuttgart: Union Deutsche Verlagsgesellschaft, 1930), 188. Seeber's 1912 trick photograph is reproduced in *Die Kinotechnik* 1 (September 1919): 13.

19 Spektator, "Autorenkünstler [sic] und Riesenfilms," *Der Kinematograph* 349 (3 September 1913).

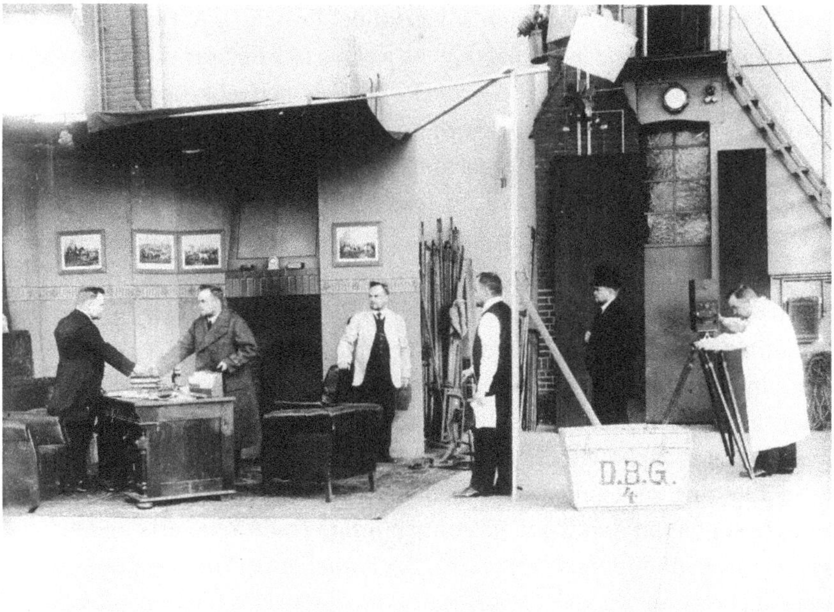

Fig. 3.1. Guido Seeber in six different functions in the studio of Deutsche Bioscop. Photograph by Guido Seeber, 1912. Source: Deutsche Kinemathek.

Alraune (1911), and Gustav Meyrink's *Der Golem* (1914). Simultaneously, nineteenth-century fantastic literature attracted renewed attention as well. The first German edition of Poe's collected works appeared in 1901 and in 1905 Ewers published a small but influential book that celebrated Poe as an artist in the "truest sense," as "a pioneer of culture in the newly discovered land of the unconscious."[20] In 1908, the first critical-historical edition of E. T. A. Hoffmann's works was launched, and around 1912 Ewers began editing a series titled "gallery of fantasts," dedicating the first volumes to Hoffmann and Poe.[21]

While capturing mass audiences with supernatural thrills, *Der Student von Prag* sought to flatter the intelligentsia with a plethora of literary references. The theme of the detached mirror image traces back to the episode

20 Hanns Heinz Ewers, *Edgar Allan Poe*, trans. Adèle Lewisohn (New York, NY: B.W. Huebsch, 1917), 11; For the first German edition of Poe's work see Edgar Allen Poe, *Edgar Allen Poes Werke in zehn Bänden*, eds. Heda and Arthur Moeller-Bruck, trans. Hedda Moeller-Bruck and Hedwig Lachmann, 10 vols. (Minden: J.C.C. Bruns Verlag, 1901-1904).

21 E. T. A. Hoffmann, *Sämtliche Werke: Historisch-kritische Ausgabe*, ed. Carl Georg von Maassen, 10 vols. (Munich: Georg Müller Verlag, 1908-1928); E. T. A. Hoffmann, *Phantastische Geschichten*, ed. Hanns Heinz Ewers (Munich: Georg Müller Verlag, n.d.); Edgar Allen Poe, *Nebelmeer*, ed. Hanns Heinz Ewers, trans. Gisela Etzel (Munich: Georg Müller, 1914).

"Die Geschichte vom verlornen Spiegelbilde" from E. T. A. Hoffmann's *Die Abenteuer der Sylvester-Nacht* (1815) as well as to Adelbert von Chamisso's *Peter Schlemihls wundersame Geschichte* (1814), which tells the story of a man selling his shadow to the devil, which in turn recalls the deal with the devil in Goethe's *Faust* (1808). Other important inspirations were E. T. A. Hoffmann's doppelganger tales *Die Elixiere des Teufels* (1815/16) and *Die Doppeltgänger* (1821) as well as Hans Christian Andersen's *Skyggen* (1847), where an escaped shadow dehumanizes and eventually assassinates his former owner. The motif of the protagonist killing himself in his double is derived from Poe's *William Wilson* (1839) and Oscar Wilde's *The Picture of Dorian Gray* (1890). The film's final shot of the double sitting on the protagonist's grave replicates imagery from Alfred de Musset's poem "La Nuit de décembre" (1835), which is also quoted repeatedly in the film's intertitles.

In an effort to demonstrate cinema's artistic potential, *Der Student von Prag* first of all underscored its connection to traditional arts and artists. Like many contemporary "art" films, *Der Student von Prag* was promoted as the joint creation of a poet and a stage actor, the first production in Deutsche Bioscop's upcoming "great artists' series."[22] Advertisements promoted the film as a "fantastical drama in four acts by Hanns Heinz Ewers. In the leading role Paul Wegener, Germany's foremost character actor."[23] The film's prologue shows them on location, looking out over the Vltava river up to the Prague castle, seemingly immersed in a conversation about the golden city of a hundred spires.[24] Besides Wegener, additional ensemble members of the renowned Deutsches Theater, including Grete Berger, John Gottowt, and Wegener's then-wife Lyda Salmonova appeared in leading roles.[25] In the opening credits they are presented in a fashion reminiscent of *tableaux vivants*: A black curtain lifts, revealing a box set draped in black and an actor in pictorial, character-specific attire, posing behind a large sign announcing his or her name and the role.[26] Representative movements like the lifting and lowering of a prop animate these "tableaux," which

22 See for instance the advertisement in *Der Kinematograph* 339 (25 June 1913).

23 Advertisement in *Der Kinematograph* 345 (6 August 1913).

24 Extradiegetic prologues that introduce the filmmakers were common throughout the 1910s. German examples include *Tragödie eines Streiks* (Messter's Projection, 1911, dir. Adolf Gärtner), *Wo ist Coletti?* (Deutsche Vitascope, 1913, dir. Max Mack); *Hoffmanns Erzählungen* (Lothar Stark, 1916, dir. Richard Oswald); *Die Puppe* (PAGU, 1919, dir. Ernst Lubitsch)

25 See premiere programme of *Der Student von Prag*, ed. Deutsche Bioscop Ges.m.b.H. Berlin [1913] in the Scripts, Gray Literature and Audio Documents Archive of the Deutsche Kinemathek Berlin.

26 For a detailed examination of tableaux vivants see Daniel Wiegand, *Gebannte Bewegung: Tableaux vivants und früher Film in der Kultur der Moderne* (Marburg: Schüren-Verlag, 2016).

appear to draw on well-established iconography. For instance, Grete Berger's attitude resembles Francisco de Goya's *La maja vestida* (1800-1805) and Lyda Salmonova's appearance approximates orientalist representations of the "gypsy" woman popularized by Ambroise Thomas's hit opera *Mignon* (1866). The opening titles signal "upscale amusement," while involvement of composer Josef Weiss, a student of Franz Liszt's, and of painter Klaus Richter, a student of Lovis Corinth's, in the film's production highlights the film's artistic ambitions.[27]

Der Student von Prag drew on the traditional arts not only for kudos, but also to expand cinema's means of expression. One of the most striking examples can be found in the scene in which The Other confronts Balduin at the card table (Figure 3.2.). Ewers's scenario specifies "lighting from above. (Like an old Rembrandt painting, so that one sees hardly more than the heads.)"[28] It is noteworthy that Ewers explicitly mentioned Rembrandt as a model given that his scenario was not intended for public view and therefore had no role in enhancing the film's artistic reputation.[29] Instead, Ewers's reference suggests that he envisioned a certain painterly look in pursuit of specific dramatic goals. As a matter of fact, much of the scene's meaning depends on its chiaroscuro illumination. The confrontation between Balduin and his other self is literally spotlighted and causes his social relations to fade into darkness.

Seeking to make a case for film art, the creators of *Der Student von Prag* did not confine themselves to paying tribute to the traditional arts, however. Instead, they took pains to draw on cinema's unique means of expression. To begin with, Ewers evidently wanted to demonstrate that it was possible to tell a captivating story purely visually and his nineteen-page scenario for *Der Student von Prag* consequently only contained nine intertitles.[30]

27 It has been claimed that *Der Student von Prag* was the first German production to feature an original score or the first to be designed by a professional visual artist. However, *Richard Wagner* (Messter Film, 1913, dir. William Wauer and Carl Froelich, score by Giuseppe Becce) and *Eva* (Messter Film, 1913, dir. Curt A. Stark, score by Julius Einödshofer) premiered before *Der Student von Prag* and Deutsche Vitascope, Messter Film, and PAGU employed designers such as Kurt Dürnhöfer, Gustav Lütkemeyer, and Hermann Warm (of future *Caligari* fame).

28 Hanns Heinz Ewers, "*Der Student von Prag*," in Helmut H. Diederichs, *Der Student von Prag: Einführung und Protokoll* (Stuttgart: Focus Verlagsgemeinschaft, 1985), 89-98, 96.

29 As Barry Salt points out, such lighting effects had previously been seen in Danish films like *Den hvide Slavehandels sidste Offer* (1911). See *Film Style and Technology: History and Analysis* (London: Starwood, 1983), 71.

30 In 1915, Deutsche Bioscop sold *Der Student von Prag* to the company Robert Glombeck, which subsequently released a re-edited version of the film with ninety intertitles. See Wilfried Kugel, "Die Rekonstruktion des Stummfilms *Der Student von Prag* (Deutsche Bioscop GmbH

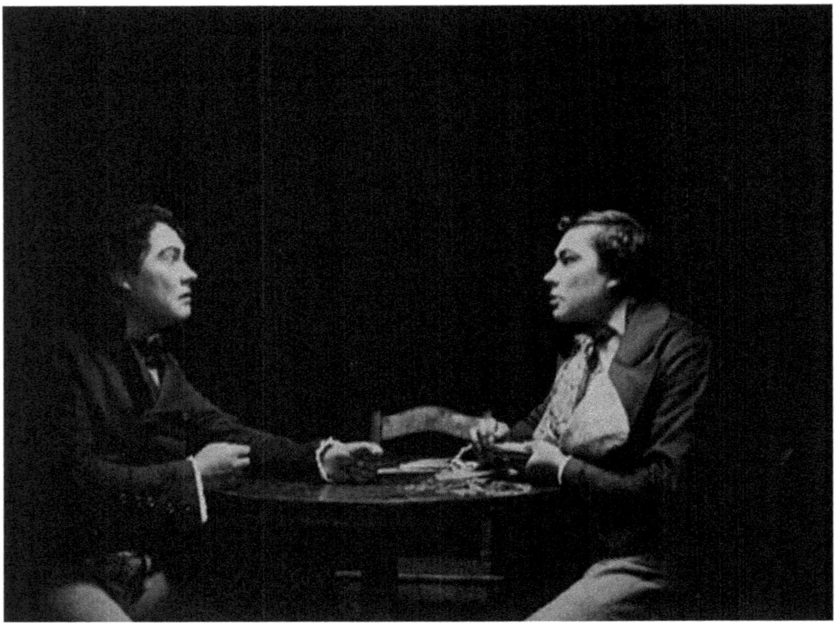

Fig. 3.2. Split screen composite with "Rembrandt" lighting in *Der Student von Prag* (1913): Playing cards with the doppelganger.

Incidentally, several contemporary "art" films, including *Das schwarze Los* (Deutsche Bioscop, 1913, dirs. John Gottowt and Emil Albes), *Die Landstraße* (Deutsche Mutoskop und Biograph, 1913, dir. Paul von Woringen), and *Der letzte Tag* (Vitascope, 1913, dir. Max Mack), likewise attempted to minimize the use of intertitles as much as possible, anticipating what has often been characterized as the pioneering achievement of *Der letzte Mann* a decade later (Union-Film/Ufa, 1924, dir. F. W. Murnau).[31]

In keeping with the assumption that cinema's artistic potential lay in areas where it had a clear advantage over spoken, "legitimate" theatre, Ewers's scenario for *Der Student von Prag* emphasized features that, as I have shown in Chapter 1, many critics regarded as "genuinely cinematic."

Berlin, 1913)." Unpublished manuscript, Berlin, 1988. The Munich Film Museum's 2016 restoration contains twenty-five intertitles.

31 See Corinna Müller, "Das 'andere' Kino? Autorenfilme in der Vorkriegsära," in *Die Modellierung des Kinofilms: Zur Geschichte des Kinoprogramms zwischen Kurzfilm und Langfilm 1905/06-1918, Mediengeschichte des Films*, vol. 2, ed. Harro Segeberg, 153-192 (Munich: Wilhelm Fink Verlag, 1998), 169; 180-181; 184. For critics' rejection of intertitles see for instance Hans Pander, "Intertitles [1923]," in *The Promise of Cinema: German Film Theory 1907-1933*, eds. Anton Kaes, Nicholas Baer, and Michael Cowan, trans. Alex H. Bush, 489-492 (Berkeley, CA: University of California Press, 2016), 489.

Simultaneously, his conception of technology as a means to materialize ideas bespeaks a techno-romantic outlook. Ewers characterized his objectives in the following way:

> From the outset I was aware that above all cinema's means of expression had to be taken into account in order to create a new art that would be on par with the theatre. [...] What one owes to the ear, one has to offer to the eye to a greater degree. Thus I first of all chose settings of extraordinary beauty that theatre scenery is hardly able to present to the viewer. Then I created in the figure of the hero's double a means of expression that is out of question for the theatre. I was thus able to play the hero's past in concrete form against the present man. In this way, I could give the drama something that has arguably never been accomplished in the cinema – a deep philosophical content.[32]

As he acknowledged here, Ewers consciously brought into effect the two domains that critics found most consistent with cinema's "technological essence." Surpassing the theatre, location shots and trick effects were the areas from which film art was to arise.

While German silent cinema is often regarded as studio-bound, a closer look at 1910s "artistic" filmmaking reveals this assumption to be misleading. Many films showcased picturesque nature scenes, in part because they constituted a potent attraction that opinion leaders considered both "essentially cinematic" and respectable. As a result, in 1913 alone, several films with artistic aspirations were shot on location abroad.[33] Simultaneously, the fantastic established itself as the prevalent genre of German cinema, because supernatural subject matter involved spectacles while also demonstrating cinema's creativity and medium specificity. In 1913, for instance, Reinhardt's *Venezianische Nacht* and *Wo ist Coletti?* (Deutsche Vitascope, 1913, dir. Max Mack) and the majority of Ewers-Rye collaborations featured a variety of tricks. However, *Der Student von Prag* stands out as a pioneering effort because here, likely for the first time, a feature-length film was entirely reliant on trick technology.

32 Hanns Heinz Ewers "Entwicklungsmöglichkeiten," *Das Lichtbild-Theater* 39 (25 September 1913); quoted in Wilfried Kugel, "Entstehung, Umfeld und Folgen: *Der Student von Prag.*" Unpublished manuscript, Berlin, 1988/2007: 10.

33 Besides *Der Student von Prag*, *Das schwarze Los* was partially shot in Ticino and Lombardy, the Asta Nielsen vehicle *Der Tod in Sevilla* (Deutsche Bioscop, dir. Urban Gad) in Andalusia, and Max Reinhardt's *Venezianische Nacht* (PAGU) and *Die Insel der Seligen* were made in Venice and Liguria, and the Ewers-Rye-Seeber collaboration for Deutsche Bioscop *Die Augen des Ole Brandis* likewise in Northern Italy.

Doppelganger effects

The trick effects in *Der Student von Prag* are simple, yet attention to their particular design provides clues about the filmmakers' methods, objectives, and obstacles. The doppelganger shots in *Der Student von Prag* were created by means of split screen composites. Although this illusion had long before been prevalent in both still photography and film, cinematographer Guido Seeber took lifelong pride in having created the first feature-length film based on this trick. Due to unstable camera equipment, convincing split screen composites still posed a challenge in the 1910s. Seeber used a French Prévost camera for *Der Student von Prag*, the same that is pictured in his trick photograph in Figure 3.1. Like all European cameras at the time, it relied on simple pressure pads to stop the film intermittently during exposure, which made it difficult to achieve impeccable registration and thus stable composites. Hand-cranking and wobbly tripods further jeopardized image steadiness. As I have shown in Chapter 2, to create a split screen composite, only part of the frame is exposed at a time while the rest is protected from light with the aid of a stationary matte. To this end, Seeber preferably mounted a slider on a sunshade in front of the lens. After the first take, the negative was rewound and the sunshade turned 180 degrees to mask and expose the other half.[34] Matte and counter-matte must fit together perfectly, since gaps or overlaps between them lead to over- or underexposed areas in the frame. If actors approach the zone between the image components, they appear partially transparent. To match two consecutive performances, silent cameramen would count out loud the number of metres exposed, while assistants monitored the numbers and the scene. Sometimes, darkly clad counterparts would play along in the masked portion of the frame to help actors with pacing and eyeline matches. This host of challenges notwithstanding, for Seeber it was a question of honour never to simulate a doppelganger illusion by shooting stand-ins from the back.[35] Even though the split-screen composites in *Der Student von Prag* are far from flawless, they attest to the filmmakers' remarkable fervour and ambition.

With only 85 shots in 77 minutes (not counting the twenty-five title cards in the most recent restoration), the cutting rate of *Der Student von*

34 For details see Guido Seeber, "Doppelgängerbilder im Film," *Die Kinotechnik* 1 (September 1919): 12-17.

35 Guido Seeber, *Der praktische Kameramann*, vol. 2, *Der Trickfilm in seinen grundsätzlichen Möglichkeiten* [1927] (Frankfurt am Main: Deutsches Filmmuseum, 1979), 61

Prag is extraordinarily slow—almost three times the average shot length of 1910s European films.[36] This tableaux-like staging may have been intended to foster a poetic atmosphere or to give prominence to the actors' performances, the vistas of Prague, and/or the doppelganger effects. Although *Der Student von Prag* constitutes a film historical milestone, in many ways its methods are still rooted in early cinema. This is particularly evident in the trick sequences, each of which consists of one tableau. In the fashion of early trick films, some portions of the same tableau were photographed split-screen and others regularly. Transitions between them were effected by means of substitution splices, which were already used to facilitate sudden, magical transformations in early trick films. The camera was stopped mid-shot, all action frozen, an element of the mise-en-scène added and/or removed, and then the filming was resumed. The stopping of the camera resulted in a few overexposed frames, which were later removed by means of an invisible cut, which is why the technique has also become known as "substitution splice." In *Der Student von Prag* the method was used to add or eliminate the frame splitting mid-shot to facilitate either the appearance of the double or to make the entire frame accessible to the protagonist after the double has disappeared. They are mostly unperceivable, because the double usually either enters or exits through fades or, like ordinary mortals, on foot. Because this modus operandi is not always obvious, it is all the more regrettable that the restorers at the Munich Film Museum chose to remove evidence of the substitution splices, rendering it difficult to reconstruct from the newer digital restorations precisely how these shots were achieved.

The doppelganger appears in eleven scenes: (1) Scapinelli orders the Other to step out of the mirror in Balduin's student lodgings. (2) Balduin encounters the Other when promenading on the gallery of Prague's Belvedere Palace. (3) The Other emerges behind a gravestone during Balduin's rendezvous with Margit at the Jewish cemetery. (4) Balduin meets the Other on the way to the duel with Waldis in the forest. (5) The Other joins Balduin at the card table at a party. (6) The Other appears during Balduin's nightly visit to Margit's boudoir. (7) The Other greets Balduin in front of the gate of the Schwarzenberg estate in Prague. (8) The Other materializes behind an open door in the Golden Lane inside the complex

36 Armin Jaeger's analysis of the 2013 restoration of *Der Student von Prag* calculates an average shot length of a hefty 41.7 seconds (image only: 51.4 seconds). See http://www.cinemetrics.lv/movie.php?movie_ID=13677, accessed July 30, 2018. Barry Salt assumes an average shot length of 15 seconds for European features in the period 1912-1917. See Salt, *Film Style and Technology*, 162.

of the Prague castle. (9) The Other drives the coach that picks up Balduin in the forest. (10) The Other joins Balduin in front of his house, Prague's Lobkowicz Palace. (11) The Other appears in Balduin's study, whereupon Balduin shoots at his double. As this breakdown indicates, the filmmakers made an effort to fashion the doppelganger scenes as heterogeneously as possible. To begin with, the double is placed in a variety of environments: Four doppelganger scenes take place indoors and seven outdoors. The most complex sequences were realized in the studio, but one third of doppelganger scenes were actually shot on location and feature recognizable sights of Prague. Outdoors, the creation of split screen composites was particularly tricky since a rapid shooting pace was necessary to avoid broken shadows caused by different positions of the sun.[37] The fact that the filmmakers made heavy use of real-world settings, which doubtlessly exacerbated the already considerable trick technical challenges they were facing, underscores their significance. Indeed, the vistas of Prague anchor the film's events in the real world and therefore lend eerie plausibility to the figure of the doppelganger.

Also with regard to composition, the filmmakers strove for variety: In five shots, the Other is placed in the left half of the split screen and in six in the right half. This effort to diversify settings and compositions may be traced to the fact that all doppelganger scenes employ the same three basic trick components: multiple exposure composites, fades, and substitution splices. The simplest doppelganger shots consist only of a split-screen composite (7, 9) or a split-screen composite with a one-sided fade (2, 8). More complex shots either commence with a split-screen composite, which is replaced with a regular shot by means of a substitution splice (3, 4, 10), or they start with a regular shot, which, following a substitution splice, is continued as a split-screen composite (5, 6). The first and the final (1, 11) doppelganger shot are the longest and technically most demanding and serve as bookends to the narrative. Both begin with a regular shot that is subsequently split using a substitution splice to facilitate the appearance of the double. Once the double has departed, a second substitution splice conceals the elimination of the split. The first shot additionally involves multiple substitution splices that facilitate the refilling of Scapinelli's magic purse, and the final shot contains a fade that allows the double to disappear.

In support of these trick components, the filmmakers took recourse to elaborate blocking, which reportedly required seven-hour rehearsals in some

37 See Seeber, *Der Trickfilm*, 40-67.

instances.[38] To cover up the substitution splices, Paul Wegener as Balduin commonly disappears behind a headstone, or a tree, turns his back to the camera, briefly leaves the room, or hides his face under a table. This insistence on non-technical means in support of the tricks was likely the result of the filmmakers' relative inexperience. Apart from Guido Seeber, all other members of the cast and crew were new to special-effects production—a fact that makes the overall project appear all the more daring.

At the same time, the intricate blocking was not only geared towards concealing the tricks, but also towards showcasing them. Spectacular effects often function at both diegetic and extradiegetic levels and consequently elicit a dual response from the audience. On the one hand, they facilitate fictional events. On the other, due to their astonishing nature, they call attention to themselves as technical feats. Dan North has characterized this paradox with respect to encounters between virtual and human actors in the following way: "Even as the shot asks us to accept that the two figures share the same narrative space, and thus that their spatial proximity can convey an empathetic bond, at another level we are invited to marvel at a complex technical achievement, and challenged to locate discrepancies in the illusion."[39] In *Der Student von Prag* the actors seek to leverage the dual perspective that North describes to heighten the audience's emotional involvement. On the face of it, Wegener's ostentatious reactions to the Other aim at conveying the protagonist's emotions and invite viewers to empathize. His incredulous staring and head scratching suggest the emotional state of someone doubting the reliability of their senses. At the same time, Wegener's actions also draw attention to the double as a technical feat. Although he never addresses the audience directly, Wegener behaves not unlike the illusionists of early trick films. When pointedly crossing to the other side of the split screen in search of the vanished doppelganger, he encourages the viewers to scrutinize the scene and to confirm the absence of any deceptive devices. Thus, in the fictional realm the audience is cued to empathize with Balduin and to experience uncertainty whether the doppelganger should be interpreted as a supernatural being or a figment of the protagonist's imagination. Simultaneously, in the real world, the spectators are invited to marvel at the powers of film technology, which reinforces their awareness of the doppelganger as a special effect.

38 For the rehearsal times, see Horst Emscher (= Josef Coböken), "Neue Wege in der Filmkunst," *Der Kinematograph* 363 (10 December 1913).

39 Dan North, *Performing Illusions: Cinema, Special Effects and the Virtual Actor* (London: Wallflower Press, 2008), 2.

The vistas and mysticism of old Prague

In view of the fact that *Der Student von Prag* is considered a milestone of special-effect history, it seems baffling that its trick technological accomplishments only played a subordinate role in the film's original reception. Hardly any reviewer went beyond general remarks like the double was "represented quite convincingly by means of various photographic tricks"[40] or that the filmmakers had accomplished a "technical feat of the first rank."[41] Some simply copied from studio publicity, which stated that the doppelganger was "a means of expression that only the cinema, but never the stage, can show in such perfection."[42] The lack of critical attention to the doppelganger effects attests to the deep-seated reserve towards technology in the cultural sphere, a stance that techno-romantic lines of argument sought to undermine.

For an overwhelming majority of commentators the main attraction of *Der Student von Prag* was not the double, but the city of Prague. For example, one of the few critics who commended Seeber's cinematography directly did not refer to the doppelganger effects but instead the location shots: "Anyone who has ever visited the Prague Castle will admire the clarity and focus of the exemplary cinematography, for which the cameraman was deservedly mentioned in the credits."[43] As a matter of fact, virtually all reviewers raved about the "extraordinary beautiful scenery,"[44] the "unique magic of the Golden City of Prague,"[45] and "the effects of the sublimely beautiful shots of old Prague, of the Jewish cemetery, and the labyrinthine lanes and alleys adorned with historic tarnish."[46] The studio evidently anticipated this inclination and its publicity clearly prioritized location cinematography over the doppelganger shots. For instance, a Deutsche Bioscop pre-release advertisement elaborated (in bold type) on the display of "historical locations

40 "Der Student von Prag," *Schlesische Zeitung* (23 August 1913); reprinted in Deutsche Bioscop, "Unsere Prophezeihung über *Der Student von Prag*," *Der Kinematograph* 349 (3 September 1913).
41 Alfred Richard Meyer, "Der Student von Prag," *Die Bücherei Maiandros* (1 October 1913): 11.
42 Premiere programme of *Der Student von Prag*, ed. Deutsche Bioscop Ges.m.b.H. Berlin [1913] in the Scripts, Gray Literature, and Audio Documents Archive of the Deutsche Kinemathek Berlin. See for example "Der Student von Prag," *Das kleine Journal* (25 August 1913).
43 Karl Bleibtreu, "Filmkritik [1914]," in *Kein Tag ohne Kino: Schriftsteller über den Stummfilm*, ed. Fritz Güttinger, 272 (Frankfurt am Main: Deutsches Filmmuseum, 1984).
44 G. K., "Künstlerfilms," *Der Kinematograph* 347 (20 August 1913).
45 Alfred Richard Meyer, "Der Student von Prag," *Die Bücherei Maiandros* (1 October 1913): 11.
46 *Berliner Tageblatt* (23 August 1913), reprinted in Deutsche Bioscop, "Unsere Prophezeihung über *Der Student von Prag*," *Der Kinematograph* 349 (3 September 1913).

of curious charm and beauty [...] that are usually inaccessible to the public,"[47] while the figure of the double was treated as secondary.

Prague was indeed an apt setting for culturally sophisticated photoplay with a supernatural aura. The picturesque Golden City with its mighty castle, its palaces and gardens, many of which were featured in the film, looks back on a compelling past. Prague possesses one of Europe's oldest universities with a strong presence of German nationalist duelling fraternities.[48] In the *Der Student von Prag*, Balduin's initial attire—and that of the double throughout the film—is recognizable as those of corps students. He is described as the "best fencer of Prague," which references the traditional fencing practiced by some right-wing student corporations in the German-speaking world. Finally, Prague was, in André Breton's words, the "magical capital of old Europe."[49] The city's mysterious ambiance corresponded to the film's subject matter as well as Ewers's, Wegener's, and Rye's personal predilection for the occult—an issue that I will explore in the context of *Nosferatu* in Chapter 4. As author Walter Mehring recalled, Prague was known as "an occult metropolis, because it lay nowhere in Europe, because, as the theosophists say, it lay in an astral plane."[50] In the sixteenth century, Holy Roman emperors Maximilian II and Rudolf II assembled at their courts a remarkable group of magicians, alchemists, cabbalists, and astrologers, and Jewish mystic Rabbi Judah Loew ben Bezalel, known as the Maharal of Prague, reportedly created a golem. In the twentieth century, prominent authors of the mystical, fantastic, and grotesque like Rainer Maria Rilke, Franz Kafka, Karl Hans Strobl, and Gustav Meyrink were residents of Prague.

Even if they rarely singled out trick effects, reviewers of *Der Student von Prag* recognized that Prague's mystery and beautiful vistas were more than simply decorative, that they served expressive goals and thus substantiated the film's artistic value. For instance, when complimenting the director for "revealing not only the picturesque, but all that is subjectively poetic in the natural ambience and the architecture of old Prague," author Karl Bleibtreu asserted cinema's ability to capture something beyond the scope

47 Advertisement in *Der Kinematograph* 345 (6 August 1913).

48 Ewers was a member and life-long supporter of three duelling fraternities. See Kugel, *Der Unverantwortliche*, 34, 36, 313, 330.

49 André Breton, "Surrealist Situation of the Object: Situation of the Surrealist Object [1935]," in *Manifestoes of Surrealism*, trans. Richard Seaver and Helen Lane, 255 (Ann Arbor, MI: University of Michigan Press, 1972).

50 Walter Mehring, *Die verlorene Bibliothek: Autobiographie einer Kultur* (Hamburg: Rowohlt-Verlag, 1952), 198.

of mechanical reproductions.[51] Similarly, the militant cinema reformer Adolf Sellmann believed that *Der Student von Prag* achieved a unity between spectacle and substance, praising the fact "that "magnificent location shots were joined with the mystical-romantic content to form a true work of art."[52] According to Horst Emscher, films like *Der Student von Prag* "require a cameraman who is not only a master of his métier, but also possesses refined artistic sensibility and thus becomes an artistically equal factor alongside the director and the author."[53] Just as Ewers had hoped, *Der Student von Prag* demonstrated to many critics the validity of the techno-romantic lines of argument, testifying to the camera's efficacy as a creative agent.

The horrific mirror image

Beyond offering stunning views, *Der Student von Prag* impressed critics especially with its ability to control audience emotions. Reviewers described the film as "suspenseful,"[54] "gripping,"[55] "giving you the creeps,"[56] or even as "seizing viewers with a frigid shock to the nerves."[57] In *Der Student von Prag*, "the ghastly fantastic comes alive, through vivid representations the ghostly gains blood heat without losing its eeriness in the least."[58] These responses bear witness to the fact that the film succeeded in filling its audiences with terror, which was highly unusual at the time. In contrast to the *féeries*, biblical or trick films of early cinema, the intended audience reaction to the supernatural was not surprise, astonishment, or amusement, but fear. Supernatural horror first emerged around 1910 when shorts like *Rusalka* (Khanzhonkov, 1910, dir. Vasili Goncharov), *Frankenstein* (Edison Manufacturing Company, 1910, dir. J. Searle Dawley) or Abel Gance's *Le Masque d'horreur* (Le Film Français, 1912) began to frame the supernatural as

51 Bleibtreu, "Filmkritik [1914]," in *Kein Tag ohne Kino*, ed. Güttinger, 272.

52 Adolf Sellmann, *Kino und Volksbildung* (M. Gladbach: Gemeinnützige Volksbibliothek, 1914), 7.

53 Horst Emscher (= Josef Coböken), "Neue Wege in der Filmkunst," *Der Kinematograph* 363 (10 December 1913).

54 "Der Student von Prag," *Das Kleine Journal* (25 August 1913); reprinted in Deutsche Bioscop, "Unsere Prophezeihung über *Der Student von Prag*," *Der Kinematograph* 349 (3 September 1913).

55 eht, "Theater und Musik," *Vossische Zeitung* (23 August 1913).

56 "Der Student von Prag," *Breslauer Zeitung* (23 August 1913); reprinted in Deutsche Bioscop, "Unsere Prophezeihung über *Der Student von Prag*," *Der Kinematograph* 349 (3 September 1913).

57 Heinrich Wildau, "Films der Woche," *Die Zeit am Montag* (25 August 1913); partially reprinted in Deutsche Bioscop, "Unsere Prophezeihung über *Der Student von Prag*," *Der Kinematograph* 349 (3 September 1913).

58 Bleibtreu, "Filmkritik [1914]," in *Kein Tag ohne Kino*, ed. Güttinger, 272.

menacing. It was not until *Der Student von Prag*, however, that paranormal horror became an element of feature-length film.

Noël Carroll has defined the horror genre as a group of films that are designed to provoke fear in the audience by confronting protagonists with a lethal danger, which is embodied in a supernatural monster.[59] In *Der Student von Prag*, the Other constitutes such a monster. Unaffected by the laws of nature, Balduin's double threatens and eventually destroys him. At the same time, with its simple plot, primacy of action, and lack of character psychology, *Der Student von Prag* abides by silent narrative cinema's presumed affinity to folk stories. We learn nothing about the protagonist Balduin except that he is a poor student, great fencer, and in love with an unattainable heiress. While this may seem like a shortcoming, as folklorist Maria Tatar has argued with regard to fairy tales, lack of character psychology may in fact intensify a narrative's symbolic potency: "By rigorously avoiding psychological analysis, the plots of fairy tales become charged with symbolic meaning. The physical descriptions and outer events of the tale serve not only to further the plot, but also to fashion ciphers of psychological realities."[60] In important ways, Tatar's observation also pertains to silent cinema. Unable to relate characters' thoughts and feelings verbally, silent filmmakers sought to impart interiority indirectly by way of actions, events, and visual signs. In *Der Student von Prag*, Wegener's histrionic gesticulation roughly outlines Balduin's state of mind and visual devices such as the Rembrandt lighting or the decorative skull in Balduin's student quarters provide symbolic clues to his situation. However, bereft of spoken language, the film's capacity to convey meaning by such means necessarily falls short of that of the theatre. *Der Student von Prag* overcomes this limitation by introducing the uncanny doppelganger as an outward manifestation of abstract concepts.

Ewers conceived of the Other as an allegory for Balduin's past, which clings to him and becomes his undoing.[61] Yet already psychologist Otto Rank, in his 1914 scholarly analysis of *Der Student von Prag*, pointed out that "this allegorical interpretation is unable either to plumb the content of the film or to justify fully the lively impression of its plot."[62] Instead of reading

59 Noël Carroll, "Film, Emotion, and Genre," in *Passionate Views*, eds. Carl Plantinga and Greg M. Smith, 21-47 (Baltimore, MD: Johns Hopkins University Press, 1999), 38.

60 Maria Tatar, *The Hard Facts of the Grimms' Fairy Tales* (Princeton, NJ: Princeton University Press, 2003), 79.

61 See advertisement in *Der Kinematograph* 345 (7 August 1913).

62 Otto Rank, *The Double: A Psychoanalytic Study* [1914], ed. and trans. Harry Tucker, Jr. (Chapel Hill, NC: The University of North Carolina Press, 1971), 6.

the double as a representation of the hero's past, Rank concluded that "the interesting and meaningful problems of man's relation to himself—and the fateful disturbance of this relation—finds [sic] here an imaginative representation."[63] Trick technology makes palpable the identity of protagonist and monster and thus allows the doppelganger to become a material emblem for the ideal. The techno-romantic paradigm thus enabled *Der Student von Prag* to non-verbally address philosophical issues and articulate universal, deep-rooted fears and desires.

The example of *Der Student von Prag* convinced Otto Rank that the representation of psychic processes on film could in fact surpass verbal communication: "It may perhaps turn out that cinematography, which in numerous ways reminds us of the dream-work, can also express certain psychological facts and relationships—which the writer often is unable to describe with verbal clarity—in such clear and conspicuous imagery that it facilitates our understanding of them."[64]

Most intellectuals at the time would have winced at this assertion. After all, a prevalent view held that "cinema's silent and visual portrayals, even in their most perfected form, lack the possibility to even approximate the moods and vibrations that the word evokes in the soul of the reader."[65] Still, some critics shared Rank's sentiment. In a remarkable review that appeared in *Bild & Film*, the leading publication of the cinema reform movement, educator Ernst Lorenzen maintained that *Der Student von Prag* showcased cinema's ability to articulate the indescribable:

> The dreamlike-demonic, slumbering below the threshold of consciousness in every human soul, which suddenly awakes and then struggles with all rational thought – here [in *Der Student von Prag*] it is expanded into a plot. The dream becomes image, which speaks so potently that the viewers ask themselves again and again: is this reality or is it merely a vision? This inner conflict causes the viewers to participate in the action as though they themselves had to fight through this piece of human destiny. [...] The photographic apparatus made it believable, because it could effortlessly solve all technical difficulties without interrupting the dreamlike quality of the whole, but rather underscored it as true and possible.[66]

63 Rank, *The Double*, 7.
64 Rank, *The Double*, 4.
65 "Der Autorenfilm und seine Bewertung," *Der Kinematograph* 326 (26 March 1913).
66 Ernst Lorenzen, "Hanns Heinz Ewers: Student von Prag," *Bild & Film* III:6 (1913/1914): 141.

For Lorenzen, the significance of *Der Student von Prag* lies in the actualization of inner psychic processes. At the same time, because the filmic images appear so factual, the audience is left uncertain whether Balduin's double constitutes a supernatural phenomenon and is objectively present or whether the doppelganger originates in a subjective delusion. As Lorenzen points out, this ambiguity creates an "inner conflict" in the viewers, which prompts them to fully immerse in the fictional world and its drama. Lorenzen's "inner conflict" is of course akin to the interpretive ambiguity that characterizes both Sigmund Freud's uncanny and Tzvetan Todorov's literary fantastic. Freud's uncanny describes an unsettling, fearful affect that most frequently arises from experiences that seemingly challenge our rational worldview. Uncanny events evoke repressed fears or desires or confirm ideas that we would ordinarily dismiss as superstitions. Todorov's fantastic, in contrast, refers to a narrative mode, namely an ambiguous, potentially supernatural situation within a realistic context. For Freud, the uncanny arises from a "conflict of judgment whether things which [...] are regarded as incredible, may not, after all, be possible."[67] In a similar way, Todorov characterizes a seemingly magical incident within a realistic narrative context that cannot be classified as either of natural or supernatural origin as "fantastic."[68] In order to correlate both, we might say that the *affective* reaction to Todorov's fantastic roughly corresponds to Freud's uncanny.[69]

In *Der Student von Prag* the hesitation between natural or supernatural explanations is first evoked in the moment when Balduin's reflection steps out of the mirror, prompting viewers, like Balduin, to doubt their own visual perception – a powerful and unsettling experience that Tom Gunning has called the "optical uncanny."[70] The groundwork for this climactic scene is laid earlier in the film when we see Balduin performing fencing moves in front of a mirror in his student dwelling. The mirror's unusual size seems conspicuously at odds with his dire financial situation. Balduin launches three attacks against his reflection, allowing the audience to

67 Sigmund Freud, "The Uncanny [1919]," in *The Standard Edition of the Complete Works of Sigmund Freud*, vol. 17, eds. James Strachey et al., 219-256 (London: The Hogarth Press and the Institute of Psychoanalysis, 1953-1974), 250.

68 Tsvetan Todorov, *The Fantastic: A Structural Approach to a Literary Genre*, trans. Richard Howard (Ithaca, NY: Cornell University Press, 1975), 33.

69 Todorov, *The Fantastic*, 47.

70 Tom Gunning, "Uncanny Reflections, Modern Illusions: Sighting the Modern Optical Uncanny," in *Uncanny Modernity: Cultural Theories, Modern Anxieties*, eds. Jo Collins and John Jervis, 68-90 (New York, NY: Palgrave Macmillan, 2008).

witness both the identity and the antagonism between him and his mirror image. Two scenes later we are back in the same room, where the lovesick Balduin receives a visit from Scapinelli. The line of sight is the same, yet the camera is set back a few metres, capturing a larger field of view to facilitate the upcoming split-screen portion of the scene. Having signed the deceitful contract, Balduin follows Scapinelli around the room, pointing out some of the worthless items in his possession, until he pauses in the left portion of the frame opposite his oversized mirror. Scapinelli points to the mirror, then to the incredulous Balduin and then to the contract. As Balduin lowers his head to re-read the agreement, Scapinelli makes magic passes over the mirror's surface. At this point, the restorers at the Munich Film Museum made the decision to insert the title card with the text of the contract for a second time in order to mask a slight jump when the split screen portion of the shot begins.[71] The doppelganger part commences as Balduin and the Other simultaneously raise their heads and drop the contract to the floor. The Other, staring at Balduin, starts to approach and eventually steps through the mirror frame into Balduin's space (Figure 3.3.). Here, the Other lingers for a moment, grinning and nodding towards the protagonist, then turns towards the exit, still staring at Balduin. Followed by Scapinelli, the doppelganger leaves through the door. Balduin looks after them, with his back turned towards the camera, which helps to conceal the scene's second substitution splice. Showing a mixture of bewilderment and amusement, Balduin now crosses into the (previously inaccessible) right portion of the frame. He gazes after them through the door and the window before catching sight of his empty mirror and approaching it in horror.

Running close to seven minutes in what appears as one single shot, the scene is one of the longest and technically most complicated in the entire film. It includes two substitution splices, which delimitate the split screen part of the shot, as well as two actors, one of whom appears in a dual role. Given the complexity of the sequence, it is not surprising that it contains technical mistakes. For example, the door to Balduin's room was closed when the left side of the frame was exposed, but left open for the right side. As a result, when Scapinelli and the Other leave, they dematerialize through a transparent closed door. Evidently, the filmmakers chose to retain such

71 The Munich restoration contains other problematic title cards, for example during Balduin's encounter with his doppelganger on the gallery of the Belvedere Palace. Here, the dialog card "Who are you?" was placed before Balduin even notices the Other, but when he visibly speaks to him a few moments later, no intertitle is provided.

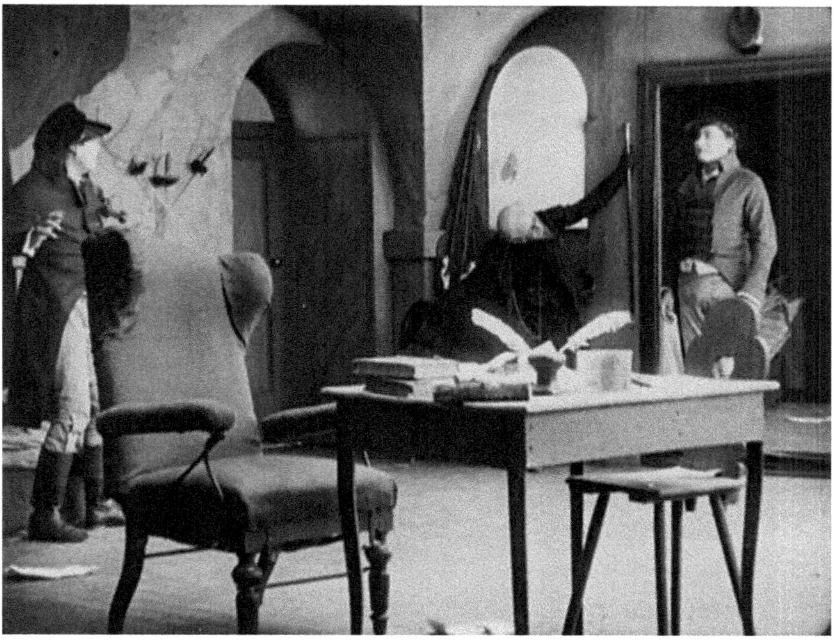

Fig. 3.3. Split screen composite in *Der Student von Prag* (1913): The doppelganger steps out of the mirror.

imperfections instead of spending additional time and expensive film stock on reshoots. However, when Seeber later recalled his work on *Der Student von Prag*, there was no mention of imperfections. Instead, he pointed out that the film "featured the first large-scale application of doppelganger shots. The execution caused a great stir back in the day. The illusion was so perfect that it was hard to convince many experts that I had exposed the respective scenes twice in succession, because in fact ... nothing could be noticed."[72] While it is certainly possible to discount this as self-delusion or boasting, it is also conceivable that the technical achievement and its emotional effectiveness were perceived compelling enough that technical flaws simply did not matter. The mirror reflection leaving its virtual existence and stepping through the frame into the real world prompts viewers to call into question not only the reliability of their senses but also the stability of their very concept of reality and consequently, the nature of the cinematic medium.

In fact, the technological double in *Der Student von Prag* repeatedly inspired critics to contemplate the essence of cinematic reproduction. In 1923, journalist Willy Haas wrote: "The problem of the double, as we

72 Guido Seeber, "Doppelgängerbilder im Film," *Kinotechnik* 1 (September 1919): 12-17.

know from Wegener's *Der Student von Prag*, is the truly demonic, the truly spiritual film problem: the film problem of all film problems."[73] Along those lines, Heide Schlüpmann and more recently Stefan Andriopoulos have given comprehensive interpretations of Balduin's doppelganger as a reflection on the dreadful disposition of the cinematic image, which, detached from its original, assumes a life of its own.[74] As Andriopoulos rightly points out, the eeriness of the process of cinematic reproduction was formulated already years before *Der Student von Prag*, when Berthold Viertel shuddered in 1910 about "this terrifying doubleness [*Doppelgängertum*] of representation."[75] Indeed, the notion that cinema produces uncontrollable, wicked duplicates of reality bespeaks the fundamental ambiguity of the techno-romantic paradigm. When construing technology as a means to access an immaterial dimension, techno-romantic lines of thought conceptualized it not necessarily as desirable, but just as easily as indeterminate or demonic and threatening. The vision of film as an uncanny doppelganger gives an eerie twist to the final shot of *Der Student von Prag*, which shows the doppelganger sitting on Balduin's grave (Figure 3.4.). According to this reading, the hero does not survive the encounter with his technologically reproduced self, but his replica proves immortal.

In addition to highlighting the terrifying quality of cinematic reproduction, *Der Student von Prag* also raises doubt regarding the assumed veracity of filmic images. The camera's mechanical nature and the chemical effects of light on the photographic emulsion suggest a direct relation between the filmic image and its referent, which does not permit any interventions. However, as Dan North has pointed out, "special effects are those interventions, subverting the integrity of the image's relation to its referent, which break down the blind trust that the viewer has been conditioned to place

73 Willy Haas, "November-Filme [1923]," in *Hätte ich das Kino! Die Schriftsteller und der Stummfilm*, eds. Ludwig Greve, Margot Pehle, and Heidi Westhoff, 172-173 (Stuttgart: Klett-Verlag, 1976), 172.

74 Heide Schlüpmann, "'Je suis la solitude:' Zum Doppelgängermotiv in *Der Student von Prag*," *Frauen und Film* 36 (1984): 10-24; Heide Schlüpmann, "The First German Art Film: Rye's *The Student of Prague*," in *German Film and Literature: Adaptations and Transformations*, ed. Eric Rentschler, 9-24 (New York, NY: Methuen, 1986); Stefan Andriopoulos, "The Terror of Reproduction: Early Cinema's Ghostly Doubles and the Right to One's Own Image," *New German Critique* 99 (Fall 2006): 151-170.

75 Berthold Viertel, "In the Cinematographic Theater" in [1910], *The Promise of Cinema: German Film Theory 1907-1933*, eds. Anton Kaes, Nicholas Baer, and Michael Cowan, trans. Eric Ames, 77-78 (Berkeley, CA: University of California Press, 2016), 78.

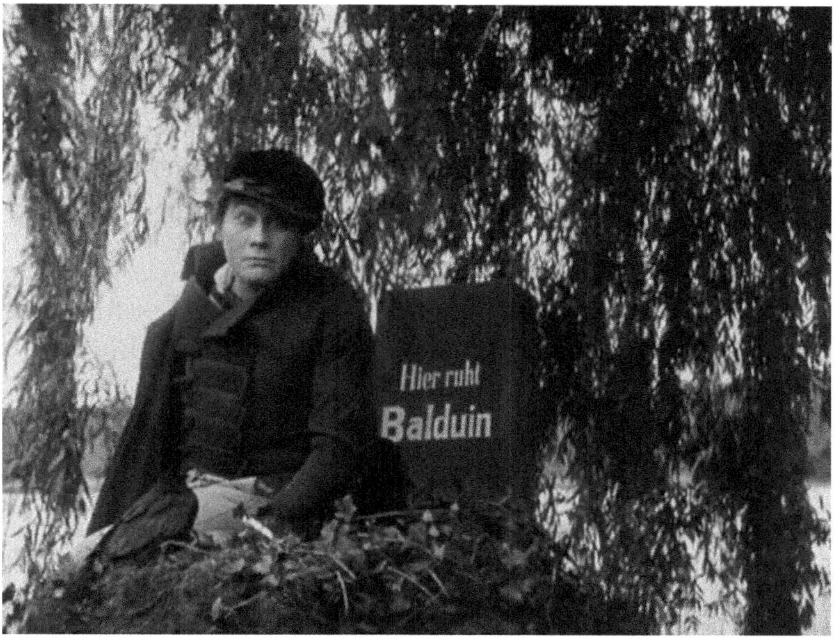

Fig. 3.4. Final shot of *Der Student von Prag* (1913): The doppelganger sits on the protagonist's grave.

in the film, and, by extension, all visual media."[76] The very existence of the doppelganger in *Der Student von Prag* provides visible evidence for the unreliability of cinema's truth claims. Ewers was fascinated by the fact that cinema's ontology encompasses both objectivity and make-believe. In his first novel, *Der Zauberlehrling* (1910), he described the cinema as a "fanatic of truth that knows no mistakes. At the same time, the cinematograph is the true alchemist; it smashes to pieces what reason preaches. It is the world's sole magician."[77]

By presenting both supernatural incidents and authentic locations, *Der Student von Prag* encompasses both of these tendencies. The intrusion of perceptibly real yet rationally inconceivable phenomena into a realistic context highlights that films, though ostensibly the result of automatic processes, are artful creations. Further, the confrontation with the supernatural in a world that viewers recognize as their own enhances their affective investment, as Béla Balázs argued in line with Freud and Todorov:

76 North, *Performing Illusions*, 23.
77 Hanns Heinz Ewers, "Schlangenfang auf Jawa [1910]," in *Kein Tag ohne Kino: Schriftsteller über den Stummfilm*, ed. Fritz Güttinger, 14-20 (Frankfurt am Main: Deutsches Filmmuseum, 1984), 15.

> It is marvellous and uncanny when unlikely events occur in this world. But in a different world, a fairy-tale world, such events are a matter of course. When the face of nature is distorted it can assume a spectral, supernatural expression. But only as long as we still recognize the face of the nature we know. [...] When, in short, the natural nature around us suddenly changes its physiognomy or its behaviour, that is when we start to feel a shiver of fear.[78]

As Balázs suggests, the affective potency of *Der Student von Prag* traces back to the juxtaposition of trick effects shots and authentic settings: The appearance of the doppelganger distorts the "face of nature," which grips the audience with "shivers of fear."

The creators of the *Der Student von Prag* pursued the ambitious goal of substantiating the feasibility of film art. Proceeding from techno-romantic assumptions, their approach relied heavily on special effects. They construed artistic filmmaking as the articulation of ideas and feelings through the imaginative application of the medium's technological assets. As I have argued in Chapter 1, contemporary critics identified those as nature scenes and trick effects. *Der Student von Prag* depicts the intrusion of an uncanny doppelganger into a real-life setting, the mystical and picturesque city of Prague. In so doing, the film bolsters medium-specific creativity with emotional thrill and visual spectacle. It combines its claim to art with mass appeal. As one of the earliest feature-length films, *Der Student von Prag* portrayed the supernatural as menacing and hence helped launch a new cinematic genre, horror. By inducing fear in the audience, it achieved unusual levels of affective viewer response. In addition, the detached mirror image invites a range of possible interpretations and it lends itself to philosophical reflections about identity and self. Finally, the confrontation between trick effects and location photography lends itself to a consideration of the eeriness of mechanical reproduction and deceptive verisimilitude of the photographic image. As a vehicle for abstract notions, the uncanny doppelganger thus bore witness to cinema's ability to convey figurative meaning and participate in the life of the mind. Popular with audiences and accommodating critics' demands for creativity and medium specificity, the juxtaposition of special effects and location work gained currency in 1910s German cinema. For

78 Béla Balázs, *Early Film Theory. Visible Man and The Spirt of Film*, ed. Erica Carter, trans. Rodney Livingstone (New York, NY/Oxford: Berghahn Books, 2010), 58. André Bazin argues along similar lines in "The Life and Death of Superimposition [1946]," in *Bazin at Work: Major Essays and Reviews from the Forties and Fifties*, ed. Bert Cardullo, trans. Alain Piette and Bert Cardullo, 73-77 (Oxford: Routledge, 1997).

instance, it became a characteristic feature of many of Paul Wegener's fantastic films. Also *Nosferatu* (Prana, 1922, dir. F.W. Murnau), the film I investigate in the next chapter, pursued this strategy. In fact, to make a case for film art, *Der Student von Prag* introduced a technological monster into a natural environment to serve as an agent of ideas and emotions. In so doing, it constituted a direct precursor to *Nosferatu*.

Bibliography

Emilie Altenloh, *Zur Soziologie des Kino: Die Kino-Unternehmung und ihre Besucher* (Leipzig: Spamerische Buchdruckerei, 1913).

Stefan Andriopoulos, "The Terror of Reproduction: Early Cinema's Ghostly Doubles and the Right to One's Own Image," *New German Critique* 99 (Fall 2006): 151-170.

Béla Balázs, *Early Film Theory. Visible Man and The Spirt of Film*, ed. Erica Carter, trans. Rodney Livingstone (New York, NY/Oxford: Berghahn Books, 2010).

André Bazin, "The Life and Death of Superimposition [1946]," in *Bazin at Work: Major Essays and Reviews from the Forties and Fifties*, ed. Bert Cardullo, trans. Alain Piette and Bert Cardullo, 73-77 (Oxford: Routledge, 1997).

Karl Bleibtreu, "Filmkritik [1914]," in *Kein Tag ohne Kino: Schriftsteller über den Stummfilm*, ed. Fritz Güttinger, 272 (Frankfurt am Main: Deutsches Filmmuseum, 1984).

André Breton, "Surrealist Situation of the Object: Situation of the Surrealist Object [1935]," in *Manifestoes of Surrealism*, trans. Richard Seaver and Helen Lane, 255 (Ann Arbor, MI: University of Michigan Press, 1972).

Noël Carroll, "Film, Emotion, and Genre," in *Passionate Views*, eds. Carl Plantinga and Greg M. Smith, 21-47 (Baltimore, MD:Johns Hopkins University Press, 1999).

"Der Autorenfilm und seine Bewertung," *Der Kinematograph* 326 (26 March 1913).

"Der Student von Prag," *Das kleine Journal* (25 August 1913).

Der Student von Prag, dir. Hanns-Heinz Ewers (1913, Edition Filmmuseum 80, Munich: film&kunst, 2016), DVD.

Deutsche Bioscop Gesellschaft, ed., *Unsere Künstler* (Promotion Booklet, 1913).

Deutsche Bioscop Ges.m.b.H. Berlin, ed., *Der Student von Prag* [1913], in the Scripts, Gray Literature, and Audio Documents Archive of the Deutsche Kinemathek Berlin.

Deutsche Bioscop, "Unsere Prophezeihung über *Der Student von Prag*," *Der Kinematograph* 349 (3 September 1913).

eht, "Theater und Musik," *Vossische Zeitung* (23 August 1913).

Horst Emscher (= Josef Coböken), "Große Kunst im Film," *Zeit im Bild* 3 (1913): 2524-2525.

Horst Emscher (= Josef Coböken), "Neue Wege in der Filmkunst," *Der Kinematograph* 363 (10 December 1913).
Marguerite Engberg, "*Studenten fra Prag* og den gådefulde Stellan Rye," in *Sekvens. Filmvidenskabelig årbog*, 161-185 (Copenhagen: Reitzel, 1982).
Hanns Heinz Ewers, "Der Kientopp," *Der Morgen* (Berlin) 18 (11 October 1907): 578.
Hanns Heinz Ewers, "*Der Student von Prag*," in Helmut H. Diederichs, *Der Student von Prag: Einführung und Protokoll* (Stuttgart: Focus Verlagsgemeinschaft, 1985).
Hanns Heinz Ewers, *Edgar Allan Poe*, trans. Adèle Lewisohn (New York, NY: B.W. Huebsch, 1917).
Hanns Heinz Ewers, "Schlangenfang auf Jawa [1910]," in *Kein Tag ohne Kino: Schriftsteller über den Stummfilm*, ed. Fritz Güttinger, 14-20 (Frankfurt am Main: Deutsches Filmmuseum, 1984).
Sigmund Freud, "The Uncanny [1919]," in *The Standard Edition of the Complete Works of Sigmund Freud*, vol. 17, eds. James Strachey et al., 219-256 (London: The Hogarth Press and the Institute of Psychoanalysis, 1953-1974).
G.K., "Künstlerfilms," *Der Kinematograph* 347 (20 August 1913).
Ludwig Greve, Margot Pehle, and Heidi Westhoff, eds. *Hätte ich das Kino! Die Schriftsteller und der Stummfilm* (Stuttgart: Klett-Verlag, 1976).
Tom Gunning, "Like Unto A Leopard: Figurative Discourse in *Cat People* (1942) and Todorov's *The Fantastic*," *Wide Angle* 3 (1988): 30-39.
Tom Gunning, "Uncanny Reflections, Modern Illusions: Sighting the Modern Optical Uncanny," in *Uncanny Modernity: Cultural Theories, Modern Anxieties*, eds. Jo Collins and John Jervis, 68-90 (New York, NY: Palgrave Macmillan, 2008).
Willy Haas, "November-Filme [1923]," in *Hätte ich das Kino! Die Schriftsteller und der Stummfilm* eds. Ludwig Greve, Margot Pehle, and Heidi Westhoff, 172-173 (Stuttgart: Klett-Verlag, 1976).
E. T. A. Hoffmann, *Phantastische Geschichten*, ed. Hanns Heinz Ewers (Munich: Georg Müller Verlag, n.d.).
E. T. A. Hoffmann, *Sämtliche Werke: Historisch-kritische Ausgabe*, ed. Carl Georg von Maassen, 10 vols. (Munich: Georg Müller Verlag, 1908-1928).
Wilfried Kugel, *Der Unverantwortliche: Das Leben des Hanns Heinz Ewers* (Düsseldorf: Grupello Verlag, 1992).
Wilfried Kugel, "Die Rekonstruktion des Stummfilms *Der Student von Prag* (Deutsche Bioscop GmbH Berlin, 1913)." Unpublished manuscript, Berlin, 1988.
Wilfried Kugel, "Entstehung, Umfeld und Folgen: *Der Student von Prag*." Unpublished manuscript, Berlin, 1988/2007.
Ernst Lorenzen, "Hanns Heinz Ewers: Student von Prag," *Bild & Film* III:6 (1913/1914): 141.
Walter Mehring, *Die verlorene Bibliothek: Autobiographie einer Kultur* (Hamburg: Rowohlt-Verlag, 1952).

Alfred Richard Meyer, "Der Student von Prag," *Die Bücherei Maiandros* (1 October 1913): 11.

Corinna Müller, "Das 'andere' Kino? Autorenfilme in der Vorkriegsära," in *Die Modellierung des Kinofilms: Zur Geschichte des Kinoprogramms zwischen Kurzfilm und Langfilm 1905/06-1918, Mediengeschichte des Films*, vol. 2, ed. Harro Segeberg, 153-192 (Munich: Wilhelm Fink Verlag, 1998).

Corinna Müller, *Frühe deutsche Kinematographie: Formale, wirtschaftliche und kulturelle Entwicklungen 1907-1912* (Stuttgart: Metzler-Verlag, 1994).

Dan North, *Performing Illusions: Cinema, Special Effects and the Virtual Actor* (London: Wallflower Press, 2008).

Hans Pander, "Intertitles [1923]," in *The Promise of Cinema: German Film Theory 1907-1933*, eds. Anton Kaes, Nicholas Baer, and Michael Cowan, trans. Alex H. Bush, 489-492 (Berkeley, CA: University of California Press, 2016).

Edgar Allen Poe, *Edgar Allen Poes Werke in zehn Bänden*, eds. Heda and Arthur Moeller-Bruck, trans. Hedda Moeller-Bruck and Hedwig Lachmann, 10 vols. (Minden: J. C. C. Bruns Verlag, 1901-1904).

Edgar Allen Poe, *Nebelmeer*, ed. Hanns Heinz Ewers, trans. Gisela Etzel (Munich: Georg Müller, 1914).

Otto Rank, *The Double: A Psychoanalytic Study* [1914], ed. and trans. Harry Tucker, Jr. (Chapel Hill, NC: The University of North Carolina Press, 1971).

Paul Rohnstein, "Beiträge zur wirtschaftlichen Entwicklung der Deutschen Film-Industrie (unter besonderer Berücksichtigung des Kinomatographentheatergewerbes [sic])," Ph.D. Dissertation, Julius-Maximilians-Universität Würzburg, 1922.

Barry Salt, *Film Style and Technology: History and Analysis* (London: Starwood, 1983).

Heide Schlüpmann, "'Je suis la solitude:' Zum Doppelgängermotiv in *Der Student von Prag*," *Frauen und Film* 36 (1984): 10-24.

Heide Schlüpmann, "The First German Art Film: Rye's *The Student of Prague*," in *German Film and Literature: Adaptations and Transformations*, ed. Eric Rentschler, 9-24 (New York, NY: Methuen, 1986).

Guido Seeber, *Der praktische Kameramann*, vol. 2, *Der Trickfilm in seinen grundsätzlichen Möglichkeiten* [1927] (Frankfurt am Main: Deutsches Filmmuseum, 1979).

Guido Seeber, "Der Student von Prag," *Lichtbild-Bühne* (15 November 1926).

Guido Seeber, "Doppelgängerbilder im Film," *Die Kinotechnik* 1 (September 1919): 12-17.

Guido Seeber, *Kamera-Kurzweil: Allerlei interessante Möglichkeiten beim Knipsen und Kubeln* (Stuttgart: Union Deutsche Verlagsgesellschaft, 1930).

Adolf Sellmann, *Kino und Volksbildung* (M. Gladbach: Gemeinnützige Volksbibliothek, 1914).

Spektator, "Autorenkünstler [sic] und Riesenfilms," *Der Kinematograph* 349 (3 September 1913).

Maria Tatar, *The Hard Facts of the Grimms' Fairy Tales* (Princeton, NJ: Princeton University Press, 2003).

Kristin Thompson, "Im Anfang war...: Some Links between the German Fantasy Films of the Teens and the Twenties," in *Before Caligari: German Cinema 1895-1920*, eds. Paolo Cherchi Usai and Lorenzo Codelli, 138-161 (Pordenone: Edizioni Biblioteca dell'Immagine, 1990).

Tsvetan Todorov, *The Fantastic: A Structural Approach to a Literary Genre*, trans. Richard Howard (Ithaca, NY: Cornell University Press, 1975).

Casper Tybjerg, "The Faces of Stellan Rye," in *A Second Life: German Cinema's First Decades*, eds. Thomas Elsaesser and Michael Wedel, 151-159 (Amsterdam: Amsterdam University Press, 1996).

Berthold Viertel, "In the Cinematographic Theater" [1910], in *The Promise of Cinema: German Film Theory, 1907-1933*, eds. Anton Kaes, Nicholas Baer, and Michael Cowan, trans. Eric Ames, 77-78 (Berkeley, CA: University of California Press, 2016).

Paul Wegener, "Warum ich für den Film spiele," *Berliner Börsen-Courier* 321 (12 July 1914).

Daniel Wiegand, *Gebannte Bewegung: Tableaux vivants und früher Film in der Kultur der Moderne* (Marburg: Schüren-Verlag, 2016).

4. Visualizing the Occult: *Nosferatu* (1922)

Abstract
Public enthusiasm for occult themes was rife in the decades around 1900 and many German filmmakers and intellectuals were occupied with esoteric concepts. Constitutive of occultist thought is a belief in secret realities beyond our perceptual abilities. The vampire tale *Nosferatu* does not advocate for specific doctrines, but many of the film's idiosyncratic aspects, particularly the appearance, behaviour, and powers of the vampire, become intelligible in novel ways when examined from an occultist perspective. The film externalizes the vampire's nature through cinematic devices and most notably special effects. The materialization of the intangible by means of technology constitutes an essentially techno-romantic project. In *Nosferatu*, it served to consolidate objectives of occultists and cinephiles for the purpose of film art.

Keywords: occultism, vampires, the invisible, camera effects, stop-motion

The period between the 1870s and the 1930s saw a surge of occultist ideas and practices across the Western world. Although the history of occultism goes back to antiquity, a sharp upturn in public interest in the last quarter of the nineteenth century suggests that the occult revival constitutes a characteristically modern phenomenon. For philosopher Richard von Coudenhove-Karlegi, enthusiasm for the occult was indicative of "modern romanticism," which also underlay other zeitgeist sentiments like nostalgia, exoticism, and faith in the future.[1] All expressed longing for other, remote, and mysterious worlds and sprang from anguish in the face of modernity's concrete, cold, and pragmatic materialism, a feature that machine technology epitomized. The techno-romantic paradigm constitutes a response to the same affliction, yet paradoxically seeks to attain the intangible, specifically

1 Richard Nikolaus Coudenhove-Kalergi, *Apologie der Technik* (Leipzig: Verlag Der neue Geist Dr. Peter Reinhold, 1922), 67-68.

Loew, Katharina, *Special Effects and German Silent Film: Techno-Romantic Cinema*.
Taylor & Francis Group, 2021
DOI 10.5117/9789463725231_CH04

through technology. What concerns me here are the intersections between techno-romantic and occultist lines of thought. Although both construe the spiritual as the essential feature of human existence, occultism's quest for invisible realities and eternal truths through the pursuit of ancient hermetic wisdom did not specifically pertain to machine technology, and techno-romantic perspectives neither constituted a systematic doctrine nor were they expressly concerned with mysterious forces. Nonetheless, in cinema, occultist and techno-romantic approaches proved highly compatible.

In the decades around 1900, fascination with the occult permeated all layers of European society. Many artists, including filmmakers, engaged with occultist thought and practices. For occultists, the visual arts held the promise of making occultist theories perceptible to the senses, as Theosophist Franz Hartmann indicated: "This spirit of light, called the soul of the world (the Astral Light), is a spiritual substance, which can be made visible and tangible by art."[2] Filmmakers and critics appreciated occult subject matter for its popular appeal and for demonstrating core functions of art, such as expressing the imagination, facilitating encounters with the mysterious, and communicating the effects of mental processes. Because visualizations of psychic and supernatural phenomena inevitably drew on special effects, occult content used to advantage cinema's unique area of competence, encouraged technical and stylistic experiments, and consequently helped expand cinema's means of expression. Asserting technology's value in the service of an idealist project, the cinematic occult thus epitomizes a techno-romantic attitude.

Compared to researchers in literature and the visual arts, film scholars have paid little attention to occultism. Although even a cursory look at film history suggests that a host of silent productions, from early trick films and serials to experimental works and late silent horror classics, can be associated with occultist ideas and motifs, the bearing of the occultist movement on silent film culture has yet to be scrutinized. The occult is an unwieldy object of investigation. Concerned with the scientifically unknowable, occultism is often seen as hokum or indicative of a reactionary relapse into pre-modern mysticism. Occultists have constructed intricate, sometimes secret knowledge and belief systems, which, to make matters worse, may vary widely from one school of thought to another. Further, an artist's engagement with the occult does not automatically result in occultist artworks, and in a commercial and collective medium like cinema occultist thought may be particularly difficult to identify.

2 Franz Hartmann, *Magic, White and Black* [1886] (New York, NY: The Path, 1895), 213.

The vampire tale *Nosferatu* lends itself to investigations of occultist traces in cinema, because it was created with the explicit objective of demonstrating the artistic treatment of such ideas on film. However, with the exception of Sylvain Exertier and Luciano Berriatúa, whose work has met with little response in Anglophone scholarship, few scholars have examined this connection in any detail.[3] Instead, research on *Nosferatu* has primarily focused on three areas. First, academics have probed the film's relation to the traditional arts, in particular to Bram Stoker's *Dracula* and Romantic painting.[4] Second, it has been read psychoanalytically with regard to issues of gender and sexuality embodied in the figure of the vampire.[5] Third, scholars have construed the film in light of its socio-historical context. For instance, they have understood *Nosferatu* as an expression of the national trauma of World War I or as a harbinger of National Socialism, confirming Siegfried Kracauer's influential interpretation of Count Orlok as one of many "tyrant" figures in German cinema that presaged Hitler.[6]

What is more, various scholars have also detected antisemitic imagery in the film.[7] The vampire's hooked nose and long claw-like fingers, his association with rats and the plague, his alterity, origins in Eastern Europe, and clandestine invasion of Germany have been read as anti-Jewish stereotypes. Indeed, vampirism itself is a well-established antisemitic trope, evoking blood-libel accusations and the identification of "Jews" with bloodsucking

3 Sylvain Exertier, "La lettre oubliée de *Nosferatu*," *Positif* (March 1980): 47-51; Luciano Berriatúa, *Nosferatu: Un film Erótico – Ocultista – Espiritista – Metafísico* (Valladolid: Divisa Red, 2009).

4 For *Dracula*, see for instance Friedrich A. Kittler, "Dracula's Legacy," trans. William Stephen Davis, *Stanford Humanities Review* 1 (1989): 143-173; Judith Mayne, "Dracula in the Twilight: Murnau's *Nosferatu* (1922)," in *German Film & Literature: Adaptations and Transformations*, ed. Eric Rentschler, 25-39 (New York, NY: Methuen, 1986). For Romantic painting see Lotte H. Eisner, *The Haunted Screen: Expressionism in the German Cinema and the Influence of Max Reinhardt* [1952], trans. Roger Greaves (Berkeley, CA: University of California Press, 2008); Angela Dalle Vacche, *Cinema and Painting: How Art is Used in Film* (Austin, TX: University of Texas Press, 1996), 161-196; Ken Calhoon, "F. W. Murnau, C. D. Friedrich, and the Conceit of the Absent Spectator," *MLN* 3 (April 2005): 633-653.

5 See for instance Robin Wood, "Murnau's Midnight and Sunrise," *Film Comment* (May/June 1976): 4-19; Thomas Elsaesser, "No End to *Nosferatu* (1922)," in *Weimar Cinema*, ed. Noah Isenberg, 79-94 (New York, NY: Columbia University Press, 2008).

6 For the psychological legacy of World War I, see Anton Kaes, *Shell Shock Cinema: Weimar Culture and the Wounds of War* (Princeton, NJ: Princeton University Press, 2009). For premonitions of Nazism see Siegfried Kracauer, *From Caligari to Hitler: A Psychological History of the German Film* [1947] (Princeton, NJ: Princeton University Press, 2004), 77-88; Patrick Colm Hogan, "Narrative Universals, Nationalism and Sacrificial Terror from *Nosferatu* to Nazism," Film Studies 8 (Summer 2006): 93-105.

7 See for instance Rolf Giesen, *The Nosferatu Story: The Seminal Horror Film, Its Predecessor and Its Enduring Legacy* (Jefferson, NC: McFarland, 2019).

capitalists. Much of this traces back to *Dracula*, which has been repeatedly read as a compilation of antisemitic metaphors, including some that are not present in *Nosferatu*, such as the vampires' stalking of children, their aversion to Christian symbols, and the Count's greed.[8] The anti-Jewish imagery in *Nosferatu* is certainly worthy of scrutiny, but regarding the film as a quasi-precursor to *Der ewige Jude* (Terra-Film, 1940, dir. Fritz Hippler) is unwarranted. It seems unlikely that Jewish author Henrik Galeen would have penned such a propaganda effort; F. W. Murnau, who lived as the "substitute son" in the house of his partner's Jewish family, directed it; and two prominent Jewish actors, Galeen's brother-in-law John Gottowt and Alexander Granach, appeared in it. Many commentators base their case for the film's antisemitism on Orlok's supposed physiological "Jewishness." However, it is an oversimplification to read a large, aquiline nose necessarily as "Jewish," given that it allegedly also characterizes noble Roman and Native American warriors. An examination of antisemitic caricatures since the nineteenth century suggests that "Jewish" noses were not only identified by their size and "bow shape," but also by "nostrility" and downward turn of the tip.[9] Further, swarthiness, hairiness, large jug ears, bulbous thick lips, protruding beady eyes with droopy upper eyelids, and a short and puffy stature mark a stereotypically "Jewish" appearance. Orlok shares none of these characteristics and his physiognomy resembles more closely the hook-nosed mummies of Egyptian pharaohs than nineteenth-century anti-Jewish caricatures.

In this chapter I focus on the ways in which *Nosferatu* construes and visualizes the nature of the vampire according to occultist tenets. His bizarre appearance, demeanour, and powers become comprehensible in novel ways when reading them as expressions of occultist thought. Given that occultism has received little attention in film studies to date, I will begin by examining ideas and cultural contexts from which the cinematic occult emerged. Insight into the extent to which occultism percolated into the European cultural sphere not only casts a new light on German silent film culture, but also suggests areas of future research.

The creators of *Nosferatu* aspired to actualizing, in a technological visual medium, complex ideas such as esoteric philosophy's law of correspondence and analogy. Adhering to a techno-romantic mind-set, they employed film technology, most notably special effects, in the service of the ideal. The trick

8 For a reading of antisemitic tropes in *Dracula*, see Carol Margaret Davison, *Anti-Semitism and British Gothic Literature* (New York, NY: Palgrave, 2004).

9 See Sander Gilman, *The Jew's Body* (Milton Park: Taylor & Francis, 1991), 169–193.

techniques employed in *Nosferatu* are simple, yet they draw from the full spectrum of effects: Prosthetic makeup, laboratory, lighting, and editing techniques, mechanical and camera effects such as multiple exposure, stop motion, and accelerated motion. All serve the same purpose: to externalize the nature of the vampire and the occult dimension he inhabits. Technology enabled film artists to pursue the unseen, transcendental, and spiritual concealed in material and abstract phenomena, an emotional reality beyond outward appearances. In *Nosferatu*, the techno-romantic project to visualize the invisible by means of film technology consolidated occultist concerns with a cinephile zeal for film art.

The modern occult revival

"Occult" is a notoriously diffuse term without a generally accepted definition. It is often used synonymously with terms such as esoteric, arcane, or mystic. Although doctrines and practices of the different occult traditions can vary significantly, they all share a preoccupation with secret or hidden spiritual realities, with phenomena that lie beyond our normal experience or current standard explanations. The occult was already a field of inquiry in antiquity, but it was not until the nineteenth century that it became the centre of a mass movement. The modern occult revival, which reached its climax in the decades around 1900, was characterized by two major tendencies. The first stood in the tradition of Mesmerism and spiritualism and was occupied with empirical investigations of paranormal and psychic phenomena inexplicable to contemporary science. The second, represented by hermetic groups and movements like the Freemasonry, the Theosophical Society (established 1875), New Rosicrucianism (since 1875), the Hermetic Order of the Golden Dawn (established 1888), the Ordo Templi Orientis (OTO, established 1904), or the Anthroposophical Society (established 1913), was devoted to the cultivation of secret knowledge and the prospect of obtaining access to higher worlds through spiritual education. Regardless of their diverse concerns and approaches, most occultist groups proceed from similar basic premises. As Bruce Elder explains, it is generally assumed

> that the universe is a single, living being with attributes we understand as pairs of opposite (male-female, positive-negative, light-dark, active-passive, yin-yang, vertical-horizontal); that this single being undergoes evolution; that this evolution reconciles the differences between what

> we understand as the opposing features of this single being; that that a principle of analogy (signatures) interrelates all entities, and relates those on a lower plane with those on a higher plane—matters are above as they are below, and objects with similar appearances have similar spiritual powers (thus the great interest in diagrams, charts, and symbols among members of these groups); that epiphanies reveal beings which our senses are unable to present; that the imagination makes reality, hence the mind and world are one; and that spiritual beings undergo an evolutionary process through which they come into possession of ever higher states of consciousness.[10]

The fin de siècle occult revival has often been interpreted as a reactionary departure from rationality and relapse into pre-modern mysticism. Most famously, Theodor W. Adorno argued in *Minima Moralia* (1951) that occultism must be understood as a "regression in consciousness" and "metaphysic of dunces"[11] The perspective manifest in Adorno's potent contention shaped the discourse for decades. More recently, however, scholars such as Corinna Treitel and Priska Pytlik have insisted that the occultist movement should not simply be construed as a contemptible flight from reason, but rather as a cultural-historical phenomenon that was both integral to and had causes of friction with industrial modernity.[12] The proliferation of occultist thought around 1900 complicates the "secularization thesis," which posits that modernization processes rendered religion irrelevant, at least in the public sphere.[13] Occultist belief systems were very much in evidence in social and cultural life at the time and although they do not usually involve the institutional worship of supreme beings, they are fundamentally concerned with spiritual phenomena and the divine. The popularity of occultist ideas and practices also seems to contradict Max Weber's notion of "modern disenchantment," which is usually described as the condition of the world once enlightened reason and scientific methods

10 R. Bruce Elder, *Harmony and Dissent: Film and Avant-garde Art Movements in the Early Twentieth Century* (Waterloo, ON: Wilfrid Laurier University Press, 2008), 16.

11 Theodor W. Adorno, *Minima Moralia: Reflections from Damaged Life* [1951], trans. E. F. N. Jephcott, 238-244 (London: Verso, 1974), 238; 241.

12 Corinna Treitel, *A Science for the Soul: Occultism and the Genesis of the German Modern* (Baltimore, MD: Johns Hopkins University Press, 2004); Priska Pytlik, *Okkultismus und Moderne ein kulturhistorisches Phänomen und seine Bedeutung für die Literatur um 1900* (Paderborn: Ferdinand Schöningh, 2005).

13 For an overview of secularization theory see Hugh Mcleod, *Secularisation in Western Europe, 1848-1914* (New York, NY: St. Martin's Press, 2000).

have superseded transcendence and mystery.[14] However, the anachronism of spiritual impulses and their incompatibility with modern rationality, which deliberations about secularization and disenchantment tend to allege, are at least open to doubt. For instance, groups like the Theosophical Society, the most influential occultist organization at the time, drew on various spiritual traditions including Neoplatonism, Gnosticism, Jewish Kabbalah, and Hermeticism, as well as tantric Hinduism and Buddhism, while also explicitly embracing Enlightenment values. As stated by its co-founder Helena Blavatsky, the society's objective was to strengthen the spiritual element in order to unsettle the unfettered dominance of materialism—according to Blavatsky an adverse reaction to earlier religious bigotry—and to reveal eternal truths from affinities between scientific, philosophical, and religious teaching.[15] What is more, as Jason A. Josephson-Storm has argued, a more comprehensive reading of Weber reveals that he did not actually postulate the disappearance of magic and mysticism in the wake of scientific progress, but rather the denial and demonization of such ideas in favour of the illusion of a rational modern world.[16] In fact, as Josephson-Storm notes, Weber himself was active in occultist circles and, although disenchantment became modernity's prevalent ideology, it did not reflect the experience of large segments of the European population.

The occult revival attests to a profound ambivalence toward modernization processes, highlighting the simultaneous sway of and deep-seated frustration with both religion and the natural sciences. Philosopher and educator Friedrich Paulsen, for instance, criticized in 1902 the fact that the natural sciences, like religion and philosophy before them, were not living up to their promise of conveying a comprehensive understanding of the world:

> A new generation, suspicious of [Hegelian] reason like an earlier one of faith, turned to science: Exact science will put us on terra firma and provide us with a reliable worldview. But science does not accomplish this. It becomes increasingly clear that science does not provide us with a holistic worldview that satisfies imagination and soul; it only produces

14 See Max Weber, "Science as Vocation [1918]," in *The Vocation Lectures*, eds. David Owen and Tracy B. Strong, trans. Rodney Livingstone, 1-32 (Indiapolis, IN: Hackett Publishing Company, 2004), 13.

15 Helena P. Blavatsky, *Isis Unveiled: A Master-Key to the Mysteries of Ancient and Modern Science and Theology*, vol. 1 (New York, NY: J.W. Bouton, 1877), xlv.

16 Jason A. Josephson-Storm, *The Myth of Disenchantment: Magic, Modernity, and the Birth of the Human Sciences* (Chicago, IL: University of Chicago Press, 2017), 269-301.

> a thousand fragmentary insights. The result is disappointment: science does not satisfy our hunger for knowledge.[17]

The impression that the natural sciences were unable to provide convincing answers was not only the result of their perceived materialist bias, but also a consequence of a profound scientific crisis, which culminated in a collapse of the conventional physical worldview at the turn of the century.

Nineteenth-century physics had offered a comprehensive and coherent understanding of the natural world. It was assumed that all basic laws of nature were known. Newtonian mechanics conceived of time and space as absolute. The cosmos was understood as a perfectly precise and predictable mechanism. Knowledge of all the parts of this clockwork universe would allow for the precise prediction of all future events. Classical physics described the world as consisting of three interconnected entities: matter, ether, and energy. All three obeyed the newly formulated laws of thermodynamics. Matter was comprised of immutable and indivisible chemical atoms. Ether was the all-pervasive medium for the propagation of light waves and other types of electromagnetic radiation.

This unified system began to falter in 1887 when strong evidence against the ether theory was found. Several subsequent discoveries, including that of x-radiation, gamma rays, and electrons, further undermined the classical worldview. In the following decades, most of the basic ideas held by scientists at the end of the nineteenth century were overturned. A host of theoretical novelties emerged across a broad range of disciplines, which radically transformed our understanding of the world. By the 1920s classical physics, with its absolutes and its ether, had become a thing of the past. Physicists led by Albert Einstein had created a conceptually new world.

The breakdown of the stable Newtonian cosmology, in particular the discovery of the first subatomic particle in 1897, sparked an intellectual and emotional public crisis. The fact that matter did not actually consist of "solid, massy, hard, impenetrable, moveable Particles," as Newton had claimed, was experienced as a disintegration of material reality.[18] Wassily Kandinsky recalled in 1913:

17 Friedrich Paulsen, *Die deutschen Universitäten und das Universitätsstudium* (Berlin: Verlag von Asher & Co, 1902), 81-82.

18 Isaac Newton, *Opticks: Or, a Treatise of the Reflections, Refractions, Inflections and Colours of Light* [1704] (Mineola, NY: Dover Publications, 1979), 400.

> The crumbling of the atom was to my soul like the crumbling of the whole world. Suddenly, the heaviest walls toppled. Everything became uncertain, tottering and weak. I would not have been surprised if a stone had dissolved in the air in front of me and become invisible. Science seemed to me destroyed: its most important basis was only an illusion, an error of the learned, who did not build their godly structures stone by stone with a steady hand in transfigured light, but groped at random in the darkness for truth and blindly mistook one object for another.[19]

The collapse of the classical worldview not only divested materialist science of its absolute authority; the discovery of subatomic particles and invisible rays provided justification for inquiries into other intangible phenomena. Like the emerging academic discipline of experimental psychology, occultism aspired to engage with inner worlds scientifically. Both fields shared an interest in psychic phenomena, including hypnosis, telepathy, psychokinesis, or near-death experiences and many psychologists, most prominently Carl Gustav Jung, were also concerned with occultist perspectives.[20] Even Sigmund Freud, whose scepticism has often been asserted, occupied himself with the occult and, as Jason A. Josephson-Storm has shown, was far more invested than is usually assumed.[21]

In fact, many distinguished scientists, including Nobel Prize laureates Marie and Pierre Curie, Charles Richet, William Crookes, John William Strutt, and Santiago Ramón y Cajal, were part of occultist circles. Parapsychologist Carl du Prel explained the relationship between science and the occultist movement in the following way: "Occultism is nothing but unknown natural science. It will be proven by the natural sciences in the future, but even today scientists can no longer make fundamental objections. Only the crudest materialists are still struggling against it."[22] Investigations into the invisible and immaterial were in vogue, and for many contemporary observers, radioactivity, dreams, clairvoyance, and somnambulism were not only analogous, but equally worthy of scientific study.

Ideologically, the European occult revival accommodated political views of all stripes. The movement attracted numerous right-wing sympathizers

19 Wassily Kandinsky, "Reminiscences," in *Modern Artists on Art*, ed. Robert L. Herbert, trans. Eugenia W. Herbert, 19-44 (Mineola, NY: Dover Publications, 1999), 27.

20 For links between psychology and the occult, see Alex Owen, *The Place of Enchantment: British Occultism and the Culture of the Modern* (Chicago, IL: University of Chicago Press, 2004).

21 Josephson-Storm, *The Myth of Disenchantment*, 191-206.

22 Carl du Prel, *Der Spiritismus* (Leipzig: Reclam-Verlag, 1893), 15.

and had strong reactionary and fascist currents. Simultaneously, many groups embraced progressive causes like gay and women's rights, ethnic pluralism, and the abolition of capital punishment as well as non-normative behaviours including non-traditional forms of sexuality, vegetarianism, or the use of psychoactive substances. The ideological ambiguity of certain occultist positions traces back to theoretical inconsistencies, which afforded diverse interpretations. For example, the Theosophical Society managed to combine egalitarian core ideals with hierarchical teachings. According to Helena Blavatsky, the group's principal goal was "to form a nucleus of the Universal Brotherhood of Humanity without distinction of race, creed, sex, caste or colour."[23] Blavatsky explicitly championed the joint purpose and common, divine origin of all human beings. At the same time, however, her concept of succeeding prehistoric races ("root races") is hierarchically structured and thus inadvertently inspired racist-esoteric theories, which in turn left traces in aspects of Nazi ideology. The association of Nazism with occultism has attracted intense attention particularly in the Anglophone world. Such scrutiny is usually geared towards portraying Nazism as the result of arcane and demonic influence. However, as Nicholas Goodrick-Clarke has shown, although Ariosophist "ideas and symbols filtered through to several anti-Semitic and nationalist groups in late Wilhelmian Germany, from which the early Nazi Party emerged," the notion that occultist thought principally inspired National Socialism is untenable.[24]

Occulture

In the decades around 1900, enthusiasm for the occult reached all strata of European society. In the uneducated classes, fortune telling, dowsing, and conjuring spirits of the dead simultaneously functioned as a parlour game and as a substitute religion. Among the cultural elites, attending séances, reading up on Theosophy, and listening to lectures by Rudolf Steiner was fashionable. Everybody, it seemed, was exploring occultist ideas, including public figures like Henri Bergson, William Butler Yeats, Maurice Maeterlinck, Fanny zu Reventlow, Albert Schweizer, Arthur Schnitzler, Walter Gropius, Igor Stravinsky, Isadora Duncan, Victor Hugo,

23 Helena Petrovna Blavatsky, *The Key to Theosophy* (London: The Theosophical Publishing Company, 1889), 39.
24 Nicholas Goodrick-Clarke, *The Occult Roots of Nazism: The Ariosophists of Austria and Germany, 1890-1935* (Wellingborough: The Aquarian Press, 1985), 5.

Leo Tolstoy, James Joyce, and Arnold Schönberg. However, not everyone who engaged in occultist thought and practices was equally committed. Some, like Arthur Conan Doyle and August Strindberg, self-identified as occultists; others, like Thomas Mann and Hermann Hesse, were simultaneously intrigued and sceptical. As Tessel M. Bauduin and Henrick Johnsson have argued, "modernist interaction with occult currents ranges across the entire religionist spectrum from faith to agnosticism or atheism, with all possible positions, including prominently typically (post-) modernist devices such as irony and paradox, in between."[25]

Author Max Brod's intense engagement with and concurrent critical detachment from occultist ideas exemplifies the attitude of many European intellectuals, including filmmakers. According to his own account, Brod associated with author and occultist Gustav Meyrink, studied a host of occultist writings, participated in numerous spiritualist séances, and attended Rudolf Steiner's Theosophist lectures. He recalled "pleasant memories of the sweet, purely physical sensation when the animated table begins to move under one's fingers, this un-earthly pressure that one cannot resist, then the multiplicity of individual ghosts [...] I cannot rid myself of the impression that there was some truth to the mysterious messages from Zemun."[26] Brod was clearly fascinated by occultist ideas, but also maintained an analytical distance: "I am neither a spiritualist nor an anti-spiritualist, neither an anti-theosophist nor a theosophist. 'What is your worldview, then?' I am a literary man. [...] 'Higher worlds' interest me only in literary terms."[27] Art was the true means of transcendence for Brod, whereas the dogmas of a particular occultist current ultimately had little spiritual significance: "A poem by Goethe or a fugue by Reger still struck me as more surprising, more mysterious and more venerable then all occult manifestations."[28]

Even if many intellectuals stopped short of endorsing the full scope of doctrines of a specific occultist group, fundamental principles—particularly the idea of an inherent unity of all aspects of existence, belief in the pre-eminence of the spiritual over the material, interest in realities beyond human perception and cognition, and a penchant for the mysterious and supernatural—shaped a generation of European scholars and artists. Author Walter Mehring, for instance, remembered that at *Café des*

25 Tessel M. Bauduin and Henrik Johnsson, "Introduction: Conceptualizing Occult Modernism," in *The Occult in Modernist Art, Literature, and Cinema*, eds. Tessel M. Bauduin and Henrik Johnsson, 1-30 (Cham: Palgrave Macmillan, 2019), 17.

26 Max Brod, "Höhere Welten," *Pan* 16 (16 June 1911): 538-545, 542-543.

27 Max Brod, "Höhere Welten," *Pan* 16 (16 June 1911): 538-545, 538.

28 Max Brod, "Höhere Welten," *Pan* 16 (16 June 1911): 538-545, 540.

Westens, a key meeting point for Berlin artists and intellectuals before World War I, parapsychological phenomena "were unanimously accepted by all stylistic cliques and political sects, their usual feuds notwithstanding. [...] Divination, telepathy, acoustic and optical hallucinations did not seem more occult to us than for example the invisible X-rays, Madame Curie's radium or the Freudian unconscious."[29] In view of this pervasive approval, it is hardly surprising that occultist thought had a profound impact on all arts around 1900, including literature, painting, architecture, photography, sculpture, music, theatre, dance, and cinema. Bauduin and Johnsson speak of a dynamic interaction between occultism and modernism, in the course of which "occultism offers modernism new modes of expression; so does modernism for occultism. [...] Occultism transforms and is transformed by modernism."[30]

Occultist interests were particularly evident in the revival of fantastic and horror literature between approximately 1880 and 1930, which, as I have indicated in Chapter 3, also manifests in *Der Student von Prag* (Deutsche Bioscop, 1913, dir. Stellan Rye). Virtually all exponents of fantastic literature at the time—authors like Bruno Schulz, Guy de Maupassant, Henry James, George du Maurier, H. G. Wells, Karl Hans Strobl, Gustav Meyrink, and Hanns Heinz Ewers—were engrossed in occultist thought and practices. However, works by authors like Christian Morgenstern, Oskar A. H. Schmitz, D. H. Lawrence, Ezra Pound, Sylvia Plath, and Oscar Wilde, which were not concerned with supernatural forces, also bespeak their creators' engagement with the occult.[31]

It is not circumstantial that the era's predominant aesthetic tenets coincide with occultist assumptions. For example, the holistic impetus of concepts such as synaesthesia, the total work of art (*Gesamtkunstwerk*), and *Stimmung*, a term that describes both the emotional tone of an environment and the intuitive attunement of an observer's mood to that external atmosphere, not only refers back to Romantic progressive universal poetry, but also has clear affinities to a central Theosophical tenet: the fundamental oneness of all there is. Blavatsky described it in the following way: "The radical unity of the ultimate essence of each constituent part of compounds

29 Walter Mehring, *Berlin Dada* (Zurich: Verlag der Arche, 1959), 14-15.

30 Bauduin and Johnsson, "Introduction," in *The Occult in Modernist Art, Literature, and Cinema*, 8.

31 For the links between modernist literature and the occult see for instance Pamela Thurschwell, *Literature, Technology, and Magical Thinking, 1880-1920* (Cambridge: Cambridge University Press, 2001); Leon Surette, *The Birth of Modernism: Ezra Pound, T.S. Eliot, W.B. Yeats, and the Occult* (Montreal & Kingston: McGill-Queen's University Press, 1993).

in Nature—from Star to mineral Atom, from the highest Dhyan Chohan [celestial beings] to the smallest infusoria [minute aquatic creatures], in the fullest acceptation of the term, and whether applied to the spiritual, intellectual, or physical worlds—this is the one fundamental law in Occult Science."[32]

The occult preoccupation with the tensions between visible and invisible, material and immaterial is particularly evident in the work of visual artists. On the one hand, occultist views found expression in the transcendental-mystical iconography of figures like Alfred Kubin and Hugo Steiner-Prag and symbolists like Fidus and Bolesław Biegas. On the other hand, experiments with form and colour, which characterize virtually all modernist avant-garde movements including Cubism, Futurism, Suprematism, Expressionism, Cubo-Futurism, Rayonism, and Surrealism, had their roots in occultist thought, which consequently also played a vital role in the articulation of the principles of abstract art.[33] In keeping with the conception of occultism as a science of invisible realities, artists sought to free themselves from visual references in the physical world and instead pursued the unseen, transcendental, and spiritual concealed in material and abstract phenomena, an emotional reality beyond outward appearances. Not coincidentally, the pioneers of nonrepresentational art, Hilma af Klint, Kazimir Malevich, Piet Mondrian, František Kupka, Max Ernst, and Wassily Kandinsky, were all closely familiar with Theosophy, Anthroposophy, and nineteenth-century Freemasonry.

In *Concerning the Spiritual in Art* (1912), a seminal text in the history of abstract art, Kandinsky formulated a stance that many German filmmakers

32 Helena P. Blavatsky, *The Secret Doctrine* [1888], ed. Boris de Zirkoff, vol. 1 (Wheaton, IL: The Theosophical Publishing House, 1893), 120.

33 For the links between modernist painting and the occult see for instance Tessel M. Bauduin, "Abstract Art as By-Product of Astral Manifestation: The Influence of Theosophy on Modern Art in Europe," in *Handbook of the Theosophical Current*, eds. Olav Hammer and Mikael Rothstein, 429-451 (Leiden: Brill, 2013); Linda Dalrymple Henderson, "Rethinking Modern Art, Science, and Occultism in Light of the Ether of Space: Wassily Kandinsky, Umberto Boccioni, and Kazimir Malevich," in *The History of Art and 'Rejected Knowledge': From the Hermetic Tradition to the 21st Century*, ed. Anna Korndorf, 218-37 (Moscow: The State Institute of Art Studies, 2018); Linda Dalrymple Henderson, "Abstraction, the Ether, and the Fourth Dimension: Kandinsky, Mondrian, and Malevich in Context," in *Kandinsky, Malevich, Mondrian: The Infinite White Abyss*, eds. Marian Ackermann and Isabelle Malz, 233-44 (Düsseldorf: Kunstsammlung Nordrhein-Westfalen, 2014); V. Loers and P. Witsman, eds., *Okkultismus und Avant-Garde: Von Munch bis Mondrian* (Frankfurt am Main: Schirn-Kunsthalle, 1995); Friedhelm Wilhelm Fischer, "Geheimlehren und moderne Kunst," in *Fin de Siècle: Zur Literatur und Kunst der Jahrhundertwende*, ed. Roger Bauer, 344–377 (Frankfurt am Main: Klostermann, 1977).

with artistic aspirations shared at least in principle.[34] Kandinsky argued that the present materialist period had crippled humanity emotionally and spiritually, and art had become purposeless as a result. True art had to emerge from an inner need, a spiritual impulse that impelled artists to search for forms that expressed inner life rather than representing external contingencies. The impetus towards abstraction was directly linked to the move away from materiality and towards spirituality, which, as Kandinsky underscored, had been spearheaded by Blavatsky and the Theosophical Society.[35] Kandinsky's radical rejection of materialism and objective representation in art seems difficult to reconcile with a medium that produces photographic records of material objects. However, and despite working in a commercial context, many German filmmakers shared the view that the purpose of art in the modern age was to give form to emotions and ideas and to touch the soul of observers. Cinema's most obvious means to, as Kandinsky postulated, adapt form to its content, was trick technology. Special effects thus came to embody the techno-romantic paradigm and allowed filmmakers to strive for the spiritual in art.

Occult tropes were so omnipresent in the cultural sphere around 1900 that it is difficult to clearly distinguish between occultism as a philosophical belief system and a pop-cultural phenomenon. Occult themes and motifs not only shaped the work of artists like Kandinsky, who was actively invested in the teachings of the Theosophical Society, but also of those who were to varying degrees "occult-curious," and those who simply sought to profit from the pervasive occultist vogue. Christopher Partridge has called the wide circulation of ideas and practices rooted in occultism "occulture."[36] The term describes a vernacular occultism not limed to elite, underground communities but one that permeates society as a whole.

As mentioned before, in fin-de-siècle Europe, occultist ideas embodied a potent alternative to pervasive reification, reductive rationalism, mechanization, societal fragmentation, and secularization, promising a metaphysical reframing of the world from a holistic and non-normative perspective. Furthermore, the secretive, exclusive, opaque, and exotic nature of occultist teachings and practices had widespread appeal. Finally, occultism dovetailed with major societal trends. As previously indicated,

34 Wassily Kandinsky, *Concerning the Spiritual in Art*, trans. Michael T. H. Sadler (Mineola, NY: Dover Publications, 1977).
35 Kandinsky, *Concerning the Spiritual in Art*, 13-14.
36 Christopher Partridge, "Occulture is Ordinary," in *Contemporary Esotericism*, eds. Egil Asprem and and Kennet Granholm, 113-133 (New York, NY: Routledge, 2014).

the discovery of invisible forces like x-radiation or radium, the invention of "disembodied" technologies like the wireless telegraph, telephone, or phonograph, or theories like the unconscious seemed to underline the validity of occultist claims about the existence of invisible forces. Inventors also worked on techno-romantic machines and processes such as Robert Hare's Spiritoscope, psychic photography, William H. Mumler's spirit photography, or Thomas Edison's spirit telephone that were geared toward making hidden forces perceptible.[37] Like the creators of *Nosferatu*, they pursued the goal of popularizing occultist ideas. However, their chief purpose was to substantiate the objective truth of these convictions. *Nosferatu*, in contrast, sought to visualize them in the service of film art.

Cinema and the occult

Nosferatu is undoubtedly the best-known silent film with occultist connections, but it was by no means the first. Occultist and occultural subjects were a constant presence on film screens from the earliest days, not least because they afforded magical spectacles and employed to advantage cinema's trick technological tools. Although early trick films were not usually concerned with a spiritual dimension, they broached topics in keeping with occultist views, including the contingency of seemingly unambiguous dichotomies like visible/invisible, animate/inanimate, and material/immaterial. In addition, trick films regularly dealt with themes that reflected an occultist fascination with "the mystical East." Theosophists, for instance, conceived of ancient Egypt as the mysterious land of the most revered wisdom, while India and Tibet represented cultures that rejected materialism and reductive rationalism in favour of a spiritual path to enlightenment. Their search for age-old secrets and the divine origins of humanity coincided with the height of European imperialism, when stereotypes about the orient and its tantalizing otherness gained currency in Western societies at large. In cinema, Christopher Partridge's observation, "nineteenth century Western 'occulture' was Orientalist," is therefore corroborated, for instance by *féeries* inspired by *One Thousand and One* Nights like *Ali Baba et les quarante*

37 For these projects see Robert Hare, *Experimental Investigation of the Spirit Manifestations* (New York, NY: Partridge & Brittan, 1855); Arthur Brunel Chatwood, *The New Photography* (London: Downey, 1896); Thomas Edison, "The Realms Beyond [1920]," in *The Diary and Sundry Observations of Thomas Alva Edison*, ed. Dagobert D. Runes, 238-241 (New York, NY: Philosophical Library, 1948).

voleurs (Pathé Frères, 1902, dir. Ferdinand Zecca) or *Le Palais des mille et une nuits* (Star Film, 1905, dir. Georges Méliès), and films featuring oriental magicians like *Le Thaumaturge chinois* (Star Film, 1904, dir. Georges Méliès) or *Le Sorcier arabe* (Pathé Frères, 1906, dir. Segundo de Chomon).[38] However, cinema also addressed the occultist vogue directly. In early films like *A Visit to the Spiritualist* (Edison, 1899, dirs. J. Stuart Blackton and Albert E. Smith) or *Spiritisme abracadabrant* (Star Films, 1900, dir. Georges Méliès), spiritualism served as pretext for supernatural occurrences, while later productions either ridiculed or endorsed occultist beliefs and practices.[39] Occultist themes and perspectives were pervasive, but have not always been recognized. An obvious example is *Metropolis* (Ufa, 1927), a film that Fritz Lang described as "a battle between modern science and occultism, the science of the middle ages." [40]The film contains numerous occultist references, many of which play a prominent role in ceremonial magician Aleister Crowley's spiritual philosophy Thelema, such as the whore of Babylon, Moloch, ancient Egyptian iconography, inverted pentagrams, and the "black magician" Rotwang.

As mentioned above, despite its substantial bearing on film history, the occultist movement has received comparatively little attention from film scholars to date.[41] Occultism's reputation as preposterous hocus-pocus, an underestimation of its socio-cultural influence, and lack of awareness of the

38 Christopher Partridge, "Orientalism and the Occult," in *The Occult World*, ed. Christopher Partridge, 611-625 (New York, NY: Routledge, 2015), 614.

39 Examples of silent films advocating occultist beliefs and practices include *The Mysteries of Myra* (International Film Service, 1916, dirs. Leopold Wharton and Theodore Wharton), *Theophrastus Paracelsus* (Joseph Delmont Film, 1916, dir. Joseph Delmont), *The Other Person/ Onder spiritistischen dwang* (Granger-Binger Films/Hollandia, 1921, dirs. Maurits Binger and B. E. Doxat-Pratt), *One Glorious Day* (Paramount, 1922, James Cruze), *Häxan* (Aljcsha Production Company/Svensk Filmindustri, 1922, dir. Benjamin Christensen), *Gespenster* (Carl Heinz Boese & Cie, 1923, dir. Carl Heinz Boese) or *The Magician* (MGM, 1926, dir. Rex Ingram). Examples of films ridiculing them include *Is Spiritualism a Fraud?* (Robert W. Paul, 1906, dir. J. H. Martin), *Ma fille n'épousera qu'un medium* (Pathé, Frères 1909), *Bébé fait du spiritisme* (Gaumont, 1912, dir. Louis Feuillade), *Der Geisterseher* (Reinhold Scholz, 1915, dir. Waldemar Hecker), or *Are Crooks Dishonest?* (Rolin Films, 1918, dir. Gilbert Pratt).

40 Peter Bogdanovich, *Fritz Lang in America* (London: Studio Vista, 1967), 124.

41 A notable exception has been research on experimental cinema. See Elder, *Harmony and Dissent*; Kristoffer Noheden, *Surrealism, Cinema, and the Search for a New Myth* (Cham: Palgrave Macmillan, 2017); Benedikt Hjartarson, "Ghosts Before Breakfast: The Appetite for the Beyond in Early Avant-Garde Film," in *The Occult in Modernist Art, Literature, and Cinema*, eds. Tessel M. Bauduin and Henrik Johnsson, 137-162 (Cham: Palgrave Macmillan, 2019); Judith Noble, "Clear Dreaming: Maya Deren, Surrealism and Magic," in *Surrealism, Occultism and Politics: In Search of the Marvellous*, eds. Tessel M. Bauduin, Victoria Ferentinou, and Daniel Zamani, 210-226 (New York, NY: Routledge, 2018).

presence of occultist motifs are partly to blame. In addition, the theoretical substructure of film studies may have also played a role. The field has long been dominated by approaches that emphasize the critique of ideology, such as Marxist film theory, Apparatus theory, Psychoanalytic film theory, and Critical Theory. As a result, film studies may have been more reluctant than literary studies or art history to engage with a mercurial socio-cultural phenomenon operating at the fringes of the knowable.

It is particularly surprising that the links between the occultist movement and German cinema, which built a reputation on its "haunted screen," have not received attention to date. During the silent era, Germany produced dozens of films with obvious occultural inclinations, including a conspicuous number of orientalist occultural films.[42] Two subject areas of interest to occultists were particularly prevalent: psychic powers and afflictions such as hypnosis, telepathy, clairvoyance, somnambulism, hallucination, dissociation, and curses on the one hand and supernatural and/or monstrous beings including hybrids, cripples, lunatics, revenants, artificial humans, devils, and magicians on the other. Anton Kaes has rightly pointed to a surge of fascination with the occult and particularly spiritualism in the wake of the First World War, "because it held out the promise of contact with the spirits of relatives killed in battle."[43] However, by the late 1910s the occult had already played a central role in European culture for decades. As I have suggested in Chapter 3, supernatural subject matter took hold of German cinema around 1913 and remained its distinguishing feature for the remainder of the silent era. There is little doubt that producers sought to take advantage of the popularity of occulture and the emotional potency of eerie spectacles. At the same time, many leading German filmmakers were personally involved in occult matters. Fritz Lang, for instance,

42 Examples include canonical films like *Das Cabinet des Dr. Caligari* (Decla, 1920, dir. Robert Wiene), *Der müde Tod* (Decla-Bioscop, 1921, dir. Fritz Lang), and *Faust* (Ufa, 1926, dir. F. W. Murnau), and less known productions like *Die Augen des Ole Brandis* (Deutsche Bioscop, 1913, dir. Stellan Rye), which revolves around the supernatural ability to see others as they really are, *Und das Wissen ist der Tod* (Deutsche Bioscop, 1915, dir. Walter Schmidthässler), a story about precognition, *Der Yoghi* (PAGU, 1916, dir. Paul Wegner), which deals with a Hindu magician and his invisibility potion, *Nachtgestalten* (1920, dir. Richard Oswald) about a diabolical plot to drain oxygen from the atmosphere in pursuit of world domination, or *Ramper, der Tiermensch* (Defu, 1927, dir. Max Reichmann), which addresses therianthropy. Many occultural films were set in Egypt, like *Die Augen der Mumie Ma* (PAGU, 1918, dir. Ernst Lubitsch) and *Das Rätsel der Sphinx* (Ellen Richter Film, 1921, dir. Adolf Gärtner), in India like *Der Fakir im Frack* (Greenbaum-Film, 1916, dir. Max Mack) or *Das indische Grabmal* (May-Film, 1921, dir. Joe May), and in Tibet like *Die Fremde* (Decla, 1917, dir. Otto Rippert) and *Lebende Buddhas* (Paul Wegener Film, 1925, dir. Paul Wegener).
43 Kaes, *Shell Shock Cinema*, 102.

characterized his favoured subjects in the following way: "I am interested in occultism, painting and modern functional rooms."[44] Also Max Reinhardt, F. W. Murnau, Paul Wegener, Henrik Galeen, Ludwig Berger, Hans Poelzig, and of course Hanns Heinz Ewers were demonstrably preoccupied with occultist ideas.[45] In the case of filmmakers like Robert Wiene, Richard Oswald, Artur Robison, Rochus Gliese, Robert Reinert, Paul Leni, Joe May, or Hans Mierendorff (who incidentally founded a production company called "Lucifer-Film"), I have not been able to establish occultist penchants, but their repeated recourse to supernatural subject matter might suggest related concerns. Indeed, these clues suggest that the impact of the occult revival on German silent cinema represents a vast area in need of detailed research.

As I argue throughout this book, cinema became an important domain for offsetting the perceived preponderance of materialism. Itself an epitome of mechanistic materialism, film was enlisted to address concerns also at the core of the occultist movement: to make perceptible realities inaccessible to our senses, to challenge the putative opposition of matter and spirit, and to accentuate the simultaneous unity and mutability of all there is. Special effects served as the principal means to bridge the material and immaterial. Indeed, German cinema's famous penchant for stylization and technical innovation reflects tenacious efforts to affirm the pre-eminence of spirit over matter. The persistent techno-romantic calls for the valorization of a spiritual dimension in film and the conception of the medium as capable of transcending material reality—positions unmistakably entwined with the occultist cause—paved the way for the acceptance of film as art.

Vampires and film art

It is mistaken to assume that an artist's attention to occult matters automatically manifests in their work. However, occultist-inflected art necessarily

44 Paul Dubro, "Frau im Mond, Fritz Lang und ein Interview," *Ufa Feuilleton* 42 (16 October 1929): 4.

45 See for instance, for Murnau, Thomas Elsaesser, *Weimar Cinema and After: Germany's Historical Imaginary* (London: Routledge, 2000), 226; for Wegener, Heide Schönemann, *Paul Wegener: Frühe Moderne im Film* (Stuttgart: Edition Axel Menges, 2003), 64; for Ewers, Wilfried Kugel, *Der Unverantwortliche: Das Leben des Hanns Heinz Ewers* (Düsseldorf: Grupello Verlag, 1992), 411; for Galeen, Piotr Sadowski, *The Semiotics of Light and Shadows: Modern Visual Arts and Weimar Cinema* (London: Bloomsbury, 2017), 133; for Ludwig Berger, see Ludwig Berger, *Wir sind vom gleichen Stoff aus dem die Träume sind: Summe eines Lebens* (Tübingen: R. Wunderlich, 1953), 365; for Max Reinhardt see Pytlik, *Okkultismus und Moderne,* 69; for Poelzig, Hans-Peter Reichmann, *Hans Poelzig: Bauten für den Film* (Frankfurt am Main: Deutsches Filmmuseum, 1997), 21.

presupposes creators with pertinent knowledge, and *Nosferatu* was produced by filmmakers with demonstrable occult interests. Director F. W. Murnau had a lifelong fascination with the occult, screenwriter Henrik Galeen—incidentally also involved in the production of *Der Student von Prag*—was associated with New Rosicrucianism, and co-producer and art director Albin Grau was a senior officer of the Berlin Pansophical Lodge, a New Rosicrucian magical order.[46] In January 1921, Grau co-founded Prana-Film, a studio named after the vital principle in Hindu philosophy and represented by a logo comprised of the interlocking spirals of the Taoist *taijitu*. Established with the avowed intention of producing films with occult themes, Prana-Film announced an ambitious line-up of future projects, which included a production about the "devil's violinist" Niccolò Paganini, adaptations of Edward Bulwer-Lytton's Rosicrucian novel *Zanoni* (1842) and his putative occultist roman à clef *The Coming Race* (1871), as well as a four-part film series titled *Höllenträume* ("Dreams of Hell").[47] However, due to financial problems, *Nosferatu* remained its only release.[48]

However, *Nosferatu* was not simply intended as a promotional film for occultism. Instead, the rendering of occultist ideas was intertwined with the objective of creating film art. The producers intended to pay heed to cinema's nature and utilize its technological capabilities for the purpose of creating a medium-specific artwork. This ambition is evident from a promotional trade press article that was published on the occasion of the film's release and likely penned by Albin Grau:

> But there are things, though not entirely new, which have only been treated very sparsely, very superficially in film. Occultism, for instance, is one of them. [...] The Prana-Film-Corporation, of whose chief executives several have long engaged with occult studies, came up with the idea of making vampirism the basis of a film script. The external cause was minor: the idea occurred to the persons in question when they observed a spider sucking dry its prey. But the idea alone was not enough. The goal was not just to show a certain aspect of occultism, but to create a real film.

46 According to Volker Lechler, the persistent rumour that Grau was subsequently also a member of the Fraternitas Saturni (established 1926), which followed the teachings of notorious English ceremonial magician Aleister Crowley, is inaccurate. See Volker Lechler, *Die ersten Jahre der Fraternitas Saturni* (Stuttgart: Verlag Volker Lechler, 2015), 69-86.

47 For a history of Prana see Michel Bouvier and Jean Louis Leutrat, *Nosferatu* (Paris: Gallimard, 1981), 230-233.

48 Without providing references, Giesen mentions a second, unreleased Prana production, the three-reel alpine documentary *Hochtouren im Vorfrühling*. Giesen, *The Nosferatu Story*, 85.

> Cinema's requirements, its conditions, and particularly its possibilities had to be tied to the topic not somehow randomly but consistently and intrinsically.[49]

The articulation of occultist ideas in a medium-specific (and thus artistic) way prompted the filmmakers to utilize a variety of cinematic techniques to make perceptible the invisible forces and realities at the heart of occultist investigations.

The source material for this ambitious project was Bram Stoker's 1897 novel *Dracula*, which had first appeared in German translation in 1908. Vampirism constituted a matter of great interest to fin-de-siècle occultists, and Stoker's treatment attracted significant attention in occultist circles.[50] In addition, the producers of *Nosferatu* doubtlessly hoped to capitalize on the novel's sensational character and the popularity of gory and supernatural subject matter. The fact that credits and press articles openly acknowledged *Dracula* suggests that the subsequent copyright problems were simply not anticipated.[51]

The question to what extent *Dracula* should be read as occultist is contested among Stoker scholars, even though, as Christine Ferguson has argued, the novel "stands as a veritable compendium of occultist beliefs, sciences and figures, riven through with allusions to alchemy, necromancy, lycanthropy, geomancy and the Eastern fakir practices that had become familiar to late Victorian audiences through the rise of comparative religious studies and the contemporary occult revival."[52] *Nosferatu*, in contrast, makes far less use of overtly occultist motifs. Apart from the cryptographic correspondence between Knock and Orlok, which ostentatiously calls attention to arcane knowledge restricted to the initiated, the film contains few references or symbolism inaccessible to a general public.[53] Instances such as Ellen's somnambulism, the characterization of Bulwer as Paracelsian, the Gnostic amulet that adorns Hutter's book on vampires, or the astrological posters with Hebrew

49 "Wie *Nosferatu* entstand," *Film BZ* (5 March 1922).

50 Christine Ferguson, "Dracula and the Occult," in *The Cambridge Companion to* Dracula, ed. Roger Luckhurst, 57-65 (Cambridge: Cambridge University Press, 2018), 57.

51 For a well-researched account of the details of Florence Stoker's lawsuit against Prana-Film see David J. Skal, *Hollywood Gothic: The Tangled Web of Dracula From Novel to Stage to Screen* (New York, NY: Norton, 1990), 77-102.

52 Ferguson, "Dracula and the Occult," in *The Cambridge Companion to* Dracula, 57.

53 For attempts to decipher these letters, see Exertier, "La Lettre oubliée," and Berriatúa, *Nosferatu*, 75-79.

and Arabic letters on Professor Bulwer's walls are either unobtrusive or universally intelligible.[54]

The film's creators likely held eclectic views and we cannot be certain what occultist doctrines and mythologies underlie *Nosferatu*. Albin Grau and Henrik Galeen had backgrounds in New Rosicrucianism, a movement that was profoundly indebted to the teachings of Blavatsky's Theosophical Society.[55] In my discussion of *Nosferatu* I will primarily reference Theosophist perspectives, in part because they shaped Rosicrucian thinking and because Theosophists, in contrast to Rosicrucians and related groups, produced a comprehensive body of written work, which will allow me to reconstruct major occultist ideas related to vampirism.

Occultist vampires: Visualizing the invisible

The conception of vampirism in *Nosferatu* adheres to the teachings of leading occultist thinkers like Éliphas Lévi, Helena Blavatsky, and Franz Hartmann. For them, vampirism was the consequence of iniquitous character and an all-powerful attachment to physical existence. Éliphas Lévi, a pioneer of modern occultism, pointed out: "When a man has lived well the astral body evaporates like a pure incense ascending towards the upper regions; but should he have lived in sin, his astral body, which holds him prisoner, still seeks the objects of his passions and wishes to return to life."[56] Lévi presupposes a view of the human constitution that was prevalent among many occultist groups. It consists of seven, hierarchically organized aspects, which are aligned with the animal/material, the human/mental, and the divine/spiritual dimension, respectively. In the Theosophist variant of this construct, the four lower, terrestrial, "animal," and ephemeral aspects of the personality are made up of (1) a physical body, (2) the astral or etheric body, i.e., a shadowy double or spirit image of the physical body, (3) the physical vitality also known as vital principle, and (4) the animal soul, the seat of instincts, material desires, and passions. The three higher aspects

54 Luciano Berriatúa has identified the poster on the left as numerological symbols associated with Jupiter originating in Athanasius Kircher, *Arithmologia* (Rome: Varesii, 1665), 204. Berriatúa, *Nosferatu*, 72. The poster on the right shows an allegory of the moon, likely a copy of *Luna* by Hans Sebald Beham (1539) with a religious Arabic text.

55 For a comprehensive study of modern Rosicrucianism, see Harald Lamprecht, *Neue Rosenkreuzer: Ein Handbuch* (Göttingen: Vandenhoeck & Ruprecht, 2004).

56 Éliphas Lévi, *Transcendental Magic: Its Doctrine and Ritual* [1854-1856], trans. Arthur Edward Waite (York Beach, ME: Weiser Books, 2001), 120.

of the personality —mental, spiritual, and immortal— encompass (5) the human soul, seat of mental faculties and consciousness, (6) the spiritual soul, seat of wisdom and intuition, and (7) the higher self or divine essence, an emanation from the absolute eternal.

At death, the lower principles separate from the higher ones in a multi-step process. The physical and animal aspects dissolve, consciousness is purified, and the spiritual elements eventually reincarnate. As evident from Lévi's account, vampires are associated with the lower, material principles. After death, they hang onto their physical bodies, whose decomposition they stall by drawing vitality and strength from others. Helena Blavatsky explained: "The bicorporeal life begins; and these unfortunate buried cataleptics sustain their miserable lives by causing their astral bodies to rob the living of their life-blood. The ethereal form can go wherever it pleases; and so long as it does not break the link which attaches it to the body, it is at liberty to wander about, either visible or invisible, and feed on human victims. "[57]According to Franz Hartmann, an ally of Blavatsky's, "such existences may wander about earth, clinging to material life, and vainly trying to escape the dissolution by which they are threatened. Partly bereft of reason, and following their animal instincts, they may become Incubi and Succubi, Vampires, stealing life from the living to prolong their own existence, regardless of the fate of their victims."[58]

As Albin Grau revealed in a trade press article, his understanding of vampirism largely followed this conception:

> The powers of the night lie in wait in order to drag a human being, who has been driven insane by his quest for human grandeur, with greedy tentacles deeper and deeper toward the earth. They magnify all animal instincts in him until a creature of absolute earthbound animality arises. After their corporeal death, the shadow of the earth-soul attaches itself to such creatures, nurtures them and endows them with immense demonic astuteness and secret powers that serve evil. Enemies of light and good, by night such creatures must plague everything human.[59]

In keeping with Theosophical thought, Grau construes vampires as shadowy revenants of wicked human beings, whose animal desires and passions confine them post mortem to the terrestrial sphere. The film externalizes this metaphysical disposition in various ways. To begin with, the astral

57 Blavatsky, *Isis Unveiled,* vol. 1, 449.
58 Hartmann, *Magic, White and Black*, 186.
59 Albin Grau, "Vampire," *Bild und Film* 21 (1921).

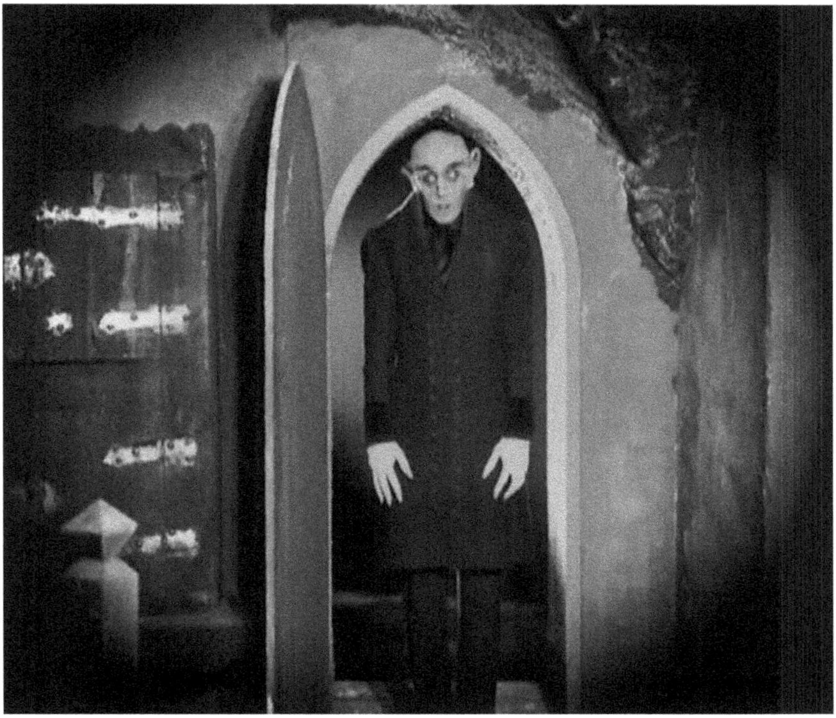

Fig. 4.1. The vampire's appearance in *Nosferatu* (1922).
Courtesy of Friedrich Wilhelm Murnau Foundation.

double must, as Blavatsky indicated, remain in proximity to the physical corpse, which may explain why Nosferatu, as contemporary critic Roland Schacht mocked, "hauls a coffin around like someone who wants to post a Christmas package shortly before seven, when the post offices close, but doesn't quite know where else he should try."[60]

Further, the vampire's bizarre appearance, aided by heavy prosthetic makeup, emblematizes his dual inner essence. On the one hand, since his physical body is deceased, his black-frocked, spindly stature, rigid movements pallid, hairless head, make him look like an emaciated corpse. On the other hand, since animal instincts dominate his nature, his physiognomy is determined by theriomorphic elements.[61] His bat-ears, rodent teeth, aquiline nose, claws, and vulturine hunch render him a hybrid of animals that elicit fear and loathing and are typically associated with death. The

60 Roland Schacht, "Nosferatu," *Das Blaue Heft* (15 April 1922).

61 Thomas Elsaesser and others have previously noted Nosferatu's animal features. See Thomas Elsaesser, "No End to *Nosferatu* (1922)," in *Weimar CInema*, ed. Isenberg, 86.

vampire's beastlike constitution is further underscored in the intertitles, which repeatedly refer to him as a "death bird," the script, which calls him "bestial," "snake," and "predatory animal,"[62] and the villagers, who link him to a werewolf, which is illustrated by cutaways to a hyena roaming the woods. A central metaphor for the vampire's disposition is the spider, an animal that sucks liquefied tissues from its prey. The film links Nosferatu to spiders in different ways. At times, slight undercranking makes the vampire's movements appear spiderlike, accelerated, and jerky, for instance in front of Hutter's house following Nosferatu's arrival in Wisborg. In a famous shot (Figure 4.2.), the vampire is framed in a low angle against the rigging of his ship like a spider in its web. In another one, he is shown clawing the window lattice of his abode, resembling, as the script specifies, "a black four-legged spider."[63] As Theosophist Franz Hartmann explained, "the astral form of an evil person may appear in an animal shape if it [sic] is so filled with brutish instincts as to become identified in his imagination with the animal which is the expression of its instincts. It may even enter the form of an animal and obsess it, and it sometimes happens that it enters such forms for its own protection against immediate decomposition and death."[64]

Although Nosferatu himself appears monstrous, vampirism is portrayed as a pervasive and—though disquieting—perfectly natural phenomenon. The film's representative of ancient scientific wisdom is Professor Bulwer, who is named after British poet Edward Bulwer-Lytton, a key inspiration to modern Theosophy. Bulwer is characterized as a Paracelsian, a follower of the famous Renaissance physician, alchemist, and mystic. In contrast to his counterpart in *Dracula*, Professor Van Helsing, Bulwer does intervene in the struggle with the vampire, but serves merely as an observer and informant of the anonymous narrator. As part of his lecture on the "mysterious workings of nature," Professor Bulwer explains to his "horrified" students the "cruelty of carnivorous plants." This is illustrated with a shot of a Venus flytrap snapping shut around a fly. In a dialog title, Bulwer compares the plant to a vampire. The lesson is intercut with a sequence at Sievers's asylum, where Knock is catching and eating flies.[65]

62 Lotte Eisner, *Murnau* [1964], trans. Martin Secker (London: Secker & Warburg, 1973), 243; 268; 270.

63 Eisner, *Murnau*, 263.

64 Hartmann, *Magic, White and Black*, 102.

65 The scene is also discussed in Thierry Lefebvre, "Les Métamorphoses de Nosferatu," *1895. Revue d'histoire du cinéma* 29 (December 1999): 61-77; and Tom Gunning, "To Scan a Ghost: The Ontology of Mediated Vision," *Grey Room* 26 (Winter 2007): 94-127.

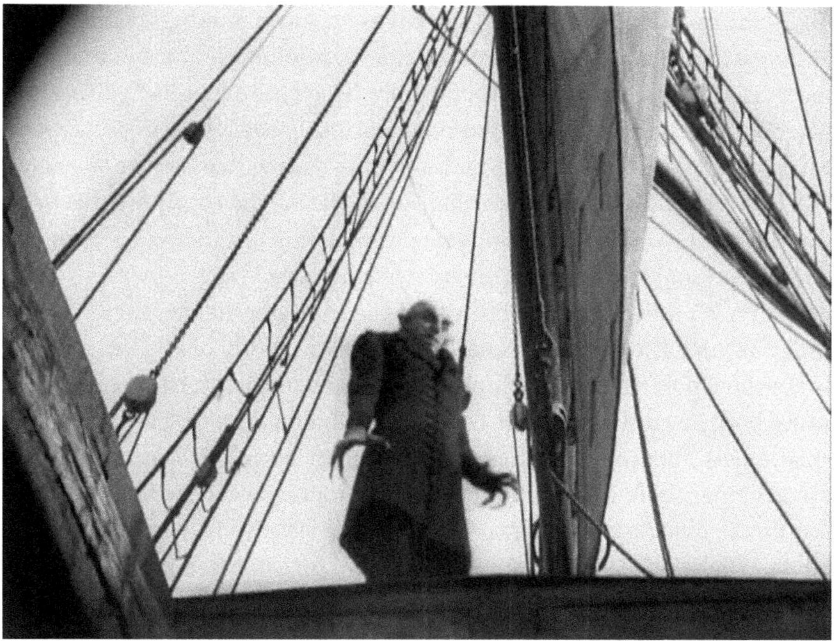

Fig. 4.2. The vampire as a spider in its web in *Nosferatu* (1922).
Courtesy of Friedrich Wilhelm Murnau Foundation.

The cross-cutting between the lecture and the asylum establishes a direct analogy between the insectivorous plant and Knock, who, as occultists would put it, is "possessed" by Nosferatu. Knock's subjugation to the vampire not only explains his servility ("The master is near!") and psychic attunement ("The master ... is ... dead."), but also his lunacy, since, according to Blavatsky, the active influence of an evil spirit may cause "the poor medium [to] become an epileptic, a maniac or a criminal."[66] Knock's outcry "blood is life!" references occultist ideas about blood as the physical manifestation of the vital principle, one of the lower aspects of the human constitution, which vampires extract from their victims to sustain their physical existence.[67]

When the scene returns to Bulwer, viewers are presented with scientific footage of a "transparent, almost disembodied" Hydra capturing an aquatic invertebrate with its tentacles. A dialog title likens the Hydra to a phantom,

66 Helena Blavatsky, "Are Chelas Mediums?" *The Theosophist* 9 (June 1884): 210-211; for a Spiritist perspective see Allan Kardec, *Experimental Spiritism: Book on Mediums*, trans. Emma A. Wood (Boston: Colby and Rich, 1874), 308-327.

67 Concurrently, Knock quotes the Book of Deuteronomy and Renfield in Stoker's *Dracula*, whose zoophagia similarly commences with the consumption of flies. See Bram Stoker, *Dracula* [1897] (New York, NY: Barnes and Noble, 2003), 155; Deuteronomy 12:23 (New International Version).

highlighting its affinity with Nosferatu as an astral or etheric being. Back at the asylum, Knock's exclamation "spiders!" is followed by a documentary insert of a spider in its web wrapping prey. By means of parallel editing and inserts, the sequence calls attention to various correspondences between vampirism in the vegetable, animal, and human kingdoms on the one hand, and in the astral sphere on the other, emphasizing the continuity between the physical world and non-physical dimensions in line with occultist doctrines about the essential oneness of existence.[68]

Not only his physiognomy, but also his bearing externalizes the vampire's inner nature. Theosophy teaches that the astral body of all living beings is composed of life force, which it transmits to the physical body. The astral body, according to then-Theosophist Rudolf Steiner, "calls inorganic substances to living existence, salvages them from lifelessness."[69] As an astral being, Nosferatu commands special kinetic powers, which extend to inanimate objects around him and which are largely visualized by special effects. Doors open for the vampire as if by an invisible hand. Some, like the castle's portal and the door of Hutter's castle bedroom, were manipulated mechanically, while others, like the barn door of Nosferatu's Wisborg dwelling or the ship hatch, were stop-motion animated. Nosferatu's coach rushes through the forest with twice the natural speed, performing jerky, mechanical movements. This effect was achieved by exposing only every other frame by undercranking the camera. The same type of accelerated, abrupt motion occurs when Nosferatu, on his departure from the castle, loads coffins and himself onto a wagon. The driverless horses, without prompting, then proceed to pull the wagon away. Here, in addition to undercranking, stop motion animation was used to stage the lid placing itself on Nosferatu's coffin. The unnatural movements associated with Nosferatu call attention to the paradoxical fact that the vampire not only causes death, but also gives, by virtue of his astral powers and film technology, an inorganic form of life.[70]

68 For different aspects of the logic of analogy that characterizes editing in *Nosferatu* see Gilbert Perez, *The Material Ghost: Films and Their Medium* (Baltimore, MD: Johns Hopkins University Press, 2000), 123-135; Dalle Vacche, *Cinema and Painting*, 188-190; Heide Schlüpmann, "Der Spiegel des Grauens: Murnaus *Nosferatu*," *Frauen und Film* 49 (December 1990): 38-51.

69 Rudolf Steiner, "Blut ist ein ganz besonderer Saft. Berlin, 25. Oktober 1906," in *Die Erkenntnis des Übersinnlichen in unserer Zeit und deren Bedeutung für das heutige Leben*, 35-65 (Dornach: Rudolf Steiner Verlag, 1983), 46.

70 Michelle Langford also stresses the inorganic nature of the vampire's movements, which she rightly links to film technology. See "Lola and the Vampire: Technologies of Time and Movement in German Cinema," in *Cinema and Technology: Cultures, Theories and Practices*,

Some of Nosferatu's kinetic powers are also pictured without special effects: The small barge that Nosferatu uses to cross a canal in Wisborg moves without any visible driving force, while the vampire stands still and erect in the centre with his coffin under his arm. He also animates his ship, the schooner Empusa, which takes its name from "a ghoul, a vampire, an evil demon taking various forms," according to Blavatsky.[71] Perverting the Biblical account of God inspiring Adam with the "breath of life" and thus causing him to become a living being, Nosferatu's "deathly breath" propels the ship with "eerie speed" towards Wisborg.[72] Murnau's copy of the script suggests that the filmmakers made plans for additional effect shots to visualize the magical forces driving the Empusa.[73] However, these tricks, which would have likely strengthened the Empusa's affinity with the Flying Dutchman's phantom ship, are not present in the surviving version of the film.

The film's most salient special effect occurs on the way to Orlok's castle, when the carriage carrying Hutter passes through a forest made up of ghostly trees set against a black sky (Figure 4.3.). It is well known that the effect was achieved by using negative rather than positive footage and by redressing carriage and driver in white so that they would still appear dark in the negative. Cinematographer Fritz Arno Wagner again undercranked the camera to achieve unnatural movements. The script describes the location as "fairy-tale forest," in which a "wise, man-sized raven [...] follows the coach with mocking eyes behind glasses."[74] In the film, the sequence is rendered more abstractly and the "coach's frenzied ride through *white forest!*," as Murnau noted, is more evocative of a passage into an alternate reality. In this context it is important to note that occultists assume the existence of seven cosmic realms that correspond to the seven aspects of the human constitution. The physical world is the lowest and densest. Immediately above is the astral plane or astral light, an intermediate domain between the physical and the spiritual, which the soul transverses after death and which the living may experience while their physical body is unconscious, for example when asleep. The white forest suggests that the carriage crosses into a sphere that appears like the ethereal double of our world. The shot

eds. Bruce Bennett, Marc Furstenau, and Adrian Mackenzie, 187-200 (New York, NY: Palgrave Macmillan, 2008), 192-193.

71 Helena Blavatsky, *The Theosophical Glossary* (London: The Theosophical Publishing Company, 1892), 114.

72 Genesis 2:7 (New International Version).

73 See scenes 105 and 110. In addition, Murnau added in writing to scene 108 and underlined "Sailing vessel navigates into the lens!" (translation amended). Eisner, *Murnau*, 257-258.

74 Eisner, *Murnau*, 242 (translation amended).

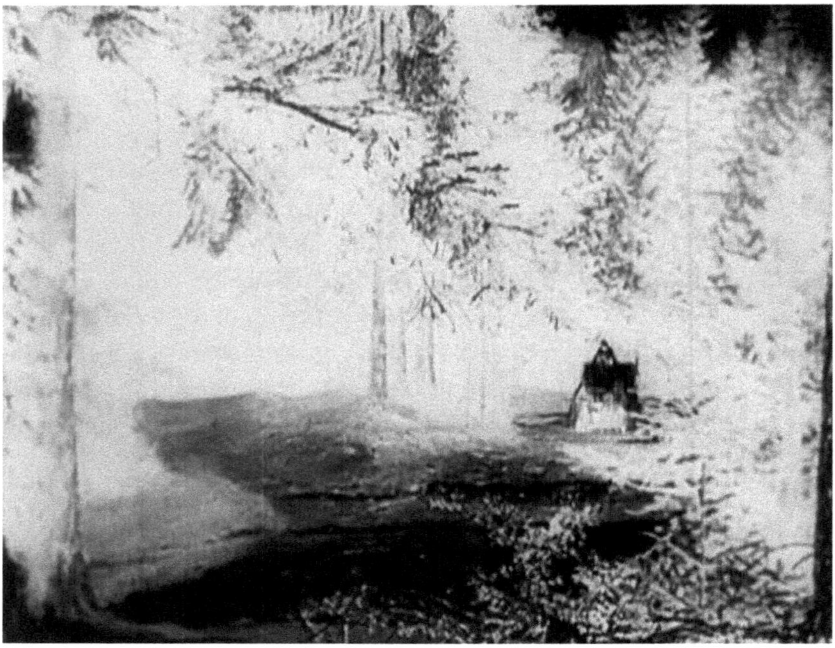

Fig. 4.3. Negative footage suggestive of the astral plane in *Nosferatu* (1922). Courtesy of Friedrich Wilhelm Murnau Foundation.

constitutes the film's only attempt to picture something akin to a different psychic or cosmic realm, a challenge it otherwise avoids.

The vampire's physiognomy, bearing, and kinetic powers call attention to the momentousness and might of hidden forces. One of the basic assumptions of occultism, this theme is central to the conception of *Nosferatu* and also underlies the treatment of one of the film's leitmotifs, the figure of shadow. Both on a visual and a linguistic level, shadows are presented as autonomous and capable of having a material impact. For example, the intertitles warn us to "take heed that his shadow not *encumber* thee like an incubus with gruesome dreams" At Orlok's castle, they recount that, "as the sun rose, the shadows of the night *withdrew* from Hutter" and later, "the spectral evening light seemed to *revive* the shadows of the castle". Emphases added. The association between shadows and the vampire may also be linked to conceptions of the vampire as a "semi-materialized astral body."[75] According to this, Nosferatu would himself be considered one of the shadows—a phantom that straddles the boundary between material and immaterial.

75 Arthur Powell, *The Astral Body and Other Astral Phenomena* (Wheaton, IL: The Theosophical Publishing House, 1927), 171.

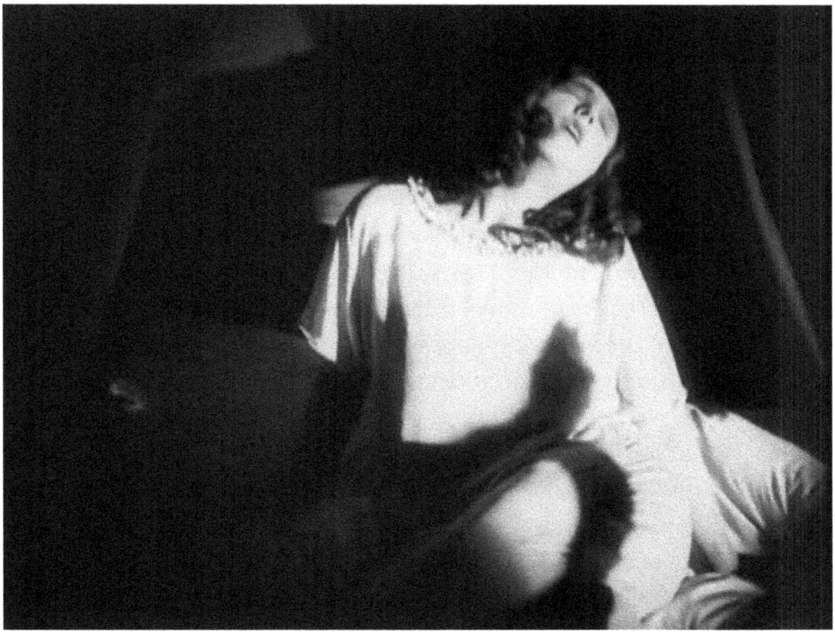

Fig. 4.4. The physical impact of the vampire's shadow in *Nosferatu* (1922).
Courtesy of Friedrich Wilhelm Murnau Foundation.

When climbing the staircase to Ellen's bedroom or attacking Hutter in his castle chamber, Nosferatu's shadow appears oversized and detached from his body. However, the most striking illustration of the corporeal force of what the intertitles call Nosferatu's "shadowy power" is the moment when, at the film's climax, Nosferatu's shadow grips Ellen's heart. Having turned her face away, she is unable to see the projection of his shadow slowly slithering up her body; yet the moment the shadow clenches its fist, her body winces.

Ellen is Nosferatu's principal antagonist. She is characterized by two attributes: selfless love and awakened psychic intuition. Both traits are embodied in the heart, which for Theosophists constitutes the centre of the spiritual consciousness within the human body. Ellen is devoted to her husband, shown to deeply care about all of creation, including animals and plants ("Why did you kill them ... the lovely flowers"), and ultimately sacrifices herself for the benefit of the entire community. Simultaneously, she is sensitive in the occultist sense, i.e., she is receptive to extrasensory perception ("her soul that night caught the cry of the Deathbird"). Her persistent anguish, trance-like states, and sleepwalking can be attributed to her knowledge of the looming dangers. She is capable of clairvoyance and spiritual communication with the vampire, which allows her to save

Hutter's life. According to the script the vampire "is listening intently as if he could feel – hear the terrified shouting in the distance."[76] As Thomas Elsaesser notes, they establish soul contact, which is visualized through cross-cutting.[77] Later, Nosferatu lies in wait at his window opposite hers ("This is how I see it – every evening ...!!!"), but because, according to occultist beliefs, "a healthy mind is a castle that cannot be invaded without the will of its master," the vampire approaches her only after having been invited.[78]

There is no doubt that the self-sacrifice of the all-loving, sinless woman for the benefit of the community brims with patriarchal gender stereotypes. At the same time, however, it is noteworthy that the struggle between Ellen and Nosferatu also inverses occultist gendered symbolism, which identifies the female principle with matter, the moon, darkness, and passivity, and the masculine principle with spirit, the sun, light, and activity.[79] Whereas Nosferatu embodies the lower animal instincts and desires, the material principle, Ellen represents the higher and immortal aspects, the spiritual principle. Symbolically, Ellen is associated with the sun, which Theosophists describe as a cosmic heart that pumps vital energy through the solar system and is itself a reflection of the (invisible) great spiritual sun.[80] It is therefore consistent that *Nosferatu* diverges from conventional formulas that stipulate the beheading, impaling, or cremation of the vampire's physical body and instead features the first death by sunlight in the history of vampire fiction.[81] In *Nosferatu*, spirituality and selfless love destroy the vampire.

The universality of these ideals stands in sharp contrast to the vampire's liminality. As Jay Johnston has observed, vampires, werewolves, and other mythological beings "cross, disrupt, and inhabit boundaries common to everyday forms of reason: animal/human; real/unreal (fantasy); actual/virtual; physical/metaphysical, culture/nature; fact/fiction. They are figures of potentially dangerous transgression (both ontologically and epistemologically): this is the heart of their appeal and abhorrence."[82] From an occultist perspective, the vampire's in-betweenness is a consequence of his belonging

76 Eisner, *Murnau*, 247.

77 Elsaesser, *Weimar Cinema and After*, 236.

78 Franz Hartmann, *The Life and Doctrines of Philippus Theophrastus, Bombast of Hohenheim Known by the Name of Paracelsus* [1891] (New York, NY: Theosophical Publishing, 1902), 143.

79 Blavatsky, *Isis Unveiled*, vol. 2, 270.

80 Blavatsky, *The Secret Doctrine*, vol. 1, 591.

81 Rosemary Ellen Guiley, *The Encyclopedia of Vampires, Werewolves, and Other Monsters* (New York, NY: Visionary Living, 2005), 272-273; Powell, *The Astral Body*, 173.

82 Jay Johnston, "Vampirism, Lycanthropy, and Otherkin," in *The Occult World*, ed. Christopher Partridge, 412-423 (New York, NY: Routledge, 2015), 412.

to the astral light or astral sphere, a liminal dimension. Occultists think of the astral light as cosmic matter that surrounds and penetrates all aspects of physical reality. It is conditioned by human thoughts and feelings, and populated by immaterial beings, both good and evil. It is the realm of illusion and hallucination, and serves as the medium through which psychic forces operate. Despite its etheric disposition, which makes it visible only to psychics, it is of material rather than spiritual nature. The astral light has two main features: it reflects and stores in pictorial form all conceptions, emotions, and incidents from the terrestrial and astral spheres. Blavatsky speaks of "photo-scenographic galleries" and Hartmann of "living mirrors."[83] In the description of Theosophist William Quan Judge, the analogies to photography and (proleptically) cinema are even more explicit: "Very much like the action of the sensitive photographic plate is this light. It takes, as Flammarion says, the pictures of every moment and holds them in its grasp. [...] As an enormous screen or reflector the astral light hangs over the earth and becomes a powerful universal hypnotizer of human beings."[84]

The striking similarities between occultist notions of the astral light and early conceptions of cinema's ontology testify to occultists' scientific imagination and acute awareness of contemporary technological developments. For the most part anticipating the invention of the Cinématographe, Theosophists construed the astral light as a visual medium that records "living images" and projects them back to the terrestrial sphere. At the same time, given the wide dissemination of occultist thought, such ideas may have also affected the initial reception of the filmic medium. Some of the seemingly contradictory features of the astral light—its association with material reality and life, secondly with immateriality, shadows, phantoms, and death, and thirdly with semblance, illusion, and interiority—were also attributed to cinema.[85]

83 Blavatsky, *Isis Unveiled*, vol. 1, 180; Hartmann, *Magic, White and Black*, 205.

84 William Q. Judge, *Echoes from the Orient: A Broad Outline of Theosophical Doctrines* (New York, NY: The Path, 1890), 59-61.

85 For cinema's association with material existence and life, see for instance Konrad Lange, "Die 'Kunst' des Lichtspieltheaters [1913]," in *Geschichte der Filmtheorie: Kunsttheoretische Texte von Méliès bis Arnheim*, ed. Helmut H. Diederichs, 75-88 (Frankfurt am Main: Suhrkamp, 2004); Carlo Mierendorff, "If I Only Had the Cinema! [1920]," in *The Promise of Cinema: German Film Theory, 1907-1933*, eds. Anton Kaes, Nicholas Baer, and Michael Cowan, trans. Jeffrey Timon, 426-433 (Berkeley, CA: University of California Press, 2016). For cinema's association with shadows, phantoms and the living dead, see for instance Herbert Tannenbaum, "Problems of the Film Drama [1913]," in *The Promise of Cinema: German Film Theory, 1907-1933*, eds. Anton Kaes, Nicholas Baer, and Michael Cowan, trans. Alex H. Bush, 192-196 (Berkeley, CA: University of California Press, 2016), 193; Willy Rath, "Künstlerische Möglichkeiten des Lichtspiels [1913],"

The ontological affinities between cinema and the astral sphere must have fascinated cinephiles with occult interests such as Grau, Murnau, and Galeen and may explain why Nosferatu also appears as a cinematic being. The vampire's dark astral powers are afforded by the cinematic apparatus, which allows viewers, as Bruce Elder has put it, to "to see, and therefore to understand, what the human eye cannot see unaided."[86] The kinship between the vampire and the filmic medium is made explicit in two narrative titles that frame the film. Following the opening title page of the Chronicle of the Great Death in Wisborg, a second title card, seemingly unrelated to the historical account, announces: "Nosferatu. Doesn't this name sound to you like the call of a death bird at midnight? Beware of uttering it, or else the images of life will pale to shadows; ghostly dreams rise up from the heart and feed on your blood." This chilling warning makes obvious references to prevailing conceptions about cinema's ontology: The evocation of the film's title at the beginning of a screening causes "the images of life" to "pale to shadows," i.e., to assume the soundless, colourless, and disembodied character that many early commentators following Maxim Gorky associated with film.[87] What is more, the audience is identified both as source and victim of cinema's "ghostly dreams" that "rise up from the heart and feed on your blood," which recalls Hanns Heinz Ewers's notion of the cinematic medium as an actualization of viewers' dreams.[88] In line with this, critic Walther Thielemann argued that *Nosferatu* presents itself as the spawn of the spectator's psyche: "The artistic effect of this strange film originates in the successful endeavour to project an inner, unformed experience outward in order to give potent expression to a folksong-like, existential dream, so powerful that it can pass as a likeness of our infantile ideas. In other

in *Geschichte der Filmtheorie: Kunsttheoretische Texte von Méliès bis Arnheim*, ed. Helmut H. Diederichs, 121-131 (Frankfurt am Main: Suhrkamp, 2004), 122. For cinema's association with semblance, illusion, and dreams, see for instance Gustav Melcher, "Die künstlerischen Vorzüge der Kinematographie," *Der Kinematograph* 116 (17 March 1909); Theodor Heinrich Mayer, "Lebende Photographien [1912]," in *Kein Tag ohne Kino: Schriftsteller über den Stummfilm*, ed. Fritz Güttinger, 119-129 (Frankfurt am Main: Deutsches Filmmuseum, 1984), 120; Robert Scheu, "Das Kino der Kommenden," *Vossische Zeitung* 117 (5 March 1914).

86 Elder, *Harmony and Dissent*, 3.

87 Maxim Gorky, "A Review of the Lumière Programme at the Nizhni-Novgorod Fair [1896]," in Jay Leyda, *Kino: A History of the Russian and Soviet Film*, trans. Leda Swan, 407-410 (London: George Allen and Unwin, 1960), 407. Several scholars, including Anton Kaes, have drawn connections between Gorky's "Kingdom of Shadows" and *Nosferatu*. See Kaes, *Shell Shock Cinema*, 125.

88 Hanns Heinz Ewers, "Vom Kinema [1910]," in *Kein Tag ohne Kino: Schriftsteller über den Stummfilm*, ed. Fritz Güttinger, 20-23 (Frankfurt am Main: Deutsches Filmmuseum, 1984), 21.

words: This film is oneirically materialized life from our subconscious."[89] At the end of the film, the final title once again alludes to the vampire's cinematic nature, adding another level of meaning to the vampire's death by sunlight. Metaphorically veiled, it suggests that *Nosferatu* is swept away when the house lights come on: "As if obliterated by the triumphant rays of the living sun, the shadow of the death bird was gone." Associating cinema with a mysterious force that deprives us of our vital energy may suggest an inimical and threatening view of the medium. However, arguably prevalent in this context is the emphasis on the medium's extensive power over our emotions, transforming a prima facie threat into an obeisance.

Nosferatu was designed with three objectives in mind: First, to exemplify the treatment of occultist ideas in a film, second, to do so in a uniquely cinematic (and thus artistic) manner, and third, to achieve wide popular appeal. The production of a "symphony of horror" with a bloodthirsty living corpse as its protagonist accommodated commercial considerations as well as occultist interests. Rather than explicitly advocating for particular occultist doctrines, *Nosferatu* employs an indirect approach. It draws on the esoteric law of correspondence and analogy, which posits that symbolic and real correlations connect all visible and invisible parts of the universe in order to call attention to invisible realities, the inner nature of things, and to establish mental connections between seemingly disparate objects and events. More than in any other aspect of the film, the physiognomy, demeanour, and powers of the vampire reveal an occultist perspective. Nosferatu's outer appearance allegorizes his inner nature as a deceased person's astral body who incorporates but the lower, animal, and material, facets of the human constitution. The film externalizes this essence through heterogeneous cinematic means of expression, most notably special effects. The reliance on trick technology for actualizing occultist thought by making the intangible sensually perceptible constitutes a fundamentally techno-romantic undertaking. The exploration and mediation of inscrutable truths is not only central to the occultist project, it represented a core concern of a broad range of idealistically inclined artists around 1900. Furthermore, the metaphorical and suggestive approach favoured by occultist thinkers also characterizes the endeavours of many artists. Finally, because so many artists engaged with occultist thought, the two groups are difficult to differentiate. *Nosferatu* provides an object lesson in the ways occultist ideas imbued silent film culture and, in conjunction with film technology, became part of the techno-romantic paradigm.

89 Walter Thielemann, "Der neue Weg im Film," *Film und Brettl* 1 (January 1922).

Bibliography

Theodor W. Adorno, *Minima Moralia: Reflections from Damaged Life* [1951], trans. E. F. N. Jephcott, 238-244 (London: Verso, 1974).

Tessel M. Bauduin, "Abstract Art as By-Product of Astral Manifestation: The Influence of Theosophy on Modern Art in Europe," in *Handbook of the Theosophical Current*, eds. Olav Hammer and Mikael Rothstein, 429-451 (Leiden: Brill, 2013).

Tessel M. Bauduin and Henrik Johnsson, "Introduction: Conceptualizing Occult Modernism," in *The Occult in Modernist Art, Literature, and Cinema*, eds. Tessel M. Bauduin and Henrik Johnsson, 1-30 (Cham: Palgrave Macmillan, 2019).

Ludwig Berger, *Wir sind vom gleichen Stoff aus dem die Träume sind: Summe eines Lebens* (Tübingen: R. Wunderlich, 1953).

Luciano Berriatúa, *Nosferatu: Un film Erótico – Ocultista – Espiritista – Metafísico* (Valladolid: Divisa Red, 2009).

Helena Blavatsky, "Are Chelas Mediums?" *The Theosophist* 9 (June 1884): 210-211.

Helena P. Blavatsky, *Isis Unveiled: A Master-Key to the Mysteries of Ancient and Modern Science and Theology*, vol. 1 (New York, NY: J. W. Bouton, 1877).

Helena Petrovna Blavatsky, *The Key to Theosophy* (London: The Theosophical Publishing Company, 1889).

Helena P. Blavatsky, *The Secret Doctrine* [1888], ed. Boris de Zirkoff, vol. 1 (Wheaton, IL: The Theosophical Publishing House, 1893).

Helena Blavatsky, *The Theosophical Glossary* (London: The Theosophical Publishing Company, 1892).

Peter Bogdanovich, *Fritz Lang in America* (London: Studio Vista, 1967).

Michel Bouvier and Jean Louis Leutrat, *Nosferatu* (Paris: Gallimard, 1981).

Max Brod, "Höhere Welten," *Pan* 16 (16 June 1911): 538-545.

Arthur Brunel Chatwood, *The New Photography* (London: Downey, 1896).

Ken Calhoon, "F. W. Murnau, C. D. Friedrich, and the Conceit of the Absent Spectator," *MLN* 3 (April 2005): 633-653.

Richard Nikolaus Coudenhove-Kalergi, *Apologie der Technik* (Leipzig: Verlag Der neue Geist Dr. Peter Reinhold, 1922).

Angela Dalle Vacche, *Cinema and Painting: How Art is Used in Film* (Austin, TX: University of Texas Press, 1996).

Carol Margaret Davison, *Anti-Semitism and British Gothic Literature* (New York, NY: Palgrave, 2004).

Paul Dubro, "Frau im Mond, Fritz Lang und ein Interview," *Ufa Feuilleton* 42 (16 October 1929): 4.

Carl du Prel, *Der Spiritismus* (Leipzig: Reclam-Verlag, 1893).

Thomas Edison, "The Realms Beyond [1920]," in *The Diary and Sundry Observations of Thomas Alva Edison*, ed. Dagobert D. Runes, 238-241 (New York, NY: Philosophical Library, 1948).

Lotte Eisner, *Murnau* [1964], trans. Martin Secker (London: Secker & Warburg, 1973).

Lotte H. Eisner, *The Haunted Screen: Expressionism in the German Cinema and the Influence of Max Reinhardt* [1952], trans. Roger Greaves (Berkeley, CA: University of California Press, 2008).

R. Bruce Elder, *Harmony and Dissent: Film and Avant-garde Art Movements in the Early Twentieth Century* (Waterloo, ON: Wilfrid Laurier University Press, 2008).

Thomas Elsaesser, "No End to *Nosferatu* (1922)," in *Weimar Cinema*, ed. Noah Isenberg, 79-94 (New York, NY: Columbia University Press, 2008).

Thomas Elsaesser, *Weimar Cinema and After: Germany's Historical Imaginary* (London: Routledge, 2000).

Hanns Heinz Ewers, "Vom Kinema [1910]," in *Kein Tag ohne Kino: Schriftsteller über den Stummfilm*, ed. Fritz Güttinger, 20-23 (Frankfurt am Main: Deutsches Filmmuseum, 1984).

Sylvain Exertier, "La Lettre oubliée de *Nosferatu*," *Positif* (March 1980): 47-51.

Christine Ferguson, "Dracula and the Occult," in *The Cambridge Companion to* Dracula, ed. Roger Luckhurst, 57-65 (Cambridge: Cambridge University Press, 2018).

Friedhelm Wilhelm Fischer, "Geheimlehren und moderne Kunst," in *Fin de Siècle: Zur Literatur und Kunst der Jahrhundertwende*, ed. Roger Bauer, 344-377 (Frankfurt am Main: Klostermann, 1977).

Rolf Giesen, *The Nosferatu Story: The Seminal Horror Film, Its Predecessor and Its Enduring Legacy* (Jefferson, NC: McFarland, 2019).

Sander Gilman, *The Jew's Body* (Milton Park: Taylor & Francis, 1991).

Nicholas Goodrick-Clarke, *The Occult Roots of Nazism: The Ariosophists of Austria and Germany, 1890-1935* (Wellingborough: The Aquarian Press, 1985).

Maxim Gorky, "A Review of the Lumière Programme at the Nizhni-Novgorod Fair [1896]," in Jay Leyda, *Kino: A History of the Russian and Soviet Film*, trans. Leda Swan, 407-410 (London: George Allen and Unwin, 1960).

Albin Grau, "Vampire," *Bild und Film* 21 (1921).

Rosemary Ellen Guiley, *The Encyclopedia of Vampires, Werewolves, and Other Monsters* (New York, NY: Visionary Living, 2005).

Tom Gunning, "To Scan a Ghost: The Ontology of Mediated Vision," *Grey Room* 26 (Winter 2007): 94-127.

Robert Hare, *Experimental Investigation of the Spirit Manifestations* (New York NY: Partridge & Brittan, 1855).

Franz Hartmann, *Magic, White and Black* [1886] (New York, NY: The Path, 1895).

Franz Hartmann, *The Life and Doctrines of Philippus Theophrastus, Bombast of Hohenheim Known by the Name of Paracelsus* [1891] (New York, NY: Theosophical Publishing, 1902).

Linda Dalrymple Henderson, "Abstraction, the Ether, and the Fourth Dimension: Kandinsky, Mondrian, and Malevich in Context," in *Kandinsky, Malevich, Mondrian: The Infinite White Abyss*, eds. Marian Ackermann and Isabelle Malz, 233-44 (Düsseldorf: Kunstsammlung Nordrhein-Westfalen, 2014).

Linda Dalrymple Henderson, "Rethinking Modern Art, Science, and Occultism in Light of the Ether of Space: Wassily Kandinsky, Umberto Boccioni, and Kazimir Malevich," in *The History of Art and 'Rejected Knowledge': From the Hermetic Tradition to the 21st Century*, ed. Anna Korndorf, 218-37 (Moscow: The State Institute of Art Studies, 2018).

Benedikt Hjartarson, "Ghosts Before Breakfast: The Appetite for the Beyond in Early Avant-Garde Film," in *The Occult in Modernist Art, Literature, and Cinema*, eds. Tessel M. Bauduin and Henrik Johnsson, 137-162 (Cham: Palgrave Macmillan, 2019).

Patrick Colm Hogan, "Narrative Universals, Nationalism and Sacrificial Terror from *Nosferatu* to Nazism," *Film Studies* 8 (Summer 2006): 93-105.

Jay Johnston, "Vampirism, Lycanthropy, and Otherkin," in *The Occult World*, ed. Christopher Partridge, 412-423 (New York, NY: Routledge, 2015).

Jason A. Josephson-Storm, *The Myth of Disenchantment: Magic, Modernity, and the Birth of the Human Sciences* (Chicago, IL: University of Chicago Press, 2017).

Anton Kaes, *Shell Shock Cinema: Weimar Culture and the Wounds of War* (Princeton, NJ: Princeton University Press, 2009).

William Q. Judge, *Echoes from the Orient: A Broad Outline of Theosophical Doctrines* (New York, NY: The Path, 1890).

Wassily Kandinsky, *Concerning the Spiritual in Art* [1912], trans. Michael T. H. Sadler (Mineola, NY: Dover Publications, 1977).

Wassily Kandinsky, "Reminiscences," in *Modern Artists on Art*, ed. Robert L. Herbert, trans. Eugenia W. Herbert, 19-44 (Mineola, NY: Dover Publications, 1999).

Allan Kardec, *Experimental Spiritism: Book on Mediums*, trans. Emma A. Wood (Boston: Colby and Rich, 1874).

Friedrich A. Kittler, "Dracula's Legacy," trans. William Stephen Davis, *Stanford Humanities Review* 1 (1989): 143-173.

Siegfried Kracauer, *From Caligari to Hitler: A Psychological History of the German Film* [1947], (Princeton, NJ: Princeton University Press, 2004).

Wilfried Kugel, *Der Unverantwortliche: Das Leben des Hanns Heinz Ewers* (Düsseldorf: Grupello Verlag, 1992).

Harald Lamprecht, *Neue Rosenkreuzer: Ein Handbuch* (Göttingen: Vandenhoeck & Ruprecht, 2004).

Konrad Lange, "Die 'Kunst' des Lichtspieltheaters [1913]," in *Geschichte der Filmtheorie: Kunsttheoretische Texte von Méliès bis Arnheim,* ed. Helmut H. Diederichs, 75-88 (Frankfurt am Main: Suhrkamp-Verlag, 2004).

Michelle Langford, "Lola and the Vampire: Technologies of Time and Movement in German Cinema," in *Cinema and Technology: Cultures, Theories and Practices*, eds. Bruce Bennett, Marc Furstenau, and Adrian Mackenzie, 187-200 (New York NY: Palgrave Macmillan, 2008).

Volker Lechler, *Die ersten Jahre der Fraternitas Saturni* (Stuttgart: Verlag Volker Lechler, 2015).

Thierry Lefebvre, "Les Métamorphoses de Nosferatu," *1895. Revue d'histoire du cinéma* 29 (December 1999): 61-77.

Éliphas Lévi, *Transcendental Magic: Its Doctrine and Ritual* [1854-1856], trans. Arthur Edward Waite (York Beach, ME: Weiser Books, 2001).

V. Loers and P. Witsman, eds., *Okkultismus und Avant-Garde: Von Munch bis Mondrian* (Frankfurt am Main: Schirn-Kunsthalle, 1995).

Theodor Heinrich Mayer, "Lebende Photographien [1912]," in *Kein Tag ohne Kino: Schriftsteller über den Stummfilm*, ed. Fritz Güttinger, 119-129 (Frankfurt am Main: Deutsches Filmmuseum, 1984).

Judith Mayne, "Dracula in the Twilight: Murnau's *Nosferatu* (1922)," in *German Film & Literature: Adaptations and Transformations*, ed. Eric Rentschler, 25-39 (New York, NY: Methuen, 1986).

Hugh Mcleod, *Secularisation in Western Europe, 1848-1914* (New York, NY: St. Martin's Press, 2000).

Walter Mehring, *Berlin Dada* (Zurich: Verlag der Arche, 1959).

Gustav Melcher, "Die künstlerischen Vorzüge der Kinematographie," *Der Kinematograph* 116 (17 March 1909).

Carlo Mierendorff, "If I Only Had the Cinema! [1920]," in *The Promise of Cinema: German Film Theory, 1907-1933*, eds. Anton Kaes, Nicholas Baer, and Michael Cowan, trans. Jeffrey Timon, 426-433 (Berkeley, CA: University of California Press, 2016).

Isaac Newton, *Opticks: Or, a Treatise of the Reflections, Refractions, Inflections and Colours of Light* [1704] (Mineola, NY: Dover Publications, 1979).

Judith Noble, "Clear Dreaming: Maya Deren, Surrealism and Magic," in *Surrealism, Occultism and Politics: In Search of the Marvellous*, eds. Tessel M. Bauduin, Victoria Ferentinou, and Daniel Zamani, 210-226 (New York, NY: Routledge, 2018).

Kristoffer Noheden, *Surrealism, Cinema, and the Search for a New Myth* (Cham: Palgrave Macmillan, 2017).

Alex Owen, *The Place of Enchantment: British Occultism and the Culture of the Modern* (Chicago, IL: University of Chicago Press, 2004).

Christopher Partridge, "Occulture is Ordinary," in *Contemporary Esotericism*, eds. Egil Asprem and Kennet Granholm, 113-133 (New York, NY: Routledge, 2014).

Christopher Partridge, "Orientalism and the Occult," in *The Occult World*, ed. Christopher Partridge, 611-625 (New York, NY: Routledge, 2015).
Friedrich Paulsen, *Die deutschen Universitäten und das Universitätsstudium* (Berlin: Verlag von Asher & Co, 1902).
Gilbert Perez, *The Material Ghost: Films and Their Medium* (Baltimore, MD: Johns Hopkins University Press, 2000).
Arthur Powell, *The Astral Body and Other Astral Phenomena* (Wheaton, IL: The Theosophical Publishing House, 1927).
Priska Pytlik, *Okkultismus und Moderne ein kulturhistorisches Phänomen und seine Bedeutung für die Literatur um 1900* (Paderborn: Ferdinand Schöningh, 2005).
Willy Rath, "Künstlerische Möglichkeiten des Lichtspiels [1913]," in *Geschichte der Filmtheorie: Kunsttheoretische Texte von Méliès bis Arnheim*, ed. Helmut H. Diederichs, 121-131 (Frankfurt am Main: Suhrkamp-Verlag, 2004).
Hans-Peter Reichmann, *Hans Poelzig: Bauten für den Film* (Frankfurt am Main: Deutsches Filmmuseum, 1997).
Piotr Sadowski, *The Semiotics of Light and Shadows: Modern Visual Arts and Weimar Cinema* (London: Bloomsbury, 2017).
Roland Schacht, "Nosferatu," *Das Blaue Heft* (15 April 1922).
Robert Scheu, "Das Kino der Kommenden," *Vossische Zeitung* 117 5 (March 1914).
Heide Schlüpmann, "Der Spiegel des Grauens: Murnaus *Nosferatu*," *Frauen und Film* 49 (December 1990): 38-51.
Heide Schönemann, *Paul Wegener: Frühe Moderne im Film* (Stuttgart: Edition Axel Menges, 2003).
David J. Skal, *Hollywood Gothic: The Tangled Web of Dracula From Novel to Stage to Screen* (New York, NY: Norton, 1990).
Rudolf Steiner, "Blut ist ein ganz besonderer Saft. Berlin, 25. Oktober 1906," in *Die Erkenntnis des Übersinnlichen in unserer Zeit und deren Bedeutung für das heutige Leben*, 35-65 (Dornach: Rudolf Steiner Verlag, 1983).
Bram Stoker, *Dracula* [1897] (New York, NY: Barnes and Noble, 2003).
Leon Surette, *The Birth of Modernism: Ezra Pound, T.S. Eliot, W.B. Yeats, and the Occult* (Montreal & Kingston: McGill-Queen's University Press, 1993).
Herbert Tannenbaum, "Problems of the Film Drama [1913]," in *The Promise of Cinema: German Film Theory, 1907-1933*, eds. Anton Kaes, Nicholas Baer, and Michael Cowan, trans. Alex H. Bush, 192-196 (Berkeley, CA: University of California Press, 2016).
Walter Thielemann, "Der neue Weg im Film," *Film und Brettl* 1 (January 1922).
Pamela Thurschwell, *Literature, Technology, and Magical Thinking, 1880-1920* (Cambridge: Cambridge University Press, 2001).
Corinna Treitel, *A Science for the Soul: Occultism and the Genesis of the German Modern* (Baltimore, MD: Johns Hopkins University Press, 2004).

Max Weber, "Science as Vocation [1918]," in *The Vocation Lectures*, eds. David Owen and Tracy B. Strong, trans. Rodney Livingstone, 1-32 (Indiapolis, IN: Hackett Publishing Company, 2004).
"Wie *Nosferatu* entstand," *Film BZ* (5 March 1922).
Robin Wood, "Murnau's Midnight and Sunrise," *Film Comment* (May/June 1976): 4-19.

5. The Technological Sublime: *Metropolis* (1927)

Abstract
Metropolis displays a deeply conflicting attitude toward industrial modernity. Conceived and marketed as a marvel of film technology, the film pursued the techno-romantic project of transcending material reality through technological means. What is more, the goal was to capture the unfathomability of technology itself. *Metropolis* simultaneously portrays technology as an agent of tyranny and dehumanization and flaunts it as spectacle. Special effects facilitate encounters with overpowering technological environments and omnipotent machines, which give rise to sentiments that are best described in terms of a "technological sublime." The sublime characterizes experiences that go beyond the earthly and finite, to attain a spiritual dimension. In attributing transcendent qualities to mechanical objects, the technological sublime embodies the techno-romantic paradigm.

Keywords: sublime, robot, science fiction, Schüfftan process, multiple exposure composites

Metropolis (Ufa, 1927, dir. Fritz Lang) exhibits an ambivalent, even paradoxical stance towards industrial modernity, which over the past century has profoundly resonated with audiences. The film's look has shaped science fiction films, music videos, video games, comics, and graphic novels. Its influence can be seen in the work of filmmakers and pop culture luminaries from Tim Burton and Ridley Scott to George Lucas and Denis Villeneuve, from Osamu Tezuka and Freddie Mercury to Beyoncé and Lady Gaga. Indeed, *Metropolis* has fashioned the way we imagine the future.

Set in a repressive urban dystopia that collapses following a human-caused disaster, *Metropolis* articulates fundamental fears about industrial modernity. The film paints a picture of the future in which nurturing

Loew, Katharina, *Special Effects and German Silent Film: Techno-Romantic Cinema*.
Taylor & Francis Group, 2021
DOI 10.5117/9789463725231_CH05

natural environments have disappeared and technology permeates all aspects of life. As people have become enslaved by god-like machines, the fusion of human and apparatus seems inevitable. Despite these horrific premonitions, however, *Metropolis* is far from taking a purely technophobic stance as it flaunts technology's splendour, magnetism, and visionary power. By confronting the audience with monstrous machines and hostile high-tech environments of hypnotic grandeur, *Metropolis* simultaneously characterizes the future as astonishing and dreadful. This apparent contradiction can be best understood in terms of a "technological sublime." The concept builds on theories by eighteenth-century philosophers like Edmund Burke, Immanuel Kant, and Friedrich Schiller, who conceived of objects that exceed our sensory capacities and result in conflicting emotions of pleasure and pain as "sublime." The sublime refers to experiences that transcend the material and finite. In *Metropolis*, stunning and unfathomable technological objects, events, and environments such as the Moloch, the robot's anthropogenesis, and the vistas of downtown Metropolis evoke the overpowering sensory impressions and concomitant emotions of terror and wonder that characterize the sublime. By construing technology as a source of transcendent experiences, the technological sublime pertains to the techno-romantic paradigm.

Like *Der Student von Prag* (Deutsche Bioscop, 1913, dir. Stellan Rye) and *Nosferatu* (Prana, 1922, dir. F. W. Murnau), *Metropolis* is engaged in the techno-romantic project of capturing truths beyond physical reality. Like them, it utilizes special effects for depicting ideas and emotions in medium-specific ways in order to assert its artistic ambitions. All three films can be read as reflections on the promise and menace of technology and the nature of the cinematic medium. However, in contrast to its predecessors, *Metropolis* addresses the subject of machine technology directly. Conceived and marketed as a film-technological marvel, *Metropolis* touts its proficiency in realizing the impossible. Special effects visualize technology's coinciding horror and fascination and thus evoke sublime experiences. By so doing, they render the unfathomability of technology itself and transform the film's central concern, the impending obliteration of humanity as a consequence of total machination, into art.

Machine-humans and human-machines

Metropolis was conceived as a multi-media package consisting of the film and a tie-in novel of the same title. A preprint of Thea von Harbou's story was serialized in *Das Illustrierte Blatt* (Frankfurter Societäts-Druckerei) starting in August 1926, and the book version was published in December 1926, shortly before the film's premiere. Both novel and film portray the relationship between human and machine as emblematic of the disastrous repercussions of industrial modernity. According to Tom Gunning, "the machine in a variety of manifestations becomes the central allegorical figure of the film."[1] In the dystopian society of Metropolis, machines are responsible for the far-reaching dehumanization and suppression of human beings. Their supremacy has advanced to the point where boundaries between human and machine, between living and non-living matter, persistently blur. For instance, the city's working class is described in the script as "human-machines" [*Mensch-Maschinen*] (scenes 5, 8, 9) with "the same faces, the same movements, the same actions, the same sorrow" (scene 5).[2] This is visualized in the film in the rigid, mechanical movements of the workers, fully in sync with the machines they operate. They perform repetitive tasks that are precisely paced by the rhythm of the mechanism and require full concentration, which eradicates their mental capacity and reduces them to debilitated human-machine hybrids. As the false Maria rants in the intertitles, the workers "lubricate the machine joints with their own blood" and "feed the machines with their own flesh." In other words, the life force of human beings is being reduced to an energy supply to machines. In the novel, Fredersen rationalizes this when arguing, "that men are used so rapidly at the machines, Freder, is no proof of the greed of the machine, but of the deficiency of the human material"(22).[3] Dispossessed of their autonomy and inner life, the workers of Metropolis begin to approximate machines. Their antagonists, the machines, on the other hand, are bestowed with animistic vitality. In the novel, the "machine brutes" (12) are attributed with humanoid mental and physical characteristics like "soft, malicious eye[s]," (33) "delicate joints" (33), "steel breast[s]" (173), and "vengeful [...] jealousy" (10). They "roar for food" (11), "gulp[] the masses down" (11), and "[throw] the masses up" (11). At any moment they could awaken from their "feigned sleep" (22) and "race [themselves] to

1 Tom Gunning, *The Films of Fritz Lang: Allegories of Vision and Modernity* (London: British Film Institute, 2000), 55.

2 Thea von Harbou, *Metropolis* [screenplay]. The scene numbers refer to the only version of the script known to survive. It is held in the Gottfried Huppertz Collection at the Filmmuseum Berlin-Deutsche Kinemathek and contains 406 numbered scenes.

3 The page numbers in the novel refer to Thea von Harbou, *Metropolis*, anonymous trans. (Mineola, NY: Dover Publications, 2015).

pieces" (22). Machines, as the novel graphically describes, are parasitic, superhuman creatures:

> And the machine, having neither head nor brain, with the tension of its [watchman's] watchfulness, sucks and sucks out the brain from the paralysed skull of its watchman, and does not stay, and sucks, and does not stay until a being is hanging to the sucked-out skull, no longer a man and not yet a machine, pumped dry, hollowed out, used up. And the machine which has sucked out and gulped down the spinal marrow and brain of the man and has wiped out the hollows in his skull with the soft, long tongue of its soft, long hissing, the machine gleams in its silver-velvet radiance, anointed with oil, beautiful, infallible – Baal and Moloch, Huitzilopochtli and Durgha. (22)

To the divinity of machines and the servitude of workers also feature prominently in the script.. Dialog titles reference "god-machines" (scenes 65 and 69) and "sacrificial slaves" (scene 65), who have to be "delivered from the machine" (scene 166). They were ultimately omitted from the film, possibly in an effort to limit bold verbal metaphors in the intertitles. The allegory of machines as gods nonetheless remains obvious in Freder's Moloch vision, in which a massive machine turns into a slave-devouring juggernaut, and in his symbolic crucifixion on Georgy's machine. The metaphorical equation of machines and gods or monsters is not unique to *Metropolis*, however, and has precursors in the work of Karl Marx, Ernst Toller, E. M. Forster, and Helmuth Plessner, among others.[4] The idea that humans might become slaves to machines became a widespread trope in the decades around 1900. Both conservatives like Oswald Spengler and Ludwig Klages and leftists like Oscar Wilde, Leopold Jacoby, Arthur Holitscher, and Wilhelm Liebknecht resorted to this metaphor when grappling with industrialization and modernity.[5]

4 See for instance Karl Marx, *Capital: A Critique of Political Economy*, vol. I: *The Process of Production of Capital* [1867], trans. Ben Fowkes (Harmondsworth: Penguin, 1990); E. M. Forster, "The Machine Stops [1909]," in *Selected Stories*, 91-124 (New York, NY: Penguin Classics, 2001); Ernst Toller, "Transformation [1919]," *Plays One*, ed. and trans. Alan Raphael Pearlman (London: Oberon Books, 2000); Ernst Toller, "The Machine Breakers [1922]," *Plays Two*, ed. and trans. Alan Raphael Pearlman (London: Oberon Books, 2019); Ernst Toller, "The German Hinkemann [1923]," *Plays Two*, ed. and trans. Alan Raphael Pearlman (London: Oberon Books, 2019); Helmuth Plessner, "Die Utopie in der Maschine [1924]," in *Gesammelte Schriften in zehn Bänden*, vol. X: *Schriften zur Soziologie und Sozialphilosophie* (Berlin: Suhrkamp, 2003).

5 See for instance Wilhelm Liebknecht, "Knowledge is Power – Power is Knowledge [1872]," in *Wilhelm Liebknecht and German Social Democracy: A Documentary History*, ed. William A. Pelz,

In *Metropolis*, both the film and the novel, the "human-machines" are confronted with a "machine-human" [*Maschinen-Mensch*] (scene 105, 108).[6] The verbal inversion of "machine-human" and "human-machine" highlights the affinity between the workers and the robot, calling attention to the film's central concern, the antagonistic symbiosis and imminent convergence of human and machine. The "machine-human" is given a female form, which conceals her machine nature. The script describes the robot as the "human of the future" (scene 108), whereas the novel calls her equally "Futura" and "Parody" (48). The former name characterizes the robot as representative of a coming age, while the latter emphasizes its nature as a travesty of humanity. Like the machines, the robot is endowed with god-like qualities: In keeping with the script, which compares it to an "Egyptian pharaoh statue" (scene 105), the film initially presents the robot on a cubic block reminiscent of a throne, with legs close together, feet parallel, and eyes gazing straight ahead. In the style of ancient Egyptian art, the emphasis is on symmetry and there is no suggestion of either facial expression or movement.[7] The fact that the "machine-human" is represented as a pharaoh, whom Egyptians considered divine, resumes the theme of "god-machines," and furthermore relates the robot to Freder's Moloch vision, which resembles an Egyptian sphinx.

The conflation of bodies and machines, the animate and the inanimate, is omnipresent in *Metropolis*. As a contemporary reviewer for *The Nation* astutely observed: "Man is inanimate. Life is metronomic. It is only the machines that are alive."[8] Machines permeate all spaces with the exception of the Eternal Gardens, an artificial paradise, and the cathedral and the catacombs, outmoded places of worship. All major characters coalesce at least temporarily with machines: Joh Fredersen is cold and rational like a mechanical instrument, Rotwang's artificial hand renders him physically part apparatus, Freder merges with the machine when standing in for Georgy, and Maria shares her appearance with the robot. Humanity is palpably threatened by its subordination to urban-industrial rationality. Indeed, the simultaneously fascinating and alarming prospect of a symbiosis between human and machine

trans. Erich Hahn (Chicago, IL: Haymarket Books, 2016; Leopold Jacoby, *Die Idee der Entwicklung* [1874/1876] (Zurich: Schabelitz, 1887); Oscar Wilde, *The Soul of Man under Socialism* [1891] (New York, NY: Harper & Row, 1970); Arthur Holitscher, *Amerika heute und morgen* [1912] (Berlin: S. Fischer, 1923); Ludwig Klages, *Mensch und Erde* [1913] (Stuttgart: A. Kroener Verlag, 1956); Oswald Spengler, *The Decline of the West* [1918/1922], trans. Charles Francis Atkinson (New York, NY: Knopf, 1966).

6 The German word "Mensch" is gender-neutral.

7 Deviating from the Egyptian model, the hands of the "machine-human" are on the armrests instead of the knees.

8 Evelyn Gerstein, "*Metropolis*," *The Nation* (23 March 1927): 323-324.

was highly topical at the time of the film's creation, captivating not only the imagination of Lang and von Harbou, but also that of numerous authors and critics including Walter Rathenau, Joseph Roth, and Bertolt Brecht.[9]

The "machine-human" turns the "human-machines" against the "god-machines" they serve. Given the novel's and the film's technophobic rhetoric one might assume that the ensuing worker's revolt is depicted as an act of human liberation from the oppressive machines. Remarkably, however, the opposite is the case. The workers put themselves in the wrong when they follow the robot in a Luddite mass frenzy. Their uncontrolled rage dwarfs the threat of technology and they impetuously proceed to destroy the very foundation of their existence: the core of Metropolis, the Heart Machine, and with it their homes and children. Their previous exploitation and enslavement notwithstanding, novel and film leave no doubt that the workers' rebellion is reprehensible. Chaos and destruction are portrayed as similarly dehumanizing and even more threatening to human existence than mechanical suppression. In an extraordinary inversion, the machines of Metropolis appear as a vital force under attack by the workers, who emerge as demonic agents.

This about-face is not of lasting consequence, however. In the end, the machines have disappeared. In front of the cathedral the workers once again march in file, but now with their heads held high. The setting in front of the cathedral is highly significant: First, the cathedral represents a middle ground between the underground machine rooms and Fredersen's office on the top floor of the New Tower of Babel (as novel and script refer to the highest building of Metropolis). Secondly, the cathedral embodies tradition, community, and spirituality—the very antithesis of modern rationality. For Béla Balázs the cathedral epitomizes a pre-conceptual "visual culture," a repository of the collective spirit of the people.[10] The newly white hair of formerly cold and cerebral Fredersen testifies to a newfound humanity and wisdom. Finally, and most importantly, a *human* heart, impersonated by Freder, supersedes the destroyed machine heart of Metropolis. The fresh beginning suggested here is conspicuously technology-free. Erasing any trace of industrialization, the film advocates for class reconciliation based solely on a common humanity.

9 See Walter Rathenau, *Zur Kritik der Zeit* [1912] (Berlin: Fischer, 1929); Joseph Roth, "Der auferstandene Mensch [1923]," in *Joseph Roth Werke*, vol. 1: *Journalistische Arbeiten 1915-1923*, ed. Klaus Westermann, 936-939 (Cologne: Kiepenheuer & Witsch, 1989-1991); Bertolt Brecht, "Singing Steyr Motor Cars [1927]," in *The Collected Poems of Bertolt Brecht*, trans. and eds. Tom Kuhn and David Constantine, 361-362 (New York, NY: Liveright Publishing, 2018).

10 *Béla Balázs: Early Film Theory. Visible Man and The Spirit of Film*, ed. Erica Carter, trans. Rodney Livingstone (New York, NY/Oxford: Berghahn Books, 2010), 9.

A vast majority of contemporary critics rejected this message of post-industrial atonement and the film's depiction of technology more generally. Remarkable for a work of science fiction, the film's depiction of machinery and automation was seen as regressive across the political spectrum. *Metropolis* rehashes, a writer for the liberal *Weltbühne* remarked, "problems and hardships of the past."[11] A commentator for the right-wing *Deutsche Allgemeine Zeitung* concurred: "*Metropolis* does not depict technology of the future, nor technology of the present, it is the past that is falsely presented to us as future, because the vision for a plausible future was lacking."[12] One of Germany's leading trade journals, *Der Film*, observed that *Metropolis* reflects a "mind-set and working conditions that were overcome fifty and more years ago."[13] Overall, many critics shared the sentiment of an Austrian provincial newspaper, which called Lang's "views on the development of technology stale, most superficial parlour philosophy, untrue, dishonest, illogical, and senseless."[14] In hindsight, even Lang himself criticized the clumsiness of his treatment. At the same time, however, he also continued to underscore the topic's relevance:

> You remember the clocks and the man who worked in harmony with them? He should become, so to speak, part of the machine. Well, that seemed to be too symbolic, too simplistic in the evocation of what we call the evils of mechanization. As well, a few years ago, I had to revise my judgment again, at the spectacle of our astronauts walking around the earth. It was the experts, they were prisoners of their space capsule, nothing else—or almost, a part of the machine that carried them.[15]

The film's heavy-handed approach notwithstanding, the theme of bio-technical synthesis may well have contributed to the enduring relevance of *Metropolis*. After all, in today's age of artificial intelligence, bionics or human-computer interaction (HCI), the threat of an antagonistic symbiosis and imminent convergence of human and machine has grown ever more topical and continues to kindle our fascination and existential fears.

11 Axel Eggebrecht, "*Metropolis*," *Weltbühne* 3 (18 January 1927): 115-116.

12 Siegfried Hartmann, "*Metropolis* Technik," *Deutsche Allgemeine Zeitung* (20 January 1927): 2.

13 Max Feige, "Der Film *Metropolis*," *Der Film* (Berlin) 1 (15 January 1927): 5-6.

14 H. H. P., "Das Zerrbild der Technik im Film: Ein Nachwort zu *Metropolis*," *Tages-Post* (Linz, Austria) 52 (6 March 1927): 3-4.

15 Gretchen Berg, "The Viennese Night: A Fritz Lang Confession, Parts One and Two (1965)," in *Fritz Lang Interviews*, ed. Barry Keith Grant, 50-76 (Jackson, MS: University Press of Mississippi, 2003), 69.

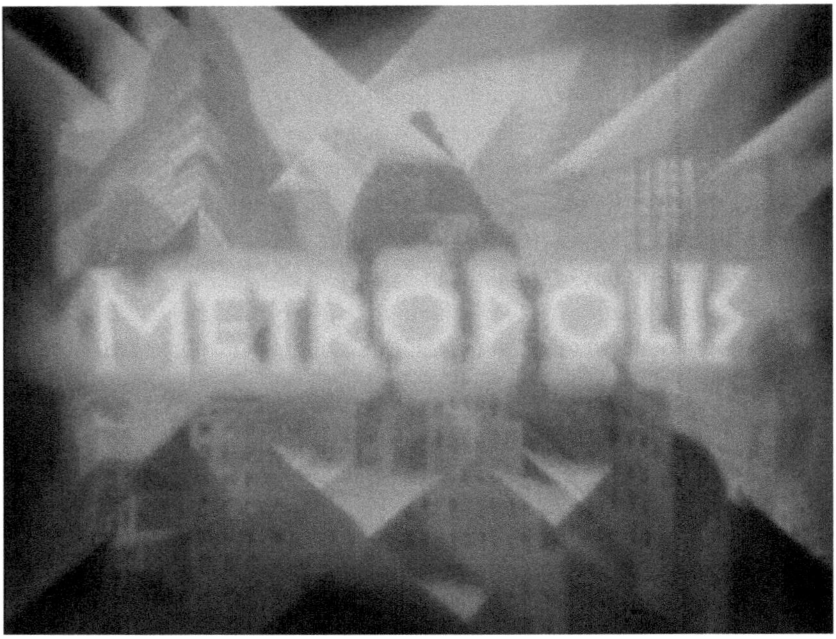

Fig. 5.1. Erich Kettelhut's animated painting of the city at daybreak in *Metropolis* (1927). Courtesy of Friedrich Wilhelm Murnau Foundation.

In *Metropolis*, not only does the human body fuse with machines, the environment is also affected by total mechanization. The city of Metropolis appears as an overpowering technological structure. The prevailing sentiment, however, is not horror, but rather an amalgam of apprehension and awe. The novel characterizes Metropolis as a "machine-city" (11), which the film visualizes in its opening sequence. The main title is constructed from intersecting diagonal and horizontal lines of light that give way to Erich Kettelhut's abstract painting of a city at daybreak (Figure 5.1.), which, as the script describes, consists of "skyscrapers, agglomerated into blocks, every block the size of cities, separated into three triangles" (scene 1). Light rays as well as bright and dark triangles move slowly and steadily up and down the facades, eventually dissolving into a montage of rhythmically moving machine parts. Their rotational and translational motions generate what Sergei Eisenstein would call "conflict[s] of graphic directions," both within and in between shots.[16] The sequence captures the majestic spectacle of the relentless, pervasive movement of mechanical work—the pulse of modernity.

16 Sergei Eisenstein, "The Cinematographic Principle and the Ideogram [1929]," in Sergei Eisenstein, *Film Form: Essays in Film Theory*, ed. and trans. Jay Leyda, 28-44 (New York, NY: Harcourt, Brace & World, 1949).

The city of Metropolis is an artificial organism with organ-like mechanical components that perform specific functions, including an encephaloid coordinating centre in the New Tower of Babel and a "heart machine" that keeps the city alive. Fritz Lang asserted that he sought in *Metropolis* to "capture the roaring rhythm of the extreme progress of civilization."[17] This aspect was not lost on reviewers. Luis Buñuel, who, like many critics, had decidedly mixed feelings about the film, nonetheless erupted into a technophile paean when praising *Metropolis* as an "exalting symphony of movement! [...] Each powerful flash of steel, the rhythmic succession of wheels, pistons, of uncreated mechanical forms is a marvellous ode, a new poetry to our eyes. Physics and Chemistry are miraculously transformed into Rhythm."[18] Contemporaries frequently contrasted the artificial rhythm of the modern city, of technology, economy, and traffic, with the organic rhythm of nature.[19] Michael Cowan has called attention to the complex role of rhythm in relation to body and technology in German modernity.[20] Indeed, in *Metropolis*, the machine city imposes its rhythm on the bodies of the workers, who move and function as its mechanical components. Organic and inorganic, animate and inanimate, machine and man have become indistinguishable. This, the film contends, constitutes an existential threat to humanity. At the same time, however, its presentation is imposing and mesmerizing.

"America" as emblem

The mechanical reshaping of the human element and ultimately its elimination is key to *Metropolis*'s conception of modernity. It can be traced, as Anton Kaes and others have shown, to contemporary debates about "Americanization" and "Americanism."[21] In contrast to the United States, where these terms were used to address aspects of a joint American identity,

17 Ludwig Spitzer, "Fritz Lang über den Film der Zukunft," *Die Filmtechnik* 2 (15 July 1925): 34-36.

18 Luis Buñuel, "*Metropolis* [1927]," in *An Unspeakable Betrayal: Selected Writings of Luis Buñuel*, trans. Garrett White, 99-102 (Berkeley, CA: University of California Press, 2000).

19 Fritz Giese, *Girlkultur: Vergleiche zwischen amerikanischem und europäischem Rhythmus und Lebensgefühl* (Munich: Delphin-Verlag, 1925), 25-31.

20 Michael Cowan, "The Heart Machine Rhythm and Body in Weimar Film and Fritz Lang's *Metropolis*," in *Modernism/modernity* 2 (April 2007): 225-248, 228.

21 Anton Kaes, *Shell Shock Cinema: Weimar Culture and the Wounds of War* (Princeton, NJ: Princeton University Press, 2009): 182-184.

in Europe they acquired a radically different meaning.[22] As early as 1877, the term "Americanization" was used to describe "the dreaded overgrowth and penetration of European culture by realism and the rapidly growing predominance of technology."[23] Fifty years later, philosopher and psychologist Richard Müller-Freienfels defined "Americanism" in remarkably similar terms as a "universal transformation of the human soul" caused by "the restructuring of the world by technology and everything associated with that."[24] Müller-Freienfels stressed that "Americanism" did not refer to actual America but instead represented an abstract idea that, although originating in actual America, had become a global phenomenon.[25] Specifically, he invoked the "quantification," "mechanization," and "standardization" of life as well as the "depersonalization of the soul" as chief characteristics of "Americanization."[26] In other words, "Americanization" and "Americanism" referred to processes of modernization that were perceived as jeopardizing essential aspects of human nature, notably spirituality and interiority. "America" became the epitome of the woes (and also the possibilities) of modern existence. *Metropolis* takes this vision to an extreme, prompting critics to describe the city of Metropolis as "hyper-American."[27]

From a European perspective, one of the most striking icons of "Americanization" was the skyscraper, which embodied both the opportunities and downsides of a highly industrialized, capitalist society. In Germany, the prospect of domestic skyscrapers the topic of a passionate public discussion since 1920.[28] On the one hand, high-rise buildings were celebrated as testaments of cutting-edge style and technology. On the other hand, they were spurned as inhumane, as "definitive emblem[s] of capitalism in audacious, but also brutally selfish form."[29] Taken together, the urban environments of *Metropolis*, specifically the spectacular skyline, the underground workers'

22 See for instance Theodore Roosevelt, "True Americanism [1894]," in *American Ideals and Other Essays, Social and Politcal*, 16-37 (Philadelphia, PA: Gebbie and Company, 1903); Royal Dixon, *Americanization* (New York, NY: Macmillan, 1916).

23 Emil du Bois-Reymond, *Reden: Erste Folge. Literatur – Philosophie – Zeitgeschichte* (Leipzig: Veit u. Comp., 1886), 240-306, 280.

24 Richard Müller-Freienfels, *Geheimnisse der Seele* (Munich: Delphin-Verlag, 1927), 292.

25 Müller-Freienfels, *Geheimnisse der Seele*, 244.

26 Müller-Freienfels, *Geheimnisse der Seele*, 249-274.

27 Willy Haas, "*Metropolis*," *Film-Kurier* (11 January 1927).

28 Jochen Meyer, "Die Hochhausdiskussion in Deutschland am Anfang der zwanziger Jahre," in *Der Schrei nach dem Turmhaus: der Ideenwettbewerb Hochhaus am Bahnhof Friedrichstrasse, Berlin 1921/22*, eds. Florian Zimmermann and Bruno Bruognolo, 187-214 (Berlin: Argon, 1988), 187.

29 Hans Schliepmann, "Wolkenkratzer," *Tägliche Rundschau* 49 (30 January 1921).

city, and the monumental machine rooms, reflect a view of modernity that is characterized by conflicting emotions. As Dietrich Neumann points out, the urban design of *Metropolis* "seems both to embrace and undermine the current notions of monumentalism, technological progress, and Americanism."[30]

Lang later often alleged that he first conceived of *Metropolis* at the sight of the Manhattan skyline.[31] This claim has been thoroughly discounted: at the time of Lang's visit to the United States in October 1924 the script of *Metropolis* was already complete.[32] Nonetheless, the sight of Manhattan was doubtlessly a transformative experience for Lang. It may well have affected the film's visual configuration and thus contributed to its deeply ambivalent attitude towards industrial modernity. Upon his return from the United States Lang fervently inquired:

> Where is the film [...] about one of these Babels of stone that call themselves American cities? The mere sight of New York at night should suffice to make this beacon of beauty the centre of a movie. [...] In the daytime, New York is the epitome of sobriety, albeit captivating on account of its movement. In the evening, New York is so beautiful, that the port entrance at night alone provides an unforgettable impression for one's whole life.[33]

Metropolis's foundational myth further underlines the significance of "America" as a key reference point, which is evident both in terms of the film's cultural criticism and its aesthetics. In keeping with the diverging perceptions of "America" in Europe at the time, the modernity of *Metropolis* is simultaneously magnificent and terrifying.

Many scholars have commented on this paradoxical treatment of modernity. In his landmark essay Andreas Huyssen distinguishes "two opposing views of modern technology." On the one hand, he identifies what he calls an "Expressionist view," which emphasizes "technology's oppressive and destructive potential." On the other hand, the film displays an "unbridled

30 Dietrich Neumann, "The Urbanistic Vision in Fritz Lang's *Metropolis*," in *Dancing on the Volcano: Essays on the Culture of the Weimar Republic*, 143-154 (New York, NY: Camden House, 1994), 154.

31 Peter Bogdanovich, *Fritz Lang in America* (London: Studio Vista, 1967), 15; Berg, "The Viennese Night," in *Fritz Lang Interviews*, ed. Grant, 68.

32 According to news reports, Fritz Lang and Thea von Harbou planned to finish the screenplay of their new film *Metropolis* during their 1924 summer vacation. See "Film – Extrablatt," *Illustriertes Wiener Extrablatt* (4 July 1924).

33 Fritz Lang, "Was ich in Amerika sah: Neuyork – Los Angeles," *Film-Kurier* (11 December 1924): 1.

confidence in technical progress and social engineering," which Huyssen associates with the "technology cult of the *Neue Sachlichkeit*."[34] Huyssen pinpoints an important antinomy, yet its association with the stylistic labels mentioned is problematic, if only because neither Expressionism nor *Neue Sachlichkeit* can be understood as inherently technophobic or technophile.[35] R. L. Rutsky seeks to reframe Huyssen's characterization in favour of a dichotomy between "rationalist, functionalist notions of technology" and "irrational, chaotic, destructive aspects."[36] Rutsky underscores that either conception of technology can be utopian or dystopian, yet in *Metropolis* both are dystopian. Indeed, as we have seen, both technology itself and human interaction with it—whether rationalist or irrational—are generally construed as dehumanizing and/or destructive. Fredersen's technological autocracy, the machines, particularly Georgy's machine and the fateful Moloch-machine, the mechanical Maria, or the workers' Luddite revolt are portrayed as harmful and dangerous. At the same time, the visual symbols associated with these conceptually adverse elements, specifically the New Tower of Babel and the city above ground, Freder's Moloch vision, the robot's anthropogenesis, and the explosion of the heart machine appear magnificent and awe-inspiring. Notably, they are often associated with spectacular light displays, including skyline and searchlights or visualized electric currents, which evoke the encounter with the Manhattan skyline that impressed Lang so deeply.

The technological sublime

The ambivalent sensation of terror and wonder associated with overpowering technological objects in *Metropolis* is evocative of theories of the sublime.[37] Originally, the term was used to describe a rhetorical style in antiquity. In the eighteenth century, it took on a constitutive role in the

34 Andreas Huyssen, "The Vamp and the Machine: Technology and Sexuality,' *New German Critique* 24/25 (Autumn 1981/Winter 1982): 221-237, 223.

35 See for instance Karlheinz Daniels, "Expressionismus und Technik," in *Technik in der Literatur*, ed. Harro Segeberg, 351-386 (Frankfurt am Main: Suhrkamp, 1987); Matthias Eberle, *Der Weltkrieg und die Künstler der Weimarer Republik. Dix, Grosz, Beckmann, Schlemmer* (Stuttgart: Belser, 1989), 22, 27.

36 R. L. Rutsky, "The Mediation of Technology and Gender: *Metropolis*, Nazism, Modernism," *New German Critique* (Autumn 1993): 3-32, 4.

37 For a comprehensive introduction see Jörg Heiniger, "Erhaben," *Ästhetische Grundbegriffe: Historisches Wörterbuch in 7 Bänden*, vol. 2, 275-310 (Stuttgart: Metzler-Verlag, 2001).

cultural imaginary, delineating the fragile limits of the emerging field of aesthetics. The term has been employed to describe both a quality of the object and the emotional response of the subject.[38] Theorists such as Edmund Burke, Immanuel Kant, Friedrich Schiller, Arthur Schopenhauer, and more recently Jean-François Lyotard conceived of the sublime in a variety of ways. Regardless of significant individual differences, however, their conceptions of the sublime have two major aspects in common: First, they invoke the sublime to characterize perceptions in face of an apparently infinite object—impressions that exceed our sensory capacity due to boundless size or overwhelming force. The sublime confronts human beings with their finitude. For Burke, it is "the strongest emotion which the mind is capable of feeling."[39] Second, they construe the subject's resulting emotions as conflicting. Kant refers to "a rapidly alternating repulsion from and attraction to one and the same object."[40] Schiller characterizes the sublime as a "mixed feeling" of pain and joy.[41] For Burke, "delightful horror" is "the most genuine effect, and truest test of the sublime."[42] Lyotard describes the Kantian sublime as "a strong and equivocal emotion: it carries with it both pleasure and pain. Better still, in it pleasure derives from pain."[43] Christiane Pries gives a detailed account of this biphasic process according to Kant:

> First phase: When encountering specific natural phenomena – those that seem exceedingly large or exceedingly powerful – our imagination, which is a sensory and finite faculty, can no longer successfully process the assailing impressions. [...] Imagination and transition fail; nature seems alienating, which results in displeasure. Second phase: This displeasure is nonetheless accompanied by pleasure, because upon further reflection it occurs to the subject that even the oversized and overpowering nature, as measured by the infinite capacity for ideas (reason), is tiny and its

38 Arthur Schopenhauer, *The World as Will and Representation* [1819], vol. 1, trans. E. F. J. Payne (New York, NY: Dover Publications, 1969), 201-202.

39 Edmund Burke, *A Philosophical Enquiry into the Sublime and Beautiful* [1757], ed. James T. Boulton (London/New York, NY: Routledge, 2008), 36.

40 Immanuel Kant, *The Critique of the Power of Judgment* [1790], ed. Paul Guyer, trans. Paul Guyer and Eric Matthews (Cambridge, UK: Cambridge University Press, 2000), 141.

41 Friedrich Schiller, *The Works of Friedrich Schiller: Aesthetical and Philosophical Essays*, ed. Nathan Haskell Dole (Boston, MA: The Wyman-Fogg Company, 1902), 127.

42 Burke, *A Philosophical Enquiry into the Sublime and Beautiful*, 73.

43 Jean-François Lyotard, *The Postmodern Condition: A Report on Knowledge*, trans. Geoff Bennington and Brian Massumi (Manchester: Manchester University Press, 1984), 77.

> power can only damage the finite, sensory side of the human, but not her intelligible capacity. As a rational being she is safe and superior to nature.[44]

While sublime experiences have most commonly been associated with natural phenomena such as mountain ranges, volcanoes, oceans, deserts, earthquakes, cataracts, avalanches, storms, or the starry vault, theorists have also applied the concept to man-made objects. Burke refers to an "artificial infinite" and calls the noise of artillery sublime.[45] Kant mentions the sublimity of the Egyptian pyramids and St. Peter's Cathedral, and also war.[46] Schopenhauer too cites the pyramids, St. Peter, and further ascribes sublime qualities to a modern water conduit.[47] Over the course of the nineteenth century, the sublime was increasingly invoked to characterize technological innovations. As Klaus Bartels and Dagmar Bellmann have emphasized, the mastering of powerful forces of nature like water, electricity, and steam allowed their sublime qualities to be transferred to technological constructs and the human beings that controlled them.[48] Considering that the technological sublime describes the evocation of the unfathomable through technology, it must be recognized as an instance of the techno-romantic paradigm.

By the second half of the nineteenth century, the sublime had exhausted its theoretical momentum and cultural currency as an aesthetic category. Simultaneously, the meaning of the German term for sublime, *erhaben,* was transforming as well. In aesthetic theory it continued to be used in the Kantian sense, but outside of aesthetics the notion of "negative pleasure"[49] faded, the word was more frequently applied to uplifting rather than overwhelming experiences, and it took on the narrower sense of "majestic, venerable, solemn." This semantic change, however, did not eradicate the Kantian sentiment itself. Nineteenth- and twentieth-century authors frequently described their experiences vis-à-vis imposing technological achievements like bridges, dams, ships, towers, factories, engines, or the telegraph in ways

44 Christiane Pries, "Einleitung," in *Das Erhabene: Zwischen Grenzerfahrung und Größenwahn*, ed. Christiane Pries, 1-30 (Berlin: Verlag Walter de Gruyter, 1989), 8-9.

45 Burke, *A Philosophical Enquiry into the Sublime and Beautiful*, 137; 82.

46 Immanuel Kant, *Observations on the Feeling of the Beautiful and the Sublime* [1764], trans. John T. Goldthwait (Berkeley, CA: University of California Press, 2003), 49; Kant, *The Critique of the Power of Judgment*, 146.

47 Artur Schopenhauer, *Metaphysik des Schönen* [1820] (Munich: Piper Verlag, 1985), 107; 110.

48 Klaus Bartels, "Über das Technisch-Erhabene," in *Das Erhabene: Zwischen Grenzerfahrung und Größenwahn*, ed. Christiane Pries, 295-316 (Berlin: Verlag Walter de Gruyter, 1989), 305; Dagmar Bellmann, *Von Höllengefährten zu schwimmenden Palästen: Die Passagierschifffahrt auf dem Atlantik (1840-1930)* (Frankfurt am Main: Campus Verlag, 2015), 92.

49 Kant, *The Critique of the Power of Judgment*, 129.

that Leo Marx famously called "the rhetoric of the technological sublime."[50] Adalbert Stifter, Karl Henckell, and Kurd Lasswitz, for instance, convey such reactions vis-à-vis the railroad, Alfred de Vigny, Theodor Fontane, and Wilhelm Bölsche expressed awe at the sight of the modern metropolis, which in the nineteenth century was still epitomized by London and Paris.[51]

Although authors like Klaus Bartels have examined the rhetoric of the technological sublime in European contexts, most scholarship, particularly following David Nye's influential *American Technological Sublime*, has been focused on the United States.[52] Nye lays out in detail the conflation of public admiration for North America's magnificent landscapes like the Grand Canyon and Niagara Falls with public awe and wonder in face of grand technological feats like the Corliss Engine, the Brooklyn Bridge, the transcontinental railroad, or the Woolworth Building.[53] In the United States, the sublime became tied to nationalist sentiment. Nye points out, "the American sublime transformed the individual's experience of immensity and awe into a belief in national greatness."[54]

From a European perspective, neither the natural nor the technological sublime was inevitably associated with "America," but "America" was associated with the sublime. The sheer size of the American continent, its vast pristine landscapes, seemingly unlimited natural resources, particularly land, water, forest, and livestock, as well as dramatic industrial development and economic expansion during the final decades of the nineteenth

50 Leo Marx, *The Machine in the Garden: Technology and the Pastoral Ideal in America* (New York, NY: Oxford University Press, 1965), 195.

51 *Adalbert Stifter Briefe*, ed. Hans Schumacher (Zurich: Manesse-Verlag, 1947), 80-81; Karl Henckell, "Viadukt," in *Gesammelte Werke*, vol. 2: *Buch des Kampfes* (Munich: J. Michael Müller Verlag, 1921), 157-158; Kurd Lasswitz, *Wirklichkeiten* (Berlin: Verlag von Emil Felber, 1900), 441; Alfred de Vigny, "Paris [1831]," in *Oeuvres complètes. Texte présenté et commenté par Ferdinand Baldensperger*, vol. 1, 164-165 (Paris: Gallimard, 1948); Theodor Fontane, *Ein Sommer in London* (Dessau: Gebrüder Katz, 1854), 2-5; Wilhelm Bölsche, "Die Poesie der Großstadt," *Das Magazin für Literatur* 40 (4 October 1890): 622-625; see also Nicholas Taylor, "The Awful Sublimity of the Victorian City: Its Aesthetic and Architectural Origins," in *The Victorian City: Images and Realities*, eds. H. J. Dyos and Michael Wolff, 431-447 (London: Routledge & Kegan Paul Books, 1973).

52 Bartels, "Über das Technisch-Erhabene," in *Das Erhabene*, ed. Pries, 295-316. For scholarship on the technological sublime in an American context see for instance Joseph Tabbi, *Postmodern Sublime: Technology and American Writing from Mailer to Cyberpunk* (Ithaca, NY: Cornell University Press, 1995); Jeremy Gilbert-Rolfe, *Beauty and the Contemporary Sublime* (New York, NY: Allworth Press, 1999); Zoltán Simon, *The Double-Edged Sword: The Technological Sublime in American Novels between 1900 and 1940* (Budapest: Akadémiai Kiadó, 2003).

53 David E. Nye, *American Technological Sublime* (Cambridge, MA: MIT University Press, 1994).

54 Nye, *American Technological Sublime*, 43.

century conveyed overwhelming immensity, even infinity. The concept of boundlessness is also reflected in the title of Ludwig Max Goldberger's 1903 travelogue *The Land of Unlimited Opportunities*, which subsequently became the principal German epithet for the United States.[55] European observers regarded monumental engineering achievements like bridges, skyscrapers, and factories as a startling and admirable testimony to American techno-enthusiasm and industrial resourcefulness. Simultaneously, they perceived them as intimidating in a number of ways: as colossal objects exceeding any human measure, as demonstration of the superior technological capabilities of a rival industrial country, and as distressing emblems of excessive and uncontrollable modernization. European responses to the "technological sublime" thus differed markedly from American ones. Instead of wholeheartedly embracing imposing technological feats, many European observers remained ambivalent. In fact, their reactions seem more closely in line with the mixed emotion of pleasure and pain theorized by eighteenth-century thinkers. For instance, in Gerhart Hauptmann's novella *Flagman Thiel* the railroad is portrayed as an overpowering, destructive "monster" (*Ungetüm*), a mystical force that causes the earth to tremble and "a veritable rain of blood [...] to descend from heaven," and whose passing is described as "sublime spectacle."[56] A similar blend of wonder and terror is also reflected in *Metropolis's* seemingly paradoxical attitude towards technology.

Powerful technological environments and omnipotent machines that humans struggle to control are of course ubiquitous in science fiction, a genre that fictionalizes the rhetoric of the technological sublime. Here, the sublime sentiment arises from a clash between the boundless technical ingenuity and the fundamental inadequacy and fragility of the human race. As Istvan Csicsery-Ronay has argued, the "science-fictional sublime" is characterized by "a sense of awe and dread in response to human technological projects that exceed the power of their human creators."[57] While Csicsery-Ronay claims that this sentiment became central to science fiction only after World War II, examples can be found already in nineteenth-century proto-science fiction. For instance, the construction of the Columbiad space gun in Jules Verne's *De la*

55 Ludwig Max Goldberger, *Das Land der unbegrenzten Möglichkeiten: Beobachtungen über das Wirtschaftsleben der Vereinigten Staaten von Amerika* (Berlin/Leipzig: F. Fontane, 1905).

56 Gerhart Hauptmann, "Flagman Thiel [1887]," in *German Novellas of Realism II*, ed. Jeffrey L. Sammons, trans. Adele S. Seltzer, 304-333 (New York, NY: Continuum, 1989): 315; 318; 348. Seltzer, in what is the most widely-used English translation, renders the original "erhabenes Schauspiel" ("sublime spectacle") as "exalted drama."

57 Istvan Csicsery-Ronay, Jr, *The Seven Beauties of Science Fiction* (Middleton, CT: Wesleyan University Press, 2008), 6-7.

Terre à la Lune (1865) or, as Klaus Bartels and Misoo Kang observe, the android in Auguste Villiers de l'Isle-Adam's *L'Ève future* are described in distinctly sublime terms.[58] Also in early literary science fiction, massive technological objects and environments like the transatlantic railroad tunnel in Bernhard Kellermann's *Der Tunnel* (1913) or Martian technology in Kurd Lasswitz's *Auf zwei Planeten* (1897) can be associated with a technological sublime. There are also numerous instances in early science fiction film, including the space rocket in Fritz Lang's *Frau im Mond* (Fritz Lang Film/Ufa, 1929), the colossal permanent air station in the Atlantic Ocean in *F.P.1 antwortet nicht* (Ufa, 1932, dir. Karl Hartl), the spectacular cityscapes in *Just Imagine* (Fox Film Corporation, 1930, dir. David Butler), or the mighty space gun in *Things to Come* (London Film Productions, 1936, dir. William Cameron Menzies). Vivian Sobchack defines science fiction as a "genre which emphasizes actual, extrapolative, or speculative science and the empirical method, interacting in a social context with the lesser emphasized, but still present, transcendentalism of magic and religion, in an attempt to reconcile man with the unknown."[59] Indeed, the genre's affinity to a sublime aesthetic originates in its ties to the unknown. The sublime, as Scott Bukatman puts it, "figures the unknown as excess."[60]

Science fiction cinema privileges technology thematically, while relying on special-effect technologies for realization. As Bukatman shows, the genre displays a range of attitudes toward technology. Examining the work of special effects supervisor Douglas Trumbull in films like *2001: A Space Odyssey* (Metro-Goldwyn-Mayer/Stanley Kubrick Productions, 1968, dir. Stanley Kubrick), *Close Encounters of the Third Kind* (Julia Phillips and Michael Phillips Productions et al., 1977, dir. Steven Spielberg), and *Blade Runner* (The Ladd Company et al., 1982, dir. Ridley Scott), Bukatman detects an ambivalence towards new technologies that is "neither unabashedly celebratory (*Star Wars*) nor darkly condemning (*Alien*)," but instead reminiscent of the conflicting emotions once attributed to the sublime.[61] The reliance on trick technology to express coexisting positive

58 Bartels counts thirty mentions of the words "sublime" and "auguste." Klaus Bartels, "Vom Erhabenen zur Simulation. Eine Technikgeschichte der Seele: Optische Medien bis 1900 (Guckkasten, Camera Obscura, Panorama, Fotografie) und der menschliche Innenraum," in *Armaturen der Sinne: Literarische und technische Medien 1870 bis 1920*, eds. Jochen Hörisch and Michael Wetzel, 17-42 (Munich: Wilhelm Fink Verlag, 1990), 38.

59 Vivian Sobchack, *Screening Space: The American Science Fiction Film* (New Brunswick, NJ: Rutgers University Press, 1987), 63.

60 Scott Bukatman, "Disobedient Machines: Animation and Autonomy," in *Beyond the Finite: The Sublime in Art and Science*, eds. Ronald Hoffmann and Iain Boyd Whyte, 128-149 (Oxford: Oxford University Press, 2011), 129.

61 Scott Bukatman, *Matters of Gravity: Special Effects and Supermen in the 20th Century* (Durham, NC: Duke University Press, 2003), 83.

and negative attitudes towards technological innovations, which characterizes Trumball's work, also describes the paradoxical treatment of technology in *Metropolis*. Here, as we will see, spectacular special-effects sequences shape a vision of the future that is simultaneously astonishing and horrific.[62]

Marketing special effects

The technophobic rhetoric of the novel and the script notwithstanding, *Metropolis* was conceived as a marvel of film technology. The production employed a broad array of innovative special effect methods, including the Schüfftan process, rear projection, miniatures in forced perspective, stop-motion animation in two and three dimensions, and highly complex multiple exposure composites. For art director Erich Kettelhut it was clear from the outset that "all trick technological finesses available were to be mobilized here."[63] Ufa aligned its elaborate marketing strategy accordingly. The publicity for *Metropolis* eclipsed that of any other Ufa picture in production at that time, including another prestige project, *Faust* (Ufa, 1925, dir. F. W. Murnau). Over the course of seventeen months of principal photography, Lang and Harbou gave frequent interviews, prominent international visitors including Sergei Eisenstein, Mae Murray, Samuel L. Rothafel, Carl Laemmle, Jack Dempsey, Grigory Aleksandrov, and Eduard Tissé toured the sets, and numerous reporters attended the shooting of spectacular crowd scenes such as the flooding of the workers' city, the construction of the Tower of Babel, and the burning of the false Maria at the stake. In addition, Ufa released regular production reports and a strikingly large number of production stills, which underscored the tremendous scale of the undertaking and the ingenuity of the filmmakers in the face of immense technological challenges. The goal was to establish, as cinematographer Günther Rittau proclaimed, "that the technology of the artistic German *Großfilm* (high-budget film production) can stand comparison with that of any other country."[64]

62 For the intersections between cinema, anxiety, science fiction, and the sublime, see Sarah Keller, *Anxious Cinephilia. Pleasure and Peril at the Movies* (New York, NY: Columbia University Press, 2020), 196-199.

63 Erich Kettelhut, *Der Schatten des Architekten*, ed. Werner Sudendorf (Munich: Belleville, 2009), 133.

64 Günther Rittau, "Special Effects in *Metropolis* [1927]," in *Fritz Lang's Metropolis: Cinematic Visions of Technology and Fear*, eds. Michael Minden and Holger Bachmann, 78-80 (Rochester, NY: Camden House, 2002), 80.

Ufa also tapped into public interest in the production process and published numerous articles under the names of cinematographer Karl Freund, art director Otto Hunte, and others that seemed to share technical insider knowledge: "How it was done," "My work on *Metropolis*," "Psst! Here we reveal -- how a fight scene is shot," "The flying camera," "1135 work hours for 9 metres 88 centimetres: From the idea to the finished scene," or "The master-builder of *Metropolis*."[65] The studio evidently aimed at striking a balance between affirming its technological pre-eminence and protecting valuable trade secrets. Photographs detailed the plans for and implementation of the Schüfftan process in the stadium of the Club of Sons and the use of a model for the Tower of Babel, and they documented the camera swing employed in the deluge sequence and the rear-projection setup for the televisual conversation between Grot and Fredersen (Figure 5.2.). These images were likely considered innocuous because the techniques in question were either fairly common or, as in the case of the Schüfftan process, patent-protected. The photos did not reveal vital technical details, as for instance how precisely projector and camera were synchronized in the case of the rear-projection setup.

At the same time, strict silence was maintained about the workings of more complex scenes like the Moloch sequence, the nightly cityscape, or the robot's anthropogenesis, leaving observers to speculate. American popular science magazine *Science and Invention*, for example, published a double-page picture exposé consisting of film stills and drawings that supposedly demonstrated how the effects of *Metropolis* had been achieved. However, the explanations given, which have recently gained much credence online, were completely fabricated.[66]

The Schüfftan process, which Fritz Lang, Günther Rittau, Otto Hunte, Erich Kettelhut, and Karl Vollbrecht had already employed in *Siegfried* (Ufa, 1924), played a key role in the production and promotion of *Metropolis*. Ufa acquired the German license to the patents one month before principal photography began on 22 May 1925 in the hopes of lowering the anticipated expenses for the monumental *Metropolis* sets. Six weeks before principal photography on *Metropolis* concluded on 30 October 1926, Ufa gave up its monopoly and made-to-order Schüfftan shots became available to the entire German film industry. In Ufa publicity, the Schüfftan process and

65 See *Metropolis Ufa Magazin Sondernummer* (Berlin: Bukwa, 1927) and *Presse- und Propagandaheft Metropolis* (Berlin: Parufamet Presse- und Propagandabteilung, 1927). Several articles were subsequently reprinted in other venues.

66 "Metropolis – A Movie Based on Science," *Science and Invention* 170 (June 1927): 106-107. See for instance http://www.smithsonianmag.com/history/1927-magazine-looks-at-metropolis-a-movie-based-on-science-4328353. Accessed 12 July 2017.

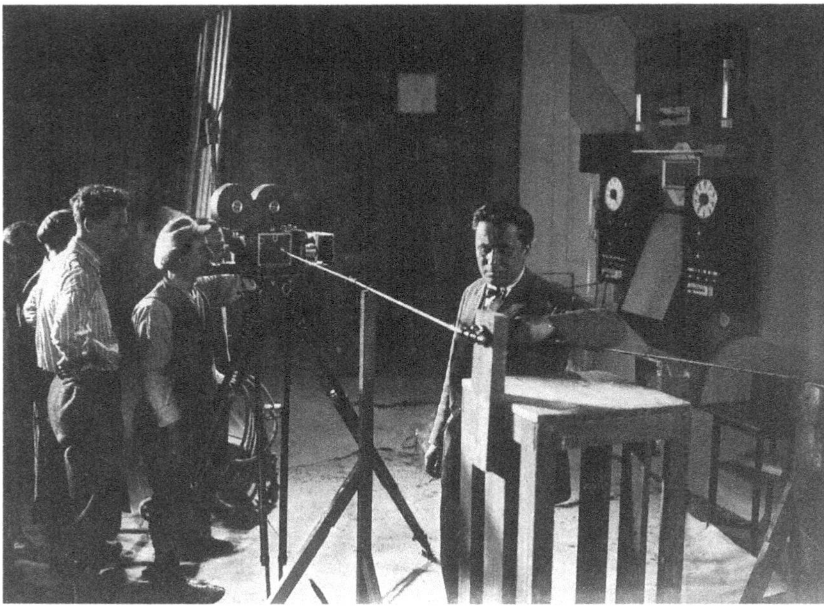

Fig. 5.2. Rear-projection setup for *Metropolis* (1927) with cinematographer Günther Rittau (centre). The Mitchell camera is connected to the projector behind the set of Fredersen's video telephone via threading rods.
Source: Deutsche Kinemathek.

Metropolis were set up to promote each other: the film's challenges confirmed the technique's unprecedented capabilities, while the technique's use demonstrated the film's technological sophistication. Together they served as evidence that the German film technology had indeed caught up with Hollywood.

Critics both endorsed and mocked *Metropolis*'s aspirations as an awe-inspiring technological wonder. In either case, they regularly highlighted the Schüfftan process's role in achieving such technological pre-eminence and resorted to language evocative of the sublime. An anonymous reviewer for *Lichtbild-Bühne*, for instance, characterized Metropolis as a visionary, even mythical undertaking. In the style of the film's Tower of Babel narrative, they marvelled, rather tongue-in-cheek: "Thus spoke Lang: Let us build a tower of technology whose top may reach unto the stars. And on the top we will write: Great is film and its Schüfftan! And great are the people who built and shot it!"[67] By referencing the Tower of Babel, the reviewer affirms the audacity of the project and ironically implies its failure. Like the Tower of Babel, *Metropolis* invokes a simultaneous sense of awe and dread in view

67 M.-l., "Technische Kritik zu Metropolis," *Lichtbild-Bühne* 10 (12 January 1927).

of its astonishing technological ambition. In fact, when discussing the Schüfftan process as used in *Metropolis*, commentators frequently resorted to language reminiscent of the sublime, regardless of whether they expressed their amazement ironically or sincerely. For author Friedrich Porges the Schüfftan process embodied technology's magical capacity to actualize the imagination, effectively elevating it to the status of a co-creator:

> Taken as a whole, *Metropolis* is not only an ingenious work of its creator Fritz Lang, without exaggeration it must be called a marvel of film technology. And this work also presents the miracles of the Schüfftan process: the "castles in the air" of fabulous proportions that Schüfftan manages to build into cinema's living pictures. Technology has once again turned a figurative notion into reality.[68]

Schüfftan's prominence in the publicity and reception of *Metropolis* is also notable given that he and his associate Ernst Kunstmann were out of the country for much of the production of *Metropolis*: Between September 1925 and November 1926, they worked at Universal Pictures. In their place, Helmar Lerski, photographer, cinematographer, and chief technician of Schüfftan's company Spiegeltechnik AG, was likely in charge of many of the Schüfftan shots in *Metropolis*.

Ufa publicity listed about a dozen shots in *Metropolis* that were realized with the aid of the Schüfftan process, several of which were set extensions.[69] The technique was used to add, among other things, contraptions to the upper portion of the machine room, top stories to the workers' apartment buildings, the bust of Hel to its pedestal, and the stadium wall in the Club of Sons.[70] The Schüfftan process also made it possible to place actors in a model environment—something no other trick technique was able to accomplish convincingly at the time. Two shots in downtown Metropolis show reflected groups of workers marching or racing over a diagonal bridge in between three-dimensional miniature skyscrapers (Figure 5.3.). The extreme contrast between the moving human dots and gigantic, sterile facades accentuates the inhumanity of the cityscape.

68 Friedrich Porges, "Luftschlösser im Film," *Die Bühne* (Vienna) 120 (25 February 1927).

69 See advertisement "Das Schüfftan'sche Kombinationsverfahren!" *Lichtbild-Bühne* 11 (13 January 1927).

70 According to the advertisement "Das Schüfftan'sche Kombinationsverfahren!" *Lichtbild-Bühne* 11 (13 January 1927), Schüfftan shots were also employed for "interior Yoshiwara," "interior cathedral," "Tower of Babel," "newsstand," and "Rotwang's house."

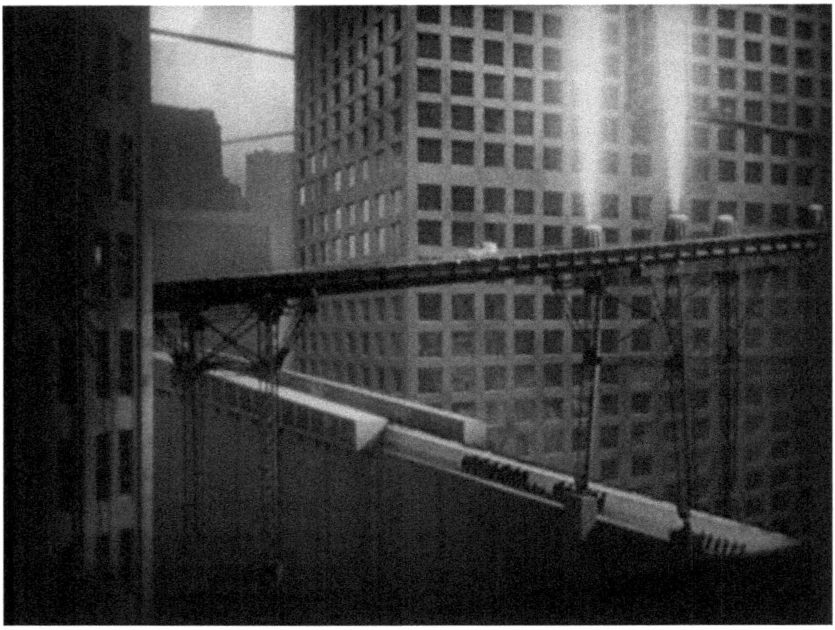

Fig. 5.3. Schüfftan process in *Metropolis* (1927): Gangs of workers marching through a miniature set. Courtesy of Friedrich Wilhelm Murnau Foundation.

Moloch

The most prominent application of the Schüfftan process in *Metropolis* is the transformation of the M-machine into Freder's Moloch vision. The apparition materializes as a hybrid mythological creature that once again blurs the boundaries between human, animal, and machine, between living and non-living matter. According to art director Erich Kettelhut, the set of the M-machine, which measured twenty metres high and wide, was designed to resemble the diesel engine of an ocean steamer (Figure 5.4). Framed in a low-angle extreme long shot, the structure is dominated by a steep stairway with a wide ramp in the middle, which is flanked by two massive bolted industrial boilers. The stairway leads up to a large cavity in which pistons operate. Already before its transformation, the M-machine appears somewhat animal-like: Its barrel-shaped roof is topped by a domed structure whose bolted eye-like covers and trunk-like shaft cause it to resemble the face of a fly. Steam is released from eight valves on top of the boilers and the roof. Twelve workers, symmetrically arranged on two levels on both sides of the stairway, are seen operating the machine on switchboards. They tilt their bodies rapidly and rhythmically from side to side, as if components of the mighty machine.

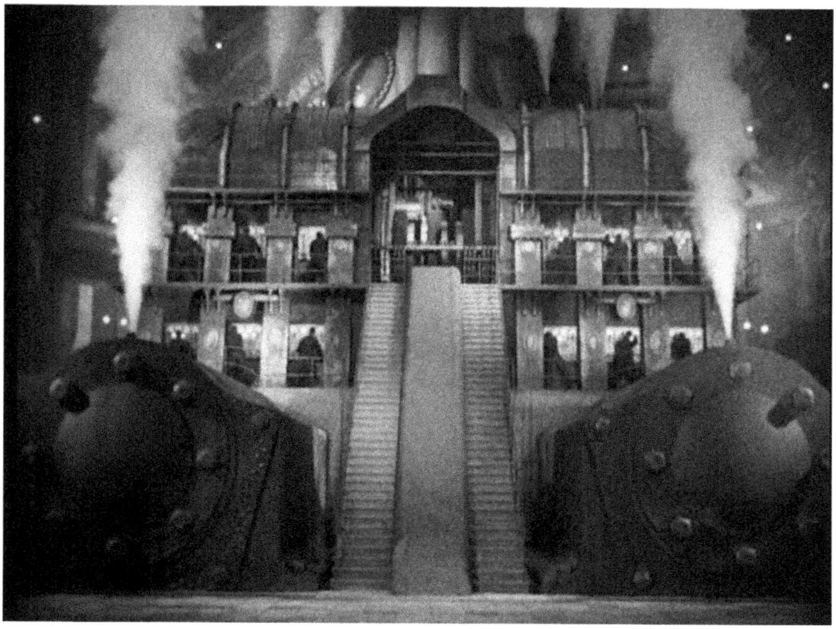

Fig. 5.4. M-machine before the explosion in *Metropolis* (1927).
Courtesy of Friedrich Wilhelm Murnau Foundation.

This technological organism is transformed into Freder's Moloch vision, which however does not resemble the bull-headed, upright bronze statue of rabbinical tradition that is famously featured in Giovanni Pastrone's *Cabiria* (Itala Film, 1914). Instead, it looks like a recumbent cat with an anthropoid head, reminiscent of an Egyptian sphinx (Figure 5.5.). Between its paws, a stairway leads up to a wide-open mouth decorated with rows of sharp teeth. Sacrificial slaves are dragged up past two priest figures and thrown into the juggernaut's throat. Subsequently, workers march up in rows of six and throw themselves into the abyss. The stairway to a sanctuary and the human sacrifice recall Mesopotamian ziggurats or Aztec houses of worship, while the mouth-shaped portal evokes doorways of Hindu and Buddhist temples or representations of the mouth of hell in Medieval European art. All aspects of its appearance highlight the machine's divine and destructive nature.

Indeed, the machine's transformation into a man-eating idol represents one of *Metropolis's* central allegories. It appears almost as a response to Béla Balázs's 1924 call for the artistic treatment of industry and factory labour in cinema, with the goal of revealing the true face of modernity:

> But the scale of its [factory work's] inhuman meaninglessness has grown to such a horrifying degree that the *machine, the man-eating machine,*

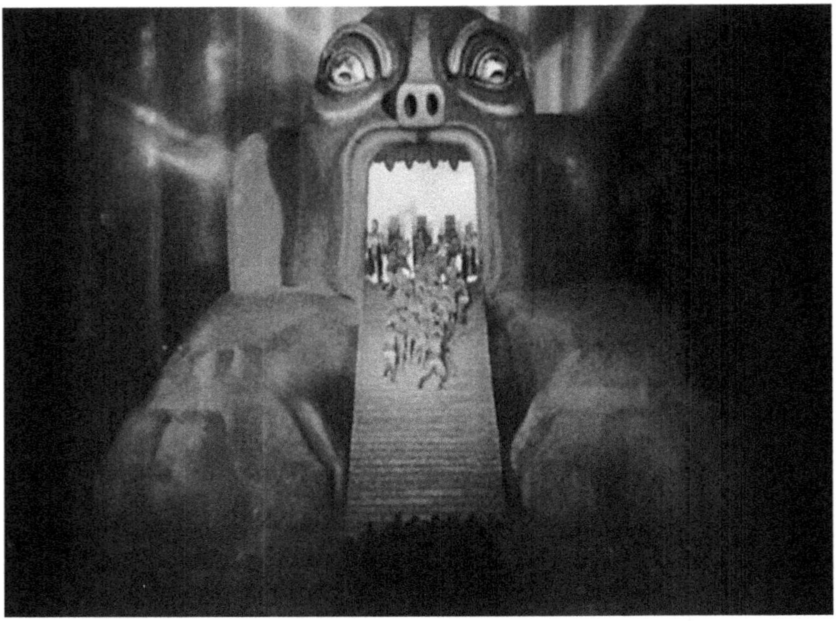

Fig. 5.5. Freder's Moloch vision in *Metropolis* (1927).
Courtesy of Friedrich Wilhelm Murnau Foundation.

> *has taken on the demonic vitality of a nightmarish monster.* [...] We see the machine *acquire a face in film*, its movements are transformed into a terrifying expressiveness.[71]

According to Balázs, it is the task of the cinematic apparatus to stage the machine's horrific nature as art. The techno-romantic sublimation of technology into art on both the thematic and the production level characterizes the fundamental ambition of *Metropolis*.

Although the transformation of the M-machine into Freder's Moloch vision constitutes one of the film's most impressive special-effect feats, the technicians did not consider it particularly challenging. Kettelhut recalled:

> Even the transformation of the entire middle part of the Moloch-machine into an oversized grotesque face with wide-open eyes and huge jaws did not cause any problems. The image of the Moloch head, which was constructed opposite the machine, was laid over the machine using the combination of the Schüfftan process and sliding mirrors, which had

71 *Béla Balázs: Early Film Theory*, ed. Erica Carter, 54. Emphases in the original.

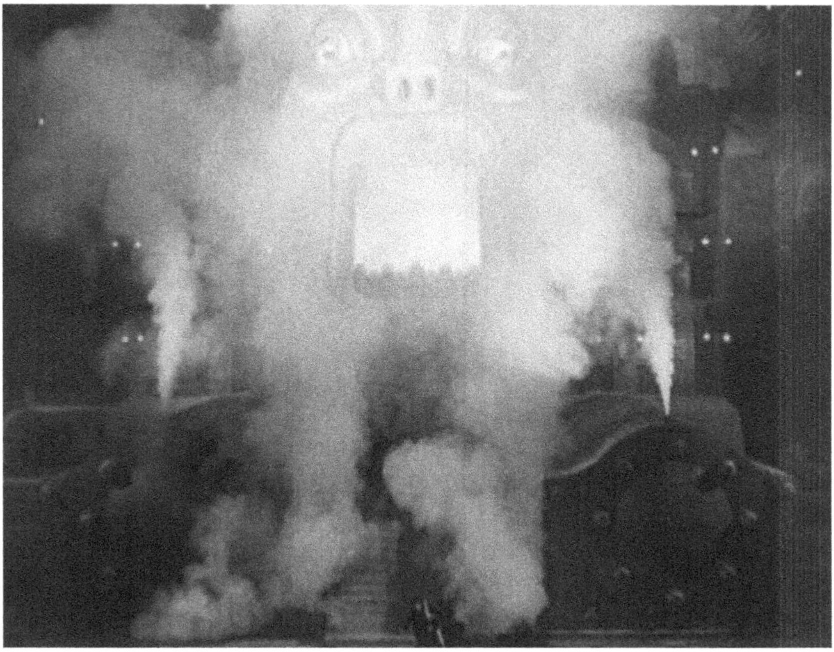

Fig. 5.6 Transformation of the M-machine into the Moloch vision facilitated by the Schüfftan process in *Metropolis* (1927).
Courtesy of Friedrich Wilhelm Murnau Foundation.

> been used in the *Nibelungen* for the transformation of the dwarfs into stone figures.[72]

On closer inspection, however, the substitution reveals itself as more complicated than Kettelhut's recollections may suggest. Following the explosion, the M-machine is enveloped in steam (which was acquired from a nearby gas plant). The brightly lit face of the Moloch appears slowly from top to bottom, seemingly materializing through the steam. At this early stage of the apparition, the Moloch face displays lower teeth, suggesting that an additional model might have been used. The steam cleverly camouflages a second step of the metamorphosis. The full Moloch figure spreads slowly from top to bottom and from the centre outwards, revealing the stairway and then the paws of the cat. In the third part of the transformation, the worker column replaces the slaves on the stairway and the left and right portions of the frame reconvert from the bottom up into the M-machine. The workers continue to march into the mouth of the juggernaut until the screen blurs. The vision has disappeared.

72 Kettelhut, *Der Schatten des Architekten*, 151.

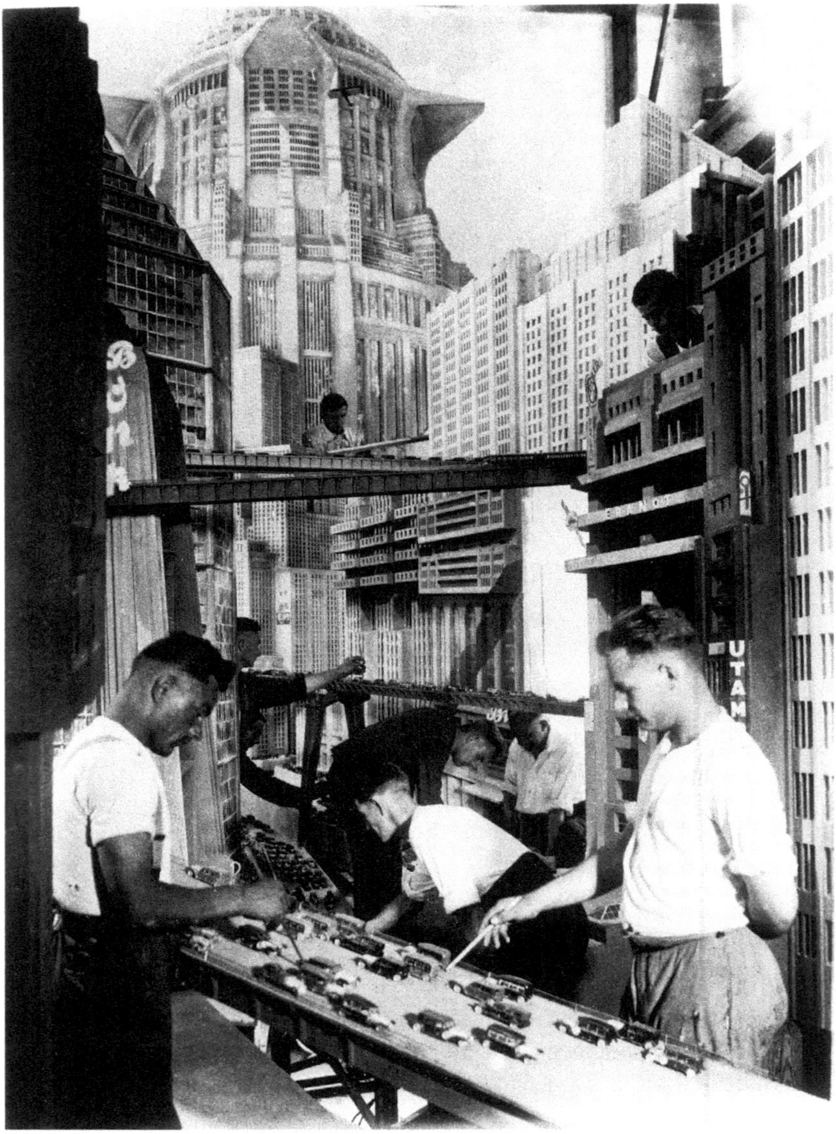

Fig. 5.7. Forced-perspective set of the great commercial street in *Metropolis* (1927) being stop-motion animated.
Source: Deutsche Kinemathek.

The Moloch represents a formidable technological monster. However, it is not only gruesome, but also mesmerizing, and thus evokes conflicting emotions of pleasure and distress. The audience is supposed to experience these, and they are also exemplified in Freder's gestures and facial expressions. In the grip of the vision, Freder raises his hands in astonishment and fear, eventually reaching out

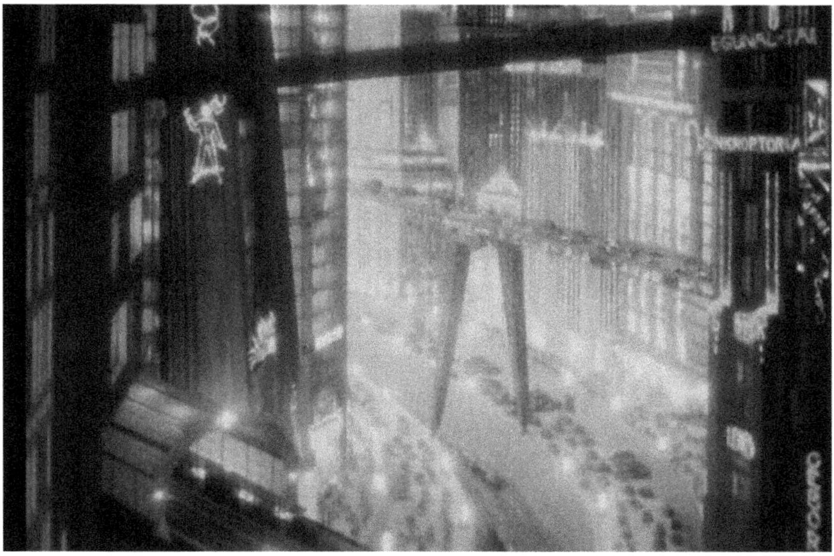

Fig. 5.8. Detail of the forced-perspective set of the great commercial street in Metropolis (1927) in night-time illumination.
Courtesy of Friedrich Wilhelm Murnau Foundation.

to the blurring apparition. His facial expressions fluctuate between awestruck and horrified, and he stares directly into the camera—aghast, yet unable to avert his gaze. In the confrontation with the Moloch, Freder externalizes paradoxical responses towards the overwhelming power of modern machines, as if putting the encounter with the technological sublime on display.

Cityscapes

Among the most iconic images of *Metropolis* are the ones showcasing immense urban environments. Compared to the Moloch vision, their symbolism is not as readily legible, yet they articulate decidedly mixed emotions about industrial modernity. This is particularly tangible in the case of the cityscape above ground, which is dominated by another momentous trick-technological achievement: the vast, stop-motion animated miniature set known as the "great commercial street." In contrast to the subterraneous workers' dwellings, which are thoroughly dismal and oppressive, the high-rises towering above appear at once boundless and claustrophobic, majestic, overbearing, and of otherworldly beauty.

The filmmakers spared no effort to bring this massive technological environment to life. By all accounts, the sequence that introduces the city above and shows traffic streaming through skyscraper canyons was among the most intricate of the entire production. For roughly one minute of screen time in the finished film, 2100 individual frames were exposed over the course of six weeks.[73] As was the case with the Moloch sequence, the members of the technical team relied heavily on their experiences with *Die Nibelungen* (Ufa, 1924, dir. Fritz Lang). Like the Rhine valley in *Siegfried*, the great commercial street of Metropolis was realized in forced perspective due to space concerns. The set was about eight metres long and relied on substantial perspective foreshortening.[74] It increased from 1:25 in the foreground to 1:100 for the New Tower of Babel in the background, which, according to art director Otto Hunte, was conceived as 500 metres high (incidentally, the first real-life building taller than 500 metres was not built until the year 2010).[75] During several months of preparation, more than three hundred model cars, trains, and airplanes in three different sizes for the fore-, middle- and background were designed and constructed. Punched-out cardboard pedestrians were hand-painted.[76] The airplanes were suspended by three hairs and guided on wires from outside of the frame.[77] For each exposure, all elements were individually moved by hand: Airplanes were advanced by one and a half centimetres, trains by one centimetre, cars by seven millimetres, and pedestrians only minimally (Figure 5.7.).[78] Even though the use of forced perspective severely limited the choice of camera setups, the technicians managed to use the same set in four, seemingly distinct shots: A panorama shot of the entire set with a back-cloth of New Tower of Babel in the background, a closer-framed shot of the centre of the set, a high-angle night shot of the lower portion with a bridge added and tracks replacing a road (Figure 5.8.), and finally a mirror-inverted daylight version of the night-time shot.[79] With the exception of the panorama, none of the shots includes a visible

73 Rittau, "Special Effects in *Metropolis* [1927]," in *Fritz Lang's Metropolis*, eds. Minden and Bachmann, 79; O. H. Schulze quoted in Rolf Giesen, "Spezialeffekte Made in Germany," *Künstliche Welten: Tricks, Special Effects und Computeranimation im Film von den Anfängen bis heute*, eds. Rolf Giesen and Claudia Meglin, 69-111 (Hamburg/Vienna: Europa Verlag, 2000), 77.

74 Berg, "The Viennese Night," in *Fritz Lang Interviews*, ed. Grant, 70.

75 Otto Hunte, "Der Baumeister von *Metropolis*," *Presse- und Propagandaheft Metropolis* (Berlin: Parufamet Presse- und Propagandabteilung, 1927).

76 Rittau, "Special Effects in *Metropolis* [1927]," in *Fritz Lang's Metropolis*, eds. Minden and Bachmann, 79; Kettelhut, *Der Schatten des Architekten*, 148.

77 Kettelhut, *Der Schatten des Architekten*, 159.

78 Kettelhut, *Der Schatten des Architekten*, 160.

79 The daylight shots appear in the sequence when Freder rushes to see his father at the New Tower of Babel and the night-time shot in a scene following the explosion of the heart machine, when the lights of Metropolis go out.

horizon, thus foregrounding the all-engulfing and oppressive nature of the great city. Simultaneously, they represent meticulously choreographed, breath-taking symphonies of movement and, in the case of the night-time shot, light, recalling the "unforgettable impression" Lang had of Manhattan.

In addition to three-dimensional models, a remarkable number of two-dimensional ones were employed as well. The most innovative kind was a type of animated painting that, because of the effort involved, was rarely, if at all, used again before the advent of CGI. Erich Kettelhut employed the technique in two cases to obviate expensive and space-consuming three-dimensional models. The first one, the city at daybreak, is located after the opening title "*Metropolis*" (Figure 5.1.). The second animated painting appears in the night-time sequence, just before the explosion of the heart machine cuts the electricity supply of Metropolis (Figure 5.9.). It shows a low angle view of the skyline of Metropolis with the New Tower of Babel rising above the roofs. Countless windows and animated advertisements illuminate the skyscrapers; trains and cars move across elevated bridges, spotlights streak across the sky. To realize the animation of the city at daybreak and of the nightly skyline, Erich Kettelhut completed two 60x45-cm photorealistic oil paintings in grey scale, which he then covered in fixative. Using glycerol-based (and therefore non-drying) paint, he then added (and erased) all moving parts shot-by-shot. For a total of ten metres of completed footage, he exposed one thousand individual frames, an undertaking that took him ten weeks to complete. Kettelhut described his process in the following terms:

> When I wanted for example a search light beam to sweep across a façade from right to left, I wiped as many millimetres from the right side of the light beam as I added on the left side. The image was exposed, and I painted, and wiped until the light beam had disappeared from the field of view. But there were three or four search light beams that darted through the image from different directions, two city trains swished past, and along with this played the illuminated advertisements. There were picture sequences with over a dozen simultaneous movements.

While the somewhat abstract picture of the city at daybreak is evidently a painting, the night-time scene is not immediately discernible as such and blends in smoothly with the next shot of the city, which makes use of the great commercial street set (Figure 5.8). Both the painting and the miniature capture the astonishing beauty of the great city at night. The low-angle perspective of the animated painting emphasizes the majestic grandeur of

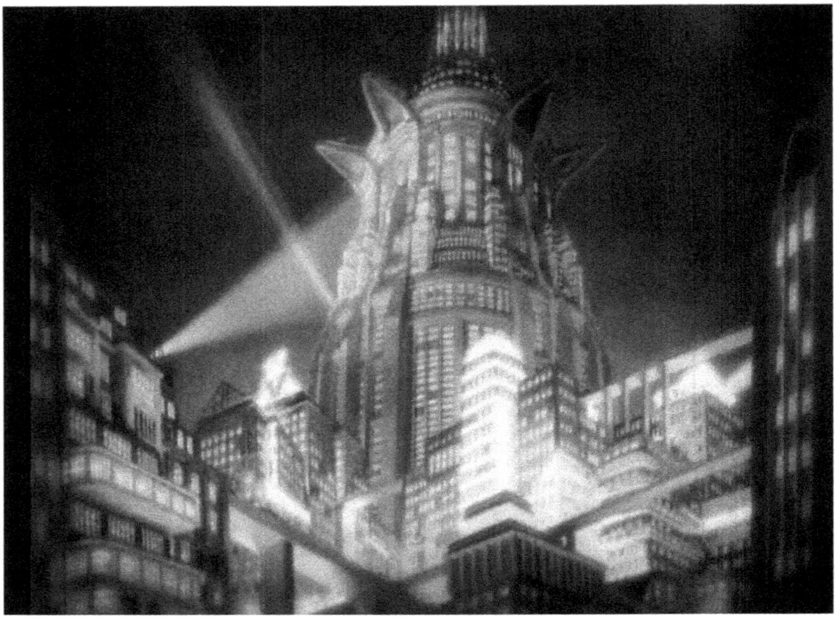

Fig. 5.9. Erich Kettelhut's animated painting of the night-time skyline in *Metropolis* (1927). Courtesy of Friedrich Wilhelm Murnau Foundation.

the city; the high-angle view into the miniature's glittering urban canyons accentuates its claustrophobic lure.

In addition to Kettelhut's two animated paintings, several still paintings were used as well. For instance, the scene of Freder's first visit to his father's office contains four paintings that, at first glance, seem motivated as views from Fredersen's office window. In fact, the first image, framed by a window lattice, can be interpreted as a plausible vista, though given the assumed height of the New Tower of Babel the angle of view is remarkably flat. In contrast, the remaining three paintings do not depict credible window views. Two juxtapose futuristic high-rises, bridges, and districts, including the city centre of Metropolis with the New Tower of Babel, in a collage-like fashion. Their multi-scale components are shown from extreme and incoherent angles, producing a captivating, hallucinatory impression of ultra-modern urbanity. The final, high angle painting shows the New Tower of Babel rising menacingly over the city, dwarfing the surrounding skyscrapers both in height and volume. The brief sequence evolves from depicting a sober external reality, a disorienting, yet alluring phantasmagoria, to an intimidating emblem of high-tech totalitarianism.

Metropolis's massive urban environments elicit an array of affective responses. The filmmakers employed the most intricate, state-of-the-art film-technological methods at an unprecedented scale to capture the "spirit" of the technological megacity in all its contradictions. It appears bleak and imposing, dazzling and oppressive, appalling and magnificent—frequently all at the same time. The city views kindle a "delightful horror" at the sight of an "artificial infinite," thus evoking the technological sublime. In keeping with the techno-romantic paradigm, trick technology in *Metropolis* serves to reveal inner truths—which include our conflicting sentiments vis-à-vis industrial modernity.

The robot's anthropogenesis

Similarly paradoxical attitudes towards technology are on display during yet another special effect sequence, the robot's anthropogenesis. One of early film history's most dazzling spectacles, the scene flaunts dramatic electric discharges and mesmerizing rings of light that glide up and down the cyborg's body. Like the Moloch and the great commercial street, the robot's anthropogenesis articulates fundamental fears about technology's impact on human existence. It depicts the dreadful juncture at which human and machine ultimately become indistinguishable. In addition, the theft of the captured woman's physiognomy encapsulates technology's threat to personal identity and our sense of reality.

The atrocious implications of the machine's conversion into a woman notwithstanding, the scene also elicits awe and wonder. Moreover, it forges associative links between the fictional world and the film's production. On both levels, technology serves as a creative agent. Within the fiction, the creator of the artificial woman is the sinister inventor Rotwang, who by virtue of his occult scientific knowledge succeeds in transmitting Maria's appearance onto the robot. He embodies the "modern magician." In the real world, on the other hand, this title should go to the cameraman, as Günther Rittau affirmed: "The cameraman is the modern magician. There is no such thing as 'impossible' for him."[80] More than any other sequence in *Metropolis*, the robot's anthropogenesis served as a showpiece for the ingenuity of German film technicians. For decades, it remained a mystery how the glowing circles surrounding the robot had been achieved. Even expert observers were at a loss to explain their creation, which contributed to endowing a technological feat with mystical ambience.[81]

80 Günther Rittau, "Special Effects in *Metropolis* [1927]," in *Fritz Lang's Metropolis: Cinematic Visions of Technology and Fear*, eds. Michael Minden and Holger Bachmann, 78-80 (Rochester, NY: Camden House, 2002), 78.

81 See for instance "Metropolis Film Seen," *New York Times* (11 January 1927): 36; Randolph Bartlett, "German Film Revision Upheld as Needed Here," *New York Times* (13 March 1927): X3; Sime, "Metropolis," *Variety* (16 March 1927): 16-17; "Metropolis – A Movie Based on Science," *Science and Invention* 170 (June 1927): 106-107.

The glowing rings appear in three distinct shots during the robot's anthropogenesis: a long shot that shows the seated robot under the pentagram and the unconscious Maria in her glass tube in the foreground, a medium long shot of the robot on its wired throne, and a close-up of its face. In addition, there is a 180° reverse angle shot of the glass tube, against a wall of Rotwang's modern alchemist kitchen, which is fitted with a hearth, twisting shelves, control indicators, and glowing induction coils and wires. The anthropogenesis begins with a medium close-up of the unconscious Maria in her glass tube, followed by shots of Rotwang moving around in his laboratory, throwing levers and activating the luminous, bubbling contents of various flasks, tubes, and retorts. He eventually positions himself at a control terminal, the relative position of which remains unclear: His straightforward stares alternating with frontal shots of robot and glass tube imply a straight line of vision, yet this is spatially impossible.

Taking a deep breath, Rotwang triggers the ceiling generator that envelops the glass tube (seen in the reverse shot) in electrostatic discharge. In the medium long shot of the seated robot, two, four, then six identical glowing circles become visible and move steadily up and down. They are complemented with spectacular lighting effects in the long shot. Cutaways reveal Rotwang hard at work controlling the transformation process and close-ups of receptacles with radiant, effervescent fluids. The electrostatic discharge from the spherical electrostatic generator on the ceiling creeps along the glass tube's conductor rails as if scanning Maria's body (Figure 5.11.). In the medium long shot, the robot is now surrounded by eight glowing rings and lit up rhythmically from the front left. A small spot on its chest begins to blink and expands into a flashing light in the form of a human heart. The metal surface of the robot's torso seems to soften and, prompted by Rotwang throwing a lever, arteries and veins become visible, and glimmer as if transporting fluid. While head and shoulders remain machine, legs, torso and arms waver between metal and flesh. The following close-up shows the robot's face encircled by four slowly moving, glowing rings and lit alternately from the left and the right. The rings disappear gradually as the robot's face dissolves into that of the false Maria. Shoulders and helmet are precisely lined up, but in contrast to the robot, which stares straight into the camera, the false Maria opens her eyes only gradually. We return to the medium close-up of Maria in the glass tube that opened the sequence. With a sigh, her head sinks to the side. The sequence ends with the false Maria staring coldly straight into the camera.

Even though, as we have seen, the marketing of *Metropolis* relied heavily on "how it was done" stories, Ufa and the crew carefully withheld all information about the production of the robot's anthropogenesis. Even in a promotional article on the film's special effects, which appeared in several

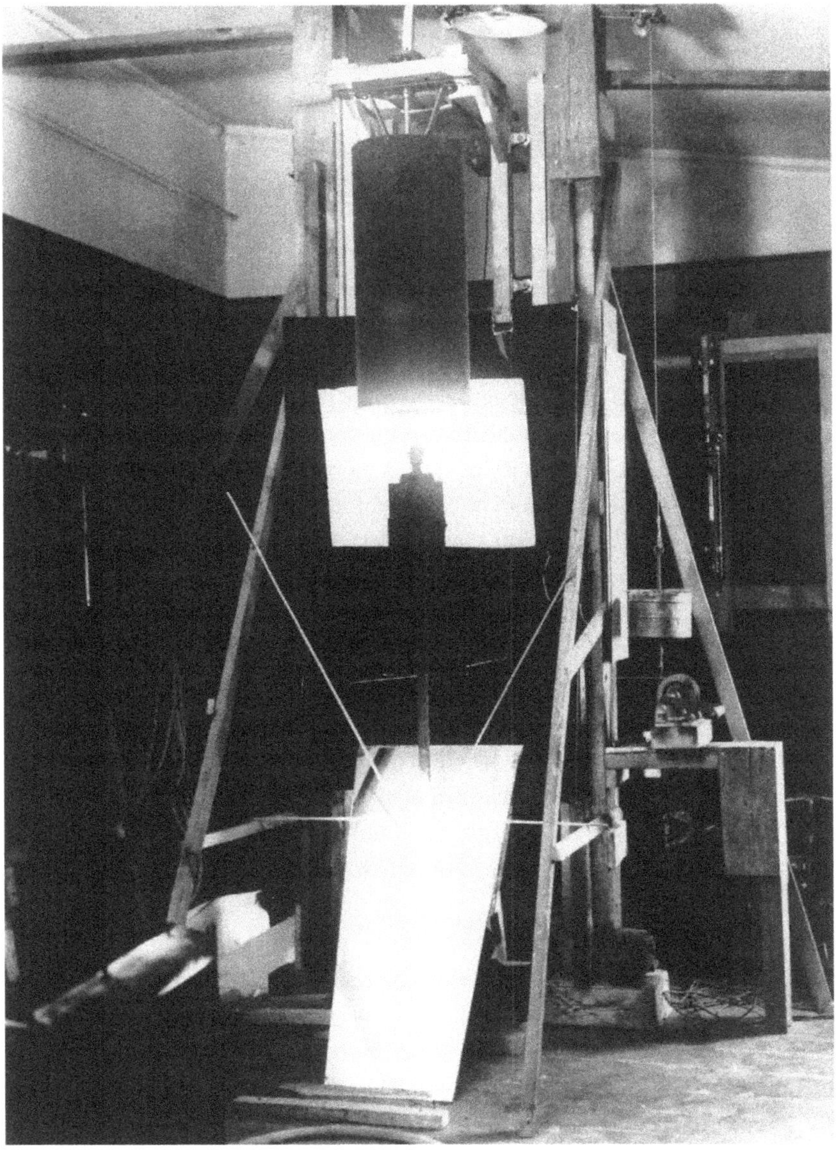

Fig. 5.10. Setup for the creation of the glowing rings during the robot's anthropogenesis in *Metropolis* (1927).
Source: Deutsche Kinemathek.

press outlets and aimed at asserting German film-technological leadership, cinematographer Günter Rittau did not reveal any workable information:

> The anthropogenesis is accomplished transferring the human form of Maria by means of electric currents. Now, electric currents have the habit of being

> invisible. But on the other hand, the fantastic-mysterious event taking place here absolutely had to be made visible. The only way of achieving this was to make the effect of the electric currents visible. Liquids in strange receptacles took on a shine and started to simmer, the electric apparatus enclosing Maria began to scintillate and was eventually completely covered by giant, shining chains of sparks moving to and fro; simultaneously, shining rings of electric current start to form on different levels around the machine woman and begin to glide up and down. The incipient anthropogenesis is illustrated by making visible a shining system of circulating blood streams. The realization of these shots required many weeks of preparatory tests in the laboratory and equally lengthy periods of practice shots. The chemistry of photography played no small role in obtaining these shots and the strangest kind of auxiliary material was used. A more detailed explanation would go too far here and would not be to the point. Suffice it to say that special frames, striae (*Schlieren*), soft soap, vignettes and extremely complicated machines especially constructed (for these shots) played a decisive role. For days, the workers had to be trained in operating the appliances, which required split-second precision. Individual pieces of film were exposed up to thirty times. Everyone who knows anything about photography will understand immediately what this means. In these shots, everything depended on the most exact calculations, on the utmost precision of the work and the appliances, and most of all on the nerves and the patience of the cameraman. I can confidently claim that shots like the ones in the Metropolis film have never been shown in any film so far.[82]

Even decades later, Fritz Lang continued to obfuscate how the effect had been done: "The concentric rings of light which surrounded her and which moved from top to bottom were, in fact, caused by a little golden ball, rapidly turning in a circle, and filmed on black velvet. We superimposed these shots in the laboratory on the shot of the sitting robot that we had already filmed."[83] Hardly surprising, Lang's fantastical account has little to do with what actually happened on the set.

It was not until the publication of Erich Kettelhut's memoirs in 2009 that a credible production report of the scene became available.[84] As Kettelhut describes in some detail, the technicians working with cinematographer Günter Rittau first created a plywood silhouette of the seated robot covered in black

82 Rittau, "Special Effects in *Metropolis* [1927]," in *Fritz Lang's Metropolis,* eds. Minden and Bachmann, 80.

83 Berg, "The Viennese Night," in *Fritz Lang Interviews*, ed. Grant, 71.

84 Kettelhut, *Der Schatten des Architekten*, 157-159.

velvet and two circular neon lights in different sizes that both fit comfortably over the plywood silhouette. One after the other, the rings were suspended horizontally on three wires from what Kettelhut calls a "fixed elevator," which was custom-built around the silhouette outside of the camera's field of view (Figure 5.10.). Due to a horizontal grease film on a small glass pane close to the lens, the neon rings, which were moved up and down at a steady pace, appear as blurred discs of moving light. For every shot in which the rings appear, Rittau, using one of Ufa's two new Mitchell cameras, recorded each ring up to six times; their pacing, starting, and end positions were meticulously planned. In the end, after countless exposures, the electrostatic discharges between Maria's glass tube, which was also substituted by a black-velvet mask, and the electrostatic generator on the ceiling were shot without the greased glass pane.

Like the city centre and the Moloch idol, the robot is a technological object that the film characterizes as malevolent and inhuman and that nonetheless appears magnificent. Spectacular lighting is a major factor in conveying this impression. What is more, in the context of the robot's anthropogenesis, light turns into a creative force. Like the artificial woman in Villiers de l'Isle-Adam's *L'Ève future*, the robot in *Metropolis* is transformed "with the sublime assistance of Light."[85] As a symbol for the divine, light has obvious ties to the sublime. In Genesis, the earth is dark, without form and void before God brings light into the world as the first act of creation. Indeed, theorists since Longinus have referred to *fiat lux* as a quintessentially sublime moment.[86] In *Metropolis* and *L'Eve future*, however, the creator is an unscrupulous scientist who, using man-made "electromagnetic power and Radiant Matter,"[87] fabricates an artificial human amidst an astonishing technological spectacle. Endowed with quasi-mystical powers, he transgresses the boundaries of the natural order and illustrates the horrific consequences of human hubris.

The robot's anthropogenesis portrays the creation of a mechanical copy of a living being, which Raymond Bellour and Tom Gunning have convincingly read as an allegory for the nature of the cinematic medium.[88] In addition, the sequence also highlights the fact that cinema fundamentally depends on light—both for film exposure and for projection. The rays emanating from

85 Auguste Villiers de l'Isle-Adam, *Tomorrow's Eve* [1886], trans. Robert Martin Adams (Chicago, IL: University of Illinois Press, 2001), 64.

86 See for instance Moses Mendelssohn, "Über das Erhabene und Naive in den schönen Wissenschaften [1758]," in *Moses Mendelssohn: Ästhetische Schriften* , ed. Anne Pollok, 216-259 (Hamburg: Felix Meinder Verlag, 2006): 244.

87 Villiers de l'Isle-Adam, *Tomorrow's Eve*, 65.

88 Raymond Bellour, "Ideal Hadaly," trans. Stanley E. Grey, *Camera Obscura* 15 (Fall 1986): 111-136; 131; Gunning, *The Films of Fritz Lang*, 67.

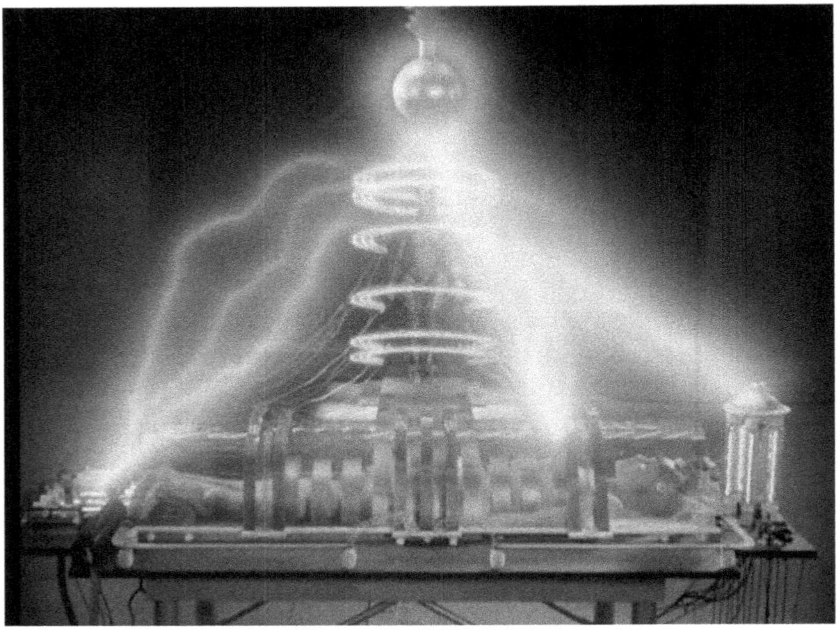

Fig. 5.11. The sublime power of light: The transmission of Maria's physiognomy onto the robot in *Metropolis* (1927). Courtesy of Friedrich Wilhelm Murnau Foundation.

the spherical electrostatic generator quite literally scan Maria's physiognomy. As seen in Figure 5.11., the long shot of the robot and the glass tube, the spatial arrangement forges links between the different components: The generator on the ceiling is located precisely at the centre of the pentagram behind the robot on the back wall, symbolically merging science and magic. What is more, the glowing rings surrounding the robot seem located directly beneath the generator, as if emerging from it. Visually, Maria, the robot, the generator, and the pentagram are connected by light, suggesting the transmission of the woman's image through the generator onto the robot. The light effects integrate the fantastic events into a coherent whole, seemingly facilitating the nefarious metamorphosis.

Thematically, the scene articulates angst about the negative effects of modern technology on notions of identity. The malicious theft of Maria's image implies the erosion of individuality, authenticity, and autonomy, and undermines our trust in the truth of our understanding of reality. This sinister purport, however, is eclipsed by the scene's sheer beauty and power, which astonishes and overwhelms the observer, both on account of the force of its imagery and the sheer magnitude and inscrutability of the film-technological accomplishments on display.

Throughout *Metropolis*, technology is presented as an agent of tyranny and dehumanization, while also appearing profoundly exciting, a modern pathway to mystical grandeur. The film's spectacular special-effects sequences highlight different aspects of industrial modernity that pose existential threats to humanity. Downtown Metropolis is an inhospitable and claustrophobic asphalt jungle. The Moloch represents the idolization of technology and the oppression resulting from mechanization. The robot's anthropogenesis addresses anxieties about the technology-induced loss of personal identity and the breakdown of our grasp on reality. Each instance traces back to the principal fears underlying the techno-romantic outlook: technology's looming obliteration of core characteristics of humanity.

In the technological megacity of Metropolis, the concept of society as living organism, in which all members contently serve their proper purpose and which the film seems to regard as ideal, has been supplanted by an antagonistic symbiosis of human and machine, a totalitarian order of masters and slaves. The concomitant blurring of boundaries between living and non-living matter threatens the very essence of what it means to be human. Humanity is identified with Christian love, as proclaimed by the prophetess Maria and embodied by Freder, the mediating "heart" personified. Rather than allowing technology to enslave humanity or engaging in dangerous and counterproductive neo-Luddite frenzies, the film contends, the human spirit must take control of technology's superior powers in the fashion stipulated by journalist Heinz Michaelis in 1923: "It is thus first necessary to disrupt the absolutism of technology and bend it to the will of mind and soul."[89] As a film-technological showcase project, *Metropolis* set out to realize this subjugation of technology in the aesthetic realm.

Like many artistically ambitious German films of the silent era, *Metropolis* pursues the techno-romantic project of transcending material reality by means of special-effect technologies. However, *Metropolis* distinguishes itself because the spiritual dimension it seeks to capture is the unfathomability of technology itself. Special effects facilitate encounters with overpowering technological environments and omnipotent machines and thus give rise to sublime sentiments. The sublime characterizes experiences that rise above the earthly and finite, and thus attain a spiritual dimension. In its attribution of transcendent qualities to mechanical objects, the technological sublime embodies the techno-romantic paradigm. By rendering technophilia and technophobia indistinguishable, it sublimates technology into art.

89 Heinz Michaelis, "Art and Technology in Film [1923]," in *The Promise of Cinema: German Film Theory, 1907-1933*, eds. Anton Kaes, Nicholas Baer, and Michael Cowan, trans. Alex H. Bush, 581-582 (Berkeley, CA: University of California Press, 2016), 581.

Bibliography

Béla Balázs: Early Film Theory. Visible Man and The Spirit of Film, ed. Erica Carter, trans. Rodney Livingstone (New York, NY/Oxford: Berghahn Books, 2010).

Klaus Bartels, "Über das Technisch-Erhabene," in *Das Erhabene: Zwischen Grenzerfahrung und Größenwahn*, ed. Christiane Pries, 295-316 (Berlin: Verlag Walter de Gruyter, 1989).

Klaus Bartels, "Vom Erhabenen zur Simulation. Eine Technikgeschichte der Seele: Optische Medien bis 1900 (Guckkasten, Camera Obscura, Panorama, Fotografie) und der menschliche Innenraum," in *Armaturen der Sinne: Literarische und technische Medien 1870 bis 1920*, eds. Jochen Hörisch and Michael Wetzel, 17-42 (Munich: Wilhelm Fink Verlag, 1990).

Randolph Bartlett, "German Film Revision Upheld as Needed Here," *New York Times* (13 March 1927): X3.

Dagmar Bellmann, *Von Höllengefährten zu schwimmenden Palästen: Die Passagierschifffahrt auf dem Atlantik (1840-1930)* (Frankfurt am Main: Campus Verlag, 2015).

Raymond Bellour, "Ideal Hadaly," trans. Stanley E. Grey, *Camera Obscura* 15 (Fall 1986): 111-136.

Gretchen Berg, "The Viennese Night: A Fritz Lang Confession, Parts One and Two (1965)," in: *Fritz Lang Interviews*, ed. Barry Keith Grant, 50-76 (Jackson, MS: University Press of Mississippi, 2003).

Peter Bogdanovich, *Fritz Lang in America* (London: Studio Vista, 1967).

Wilhelm Bölsche, "Die Poesie der Großstadt," *Das Magazin für Literatur* 40 (4 October 1890): 622-625.

Bertolt Brecht, "Singing Steyr Motor Cars [1927]," in *The Collected Poems of Bertolt Brecht*, trans. and eds. Tom Kuhn and David Constantine, 361-362 (New York, NY: Liveright Publishing, 2018).

Scott Bukatman, "Disobedient Machines: Animation and Autonomy," in *Beyond the Finite: The Sublime in Art and Science*, eds. Ronald Hoffmann and Iain Boyd Whyte, 128-149 (Oxford: Oxford University Press, 2011).

Scott Bukatman, *Matters of Gravity: Special Effects and Supermen in the 20th Century* (Durham, NC: Duke University Press, 2003).

Luis Buñuel, "*Metropolis* [1927]," in *An Unspeakable Betrayal: Selected Writings of Luis Buñuel*, trans. Garrett White, 99-102 (Berkeley, CA: University of California Press, 2000).

Edmund Burke, *A Philosophical Enquiry into the Sublime and Beautiful* [1757], ed. James T. Boulton (London/New York, NY: Routledge, 2008).

Michael Cowan, "The Heart Machine Rhythm and Body in Weimar Film and Fritz Lang's *Metropolis*," in *Modernism/modernity* 2 (April 2007): 225-248.

Istvan Csicsery-Ronay, Jr, *The Seven Beauties of Science Fiction* (Middleton, CT: Wesleyan University Press, 2008).
Karlheinz Daniels, "Expressionismus und Technik," in *Technik in der Literatur*, ed. Harro Segeberg, 351-386 (Frankfurt am Main: Suhrkamp-Verlag, 1987).
"Das Schüfftan'sche Kombinationsverfahren!" *Lichtbild-Bühne* 11 (13 January 1927).
Royal Dixon, *Americanization* (New York, NY: Macmillan, 1916).
Emil du Bois-Reymond, *Reden: Erste Folge. Literatur – Philosophie – Zeitgeschichte* (Leipzig: Veit u. Comp., 1886).
Matthias Eberle, *Der Weltkrieg und die Künstler der Weimarer Republik. Dix, Grosz, Beckmann, Schlemmer* (Stuttgart: Belser, 1989).
Axel Eggebrecht, "*Metropolis*," *Weltbühne* 3 (18 January 1927): 115-116.
Sergei Eisenstein, "The Cinematographic Principle and the Ideogram [1929]," in Sergei Eisenstein, *Film Form: Essays in Film Theory*, ed. and trans. Jay Leyda, 28-44 (New York, NY: Harcourt, Brace & World, 1949).
Max Feige, "Der Film *Metropolis*," *Der Film* (Berlin) 1 (15 January 1927).
"Film – Extrablatt," *Illustriertes Wiener Extrablatt* (4 July 1924).
Theodor Fontane, *Ein Sommer in London* (Dessau: Gebrüder Katz, 1854).
E. M. Forster, "The Machine Stops [1909]," in *Selected Stories*, 91-124 (New York, NY: Penguin Classics, 2001).
Evelyn Gerstein, "*Metropolis*," *The Nation* (23 March 1927): 323-324.
Fritz Giese, *Girlkultur: Vergleiche zwischen amerikanischem und europäischem Rhythmus und Lebensgefühl* (Munich: Delphin-Verlag, 1925).
Rolf Giesen, "Spezialeffekte Made in Germany," *Künstliche Welten: Tricks, Special Effects und Computeranimation im Film von den Anfängen bis heute*, eds. Rolf Giesen and Claudia Meglin, 69-111 (Hamburg/Vienna: Europa Verlag, 2000).
Jeremy Gilbert-Rolfe, *Beauty and the Contemporary Sublime* (New York, NY: Allworth Press, 1999).
Ludwig Max Goldberger, *Das Land der unbegrenzten Möglichkeiten: Beobachtungen über das Wirtschaftsleben der Vereinigten Staaten von Amerika* (Berlin/Leipzig: F. Fontane, 1905).
Tom Gunning, *The Films of Fritz Lang: Allegories of Vision and Modernity* (London: British Film Institute, 2000).
Willy Haas, "*Metropolis*," *Film-Kurier* (11 January 1927).
Thea von Harbou, *Metropolis* [screenplay, 1924]. Gottfried Huppertz Collection, Filmmuseum Berlin-Deutsche Kinemathek.
Thea von Harbou, *Metropolis* [novel, 1926], anonymous trans. (Mineola, NY: Dover Publications, 2015).
Siegfried Hartmann, "*Metropolis* Technik," *Deutsche Allgemeine Zeitung* (20 January 1927): 2.

Gerhart Hauptmann, "Flagman Thiel [1887]," in *German Novellas of Realism II*, ed. Jeffrey L. Sammons, trans. Adele S. Seltzer, 304-333 (New York, NY: Continuum, 1989).

Jörg Heiniger, "Erhaben," *Ästhetische Grundbegriffe: Historisches Wörterbuch in 7 Bänden*, vol. 2, 275-310 (Stuttgart: Metzler-Verlag, 2001).

Karl Henckell, "Viadukt," in *Gesammelte Werke*, vol. 2: *Buch des Kampfes* (Munich: J. Michael Müller Verlag, 1921).

H. H. P., "Das Zerrbild der Technik im Film: Ein Nachwort zu *Metropolis*," *Tages-Post* (Linz, Austria) 52 (6 March 1927): 3-4.

Arthur Holitscher, *Amerika heute und morgen* [1912] (Berlin: S. Fischer, 1923).

Otto Hunte, "Der Baumeister von *Metropolis*," *Presse- und Propagandaheft Metropolis* (Berlin: Parufamet Presse- und Propagandabteilung, 1927).

Andreas Huyssen, "The Vamp and the Machine: Technology and Sexuality," *New German Critique* 24/25 (Autum 1981/Winter 1982): 221-237.

Leopold Jacoby, *Die Idee der Entwicklung* [1874/1876] (Zurich: Schabelitz, 1887).

Anton Kaes, *Shell Shock Cinema: Weimar Culture and the Wounds of War* (Princeton, NJ: Princeton University Press, 2009).

Immanuel Kant, *Observations on the Feeling of the Beautiful and Sublime* [1764], trans. John T. Goldthwait (Berkeley, CA: University of California Press, 2003).

Immanuel Kant, *The Critique of the Power of Judgment* [1790], ed. Paul Guyer, trans. Paul Guyer and Eric Matthews (Cambridge, UK: Cambridge University Press, 2000).

Sarah Keller, *Anxious Cinephilia. Pleasure and Peril at the Movies* (New York, NY: Columbia University Press, 2020).

Erich Kettelhut, *Der Schatten des Architekten*, ed. Werner Sudendorf (Munich: Belleville, 2009).

Ludwig Klages, *Mensch und Erde* [1913] (Stuttgart: A. Kroener Verlag, 1956).

Fritz Lang, "Was ich in Amerika sah: Neuyork – Los Angeles," *Film-Kurier* (11 December 1924): 1.

Kurd Lasswitz, *Wirklichkeiten* (Berlin: Verlag von Emil Felber, 1900).

Wilhelm Liebknecht, "Knowledge is Power – Power is Knowledge [1872]," in *Wilhelm Liebknecht and German Social Democracy: A Documentary History*, ed. William A. Pelz, trans. Erich Hahn (Chicago, IL: Haymarket Books, 2016).

Jean-François Lyotard, *The Postmodern Condition: A Report on Knowledge*, trans. Geoff Bennington and Brian Massumi (Manchester: Manchester University Press, 1984).

Karl Marx, *Capital: A Critique of Political Economy*, vol. I: *The Process of Production of Capital* [1867], trans. Ben Fowkes (Harmondsworth: Penguin, 1990).

Leo Marx, *The Machine in the Garden: Technology and the Pastoral Ideal in America* (New York, NY: Oxford University Press, 1965).

Moses Mendelssohn, "Über das Erhabene und Naive in den schönen Wissenschaften [1758]," in *Moses Mendelssohn: Ästhetische Schriften*, ed. Anne Pollok, 216-259 (Hamburg: Felix Meinder Verlag, 2006).

Jochen Meyer, "Die Hochhausdiskussion in Deutschland am Anfang der zwanziger Jahre," in *Der Schrei nach dem Turmhaus: der Ideenwettbewerb Hochhaus am Bahnhof Friedrichstrasse, Berlin 1921/22*, eds. Florian Zimmermann and Bruno Bruognolo, 187-214 (Berlin: Argon, 1988).

"Metropolis – A Movie Based on Science," *Science and Invention* 170 (June 1927): 106-107.

"Metropolis Film Seen," *New York Times* (11 January 1927): 36.

Metropolis Ufa Magazin Sondernummer (Berlin: Bukwa, 1927).

Heinz Michaelis, "Art and Technology in Film [1923]," in *The Promise of Cinema: German Film Theory, 1907-1933*, eds. Anton Kaes, Nicholas Baer, and Michael Cowan, trans. Alex H. Bush, 581-582 (Berkeley, CA: University of California Press, 2016).

M.-l., "Technische Kritik zu Metropolis," *Lichtbild-Bühne* 10 (12 January 1927).

Richard Müller-Freienfels, *Geheimnisse der Seele* (Munich: Delphin-Verlag, 1927).

Dietrich Neumann, "The Urbanistic Vision in Fritz Lang's *Metropolis*," in *Dancing on the Volcano: Essays on the Culture of the Weimar Republic*, 143-154 (New York, NY: Camden House, 1994).

David E. Nye, *American Technological Sublime* (Cambridge, MA: MIT University Press, 1994).

Helmuth Plessner, "Die Utopie in der Maschine [1924]," in *Gesammelte Schriften in zehn Bänden*, vol. X: *Schriften zur Soziologie und Sozialphilosophie* (Berlin: Suhrkamp-Verlag, 2003).

Friedrich Porges, "Luftschlösser im Film," *Die Bühne* (Vienna) 120 (25 February 1927).

Presse- und Propagandaheft Metropolis (Berlin: Parufamet Presse- und Propagandabteilung, 1927).

Christiane Pries, "Einleitung," in *Das Erhabene: Zwischen Grenzerfahrung und Größenwahn*, ed. Christiane Pries, 1-30 (Berlin: Verlag Walter de Gruyter, 1989).

Walter Rathenau, *Zur Kritik der Zeit* [1912] (Berlin: Fischer, 1929).

Günther Rittau, "Special Effects in *Metropolis* [1927]," in *Fritz Lang's Metropolis: Cinematic Visions of Technology and Fear*, eds. Michael Minden and Holger Bachmann, 78-80 (Rochester, NY: Camden House, 2002).

Theodore Roosevelt, "True Americanism [1894]," in *American Ideals and Other Essays, Social and Politcal*, 16-37 (Philadelphia, PA: Gebbie and Company, 1903).

Joseph Roth, "Der auferstandene Mensch [1923]," in *Joseph Roth Werke*, vol. 1: *Journalistische Arbeiten 1915-1923*, ed. Klaus Westermann, 936-939 (Cologne: Kiepenheuer & Witsch, 1989-1991).

R. L. Rutsky, "The Mediation of Technology and Gender: *Metropolis*, Nazism, Modernism," *New German Critique* (Autumn 1993): 3-32.

Friedrich Schiller, *The Works of Friedrich Schiller: Aesthetical and Philosophical Essays*, ed. Nathan Haskell Dole (Boston, MA: The Wyman-Fogg Company, 1902).

Hans Schliepmann, "Wolkenkratzer," *Tägliche Rundschau* 49 (30 January 1921).

Artur Schopenhauer, *Metaphysik des Schönen* [1820] (Munich: Piper Verlag, 1985).

Arthur Schopenhauer, *The World as Will and Representation* [1819], vol. 1, trans. E. F. J. Payne (New York, NY: Dover Publications, 1969).

Sime, "Metropolis," *Variety* (16 March 1927): 16-17.

Zoltán Simon, *The Double-Edged Sword: The Technological Sublime in American Novels between 1900 and 1940* (Budapest: Akadémiai Kiadó, 2003).

Vivian Sobchack, *Screening Space: The American Science Fiction Film* (New Brunswick, NJ: Rutgers University Press, 1987).

Oswald Spengler, *The Decline of the West* [1918/1922], trans. Charles Francis Atkinson (New York, NY: Knopf, 1966).

Ludwig Spitzer, "Fritz Lang über den Film der Zukunft," *Die Filmtechnik* 2 (15 July 1925): 34-36.

Adalbert Stifter Briefe, ed. Hans Schumacher (Zurich: Manesse-Verlag, 1947).

Joseph Tabbi, *Postmodern Sublime: Technology and American Writing from Mailer to Cyberpunk* (Ithaca, NY: Cornell University Press, 1995).

Nicholas Taylor, "The Awful Sublimity of the Victorian City: Its Aesthetic and Architectural Origins," in *The Victorian City: Images and Realities*, eds. H. J. Dyos and Michael Wolff, 431-447 (London: Routledge & Kegan Paul Books, 1973).

Ernst Toller, "The German Hinkemann [1923]," *Plays Two*, ed. and trans. Alan Raphael Pearlman (London: Oberon Books, 2019).

Ernst Toller, "The Machine Breakers [1922]," *Plays Two*, ed. and trans. Alan Raphael Pearlman (London: Oberon Books, 2019).

Ernst Toller, "Transformation [1919]," *Plays One*, ed. and trans. Alan Raphael Pearlman (London: Oberon Books, 2000).

Alfred de Vigny, "Paris [1831]," in *Oeuvres complètes. Texte présenté et commenté par Ferdinand Baldensperger*, vol. 1, 164-165 (Paris: Gallimard, 1948).

Auguste Villiers de l'Isle-Adam, *Tomorrow's Eve* [1886], trans. Robert Martin Adams (Chicago, IL: University of Illinois Press, 2001).

Oscar Wilde, *The Soul of Man under Socialism* [1891] (New York, NY: Harper & Row, 1970).

6. “German Technique” and Hollywood

Abstract
In the mid-1920s, the innovative imagery and emotional force of German films startled American critics and filmmakers. Well-known directors like F. W. Murnau and Paul Leni were invited to Hollywood, and their American films showcased a range of unconventional camera effects, in particular moving camera feats and extreme camera angles. What galvanized American commentators about these methods was the realization that cinematic devices could be used to visualize affective content. German filmmakers proffered a novel model of cinematic immersion, which augmented the audience’s absorption in the story world with figurative levels of meaning. Prompted by objectives originating in techno-romantic thought, Hollywood began to pay increased attention to the expressive potential of technical tools, with lasting effects on American filmmaking.

Keywords: F. W. Murnau, camera angles, moving camera, immersion, camera cranes

In the mid-1920s, German films caused an upheaval in Hollywood. Critics and members of the film industry were astonished by the camera feats on display in films like *Der letzte Mann* (Ufa, 1924, dir. F. W. Murnau), *Varieté* (Ufa, 1925, dir. E. A. Dupont), and *Metropolis* (Ufa, 1927, dir. Fritz Lang). Possibly for the first time since the introduction of feature-length films, it became painfully obvious that, despite all its commercial successes, Hollywood was lagging behind technically and aesthetically. The *New York Herald Tribune* for instance proclaimed: “It is curious to note how completely Germanic this new art of the cinema seems to be. As an industry, of course, the American photoplay is supreme by at least a thousand miles, yet when it comes to the aesthetics of the screen field, even the local film makers

Loew, Katharina, *Special Effects and German Silent Film: Techno-Romantic Cinema.*
Taylor & Francis Group, 2021
DOI 10.5117/9789463725231_CH06

look to Germany for inspiration and idea."[1] In addition to aesthetic deficits, early warning signs also indicated that American market shares in Europe were slipping. As part of a multi-faceted approach to meet these challenges, studios started importing directors from Europe with expectation that they would create prestigious films with the international appeal and familiarize Hollywood with their methods.

Conventional wisdom has it that many of the impulses Hollywood received from Germany were technical in nature. Indeed, Hollywood paid close attention to German technological innovations, as producer G. A. Mincenty observed: "What interests almost every American about German films is whether he can see something technically new."[2] Scholars have often pointed to special effects, camera movement, or high-contrast lighting as examples of Germany's influence on Hollywood. However, such shorthand assertions are too inexact to be accurate. For instance, Patrick Keating has shown that Hollywood's use of high-contrast lighting and cast shadows precedes famous German examples such as *Nosferatu* (Prana, 1922, dir. F. W. Murnau).[3] In a similar vein, neither did contemporary commentators attribute innocuous tracking shots to German influence, nor is there evidence that particular special-effects techniques like the Schüfftan process were adopted from Germany. Instead, what received heightened attention in Hollywood were not so much specific methods as unusual and eye-catching effects.

In the trade press, "German technique" was typically cited in relation to "handling of light,"[4] "camera angles, the shifting of focus, 'dissolves' and simultaneous appearances on the screen,"[5] or "false perspective,"[6] as well as "gyroscopic"[7] or "acrobatic"[8] camera work, and generally "trick shots."[9] All of these refer to instances in which film technology was employed in

1 Richard Watts, "Some Additional Remarks on that New German Photoplay," *New York Herald Tribune* (11 July 1926): E3.

2 G. A. Mincenty, "Echo aus Amerika," *Filmtechnik* 12 (1925): 254.

3 Patrick Keating, *Hollywood Lighting from the Silent Era to Film Noir* (New York, NY: Columbia University Press, 2009), 73-78.

4 Randolph Bartlett, "German Film Revision Upheld as Needed Here," *New York Times* (13 March 1927): X3.

5 "Paul Fejos Praised For Broadway Direction," *Universal Weekly* (15 June 1929): 25.

6 James Mitchell Leisen, "Some Problems of the Art Director," *Transactions of S.M.P.E.* 33 (1928): 71-80.

7 "Now They'll Get 'Acting Continuity,'" *Photoplay* (June 1925): 53.

8 Hal Hall, "Cinematographers and Directors Meet Discuss Camera Trucking Problems," *American Cinematographer* (August 1932): 10.

9 "Case of Jonathan Drew," *Variety* (13 June 1928): 12.

then-unusual ways to achieve images beyond the scope of customary straight photography. The case of "German technique" underscores the difficulties in establishing special effects as a conceptual category and encourages us to think of their affinities with other cinematic devices. As Julie Turnock has argued, in studio-era Hollywood "special effects" was above all a labour classification, because certain effects were produced by a specialized work force.[10] Initially, however, tricks were the cinematographer's and the art director's domain and specialized labour was introduced only gradually over the course of the 1920s. Accordingly, boundaries were often ambiguous. Hollywood insiders spoke of "trick angles"[11] and special cameras such as the handheld Eyemo or the Akeley, which could be simultaneously panned and tilted with one hand, were said to produce "special effects."[12]

There is no doubt that "German technique" was strongly associated with technology. However, as I argue in this chapter, ultimately the significance of "German technique" for American filmmaking did not consist in the adoption of specific new devices, but rather in the establishment of cinematography as a tool to visualize ideas and emotions. Cinematographer Carl Louis Gregory observed in 1928:

> *The Last Laugh*, *Variety*, and *Metropolis*, all German made pictures started a tremendous competition among our producers. These pictures gave a new meaning to camera technique and caused a change in the methods of picture construction which are nothing less than revolutionary. In all probability there is not a single trick or camera effect in any of the three pictures mentioned that had not already been employed by American cinematographers yet only a few were able to discern the true inwardness of the almost continuous succession of effects and camera tricks employed in the making of these productions. This new technique utilizes every available device of trick photography, not as a thing in itself, but with a very definite idea of the emotional and psychological reaction which will result from the employment of the proper photographic effect.[13]

10 Julie Turnock, *Plastic Reality: Special Effects, Technology and the Emergence of 1970s Blockbuster Aesthetics* (New York, NY: Columbia University Press, 2015), 8.

11 See for instance Carroll Graham, "Ah, Those Were The Days," *Picture Play* 6 (August 1929): 24-25.

12 See for instance Carl Louis Gregory, *Motion Picture Photography* (New York, NY: Falk Publishing, 1927), 345; "Bell & Howell Pioneer Professional Standard," *American Cinematographer* 1 (January 1927): 12.

13 Carl Louis Gregory, "An Optical Printer for Trick Work," *Transactions of SMPE* 34 (1928): 419-426.

The emotional potency German pictures were able to achieve has its roots in German filmmakers' quest for film art. Although filmmaking in Germany was conditioned by the same commercial considerations as elsewhere, in contrast to the United States aesthetic concerns held disproportionally greater sway over the film business. Unlike most of their American colleagues, leading German film practitioners saw themselves as part of the cultural elites and shared the intelligentsia's prevalent techno-romantic convictions. While in the United States cinema's social merit depended on its success as an industry, in Germany it was above all cultural cachet, i.e., its status as an art.

Artistic filmmaking required the use of film technology to communicate something beyond material reality, which was precisely the aspect that for German observers was missing in American cinema. Fritz Lang claimed: "Americans have still not understood how to use their magnificent equipment to elevate the miracle of photography into the realm of the spirit."[14] In a similar vein, G. A. Mincenty commented about psychology on screen: "American film lacks this aspect altogether; and that is one of the strongest weapons of German film."[15]

In contrast to the American mode of cinematic narration, which at the time relied predominantly on the mise-en-scène and editing, German-style filmmaking was noted for its ability to convey meaning cinematographically. This aptitude was the result of long-standing experiments with the cinematic apparatus that were motivated by filmmakers' artistic aspirations and techno-romantic convictions. As the embodiment of cinema's medium specificity, the camera was construed as its principal artistic agency. As such, its purpose was the moulding of cinema's raw material—the profilmic—in accordance with the artists' conceptions.

The evolution of expressive cinematography commenced with the use of cinematography as a special effect. In keeping with techno-romantic tenets, expressive cinematography was inextricably linked with the goal of subjecting the machine to the primacy of the human mind. As a result, moving camera feats and extreme camera angles were employed in much the same way as other camera devices that are traditionally considered special effects, such as multiple exposure composites. They likewise served the purpose of evoking ideas and emotions in the audience and emotionally

14 Fritz Lang, "The Future of the Feature Film in Germany [1926]," in *The Weimar Republic Sourcebook*, eds. Anton Kaes, Martin Jay, and Edward Dimendberg, trans. Don Reneau (Berkeley, CA: University of California Press, 1994), 623.

15 G. A. Mincenty, "Echo aus Amerika," *Filmtechnik* 12 (1925): 254.

immersing them in a reality that had been moulded by the human mind. Hollywood greatly admired the originality and emotional effectiveness of German films and quickly seized on German-style camera effects. Under the influence of German filmmaking, cinematography became a core component of filmic narration and a means to depict ideas and emotions, which dramatically changed American cinema.

The "unchained camera"

When discussing the international development of cinematic expressivity, scholars often point to the "unchained camera" (*entfesselte Kamera*) as Germany's principal contribution. According to Kristin Thompson, the "unchained camera" was one of the distinctly German devices that left strong marks on American filmmaking.[16] However, the term's significance is illusive. 1920s Germany had a penchant for ill-defined buzzwords such as *Kammerspielfilm* or *Autorenfilm*. Some of these labels have since become part of the film historical discourse, but their unclear meaning has helped to perpetuate significant misconceptions. Contemporaries often defined the "unchained camera" as camerawork that had been "liberated from the tripod."[17] This seemingly clear-cut characterization is not without problems, however. For tracking shots, for instance, tripod and camera were usually placed on a mobile platform, a setup that does not differ in principle from the phantom rides of early cinema. Strictly speaking, tracking shots therefore are not "tripod-less," a condition that essentially only applied to handheld shooting. Nonetheless, scholars and even some contemporary observers have referred to any camera movement, including pans, tilts, and following shots, as "unchained camera." This has obscured our understanding of the innovations in 1920s German camerawork.

Pans and tilts on the one hand and tracking and craning shots on the other gained currency at different historical moments and should consequently be considered separately. Pans and tilts became standard tools early on in film history, also in the German context. In 1900 Guido Seeber used a protracted pan to flaunt a German warship on its way to suppress the Yihetuan Movement in *Ausfahrt der Chinakrieger von Bremerhaven mit der Straßburg am 31.7.1900*. Several German films of the 1910s, including *Das Ende*

16 Kristin Thompson, *Herr Lubitsch Goes to Hollywood: German and American Film after World War I* (Amsterdam: Amsterdam University Press, 2005), 118-124.

17 See for instance "Die entfesselte Kamera," *Ufa-Magazin* (1927): 216-217.

des Homunculus (Deutsche Bioscop, 1916, dir. Otto Rippert) and *Hoffmanns Erzählungen* (Lothar Stark, 1916, dir. Richard Oswald), feature extended pans that fulfil expressive functions. Occasional tracking shots can be found as well, as for instance in Ernst Lubitsch's *Die Augen der Mumie Mâ* (PAGU, 1918), where the camera recoils from the advancing murderous Radu. Lotte Eisner moreover notes that Hans Janowitz's script for Murnau's lost 1920 film *Der Januskopf* (Lipow-Film/Decla-Bioscop) contains the direction "camera follows him up the stairs," which would have required an elevator or handheld setup.[18] Still, prior to *Sylvester* (Rex-Film, 1924, dir. Lupu Pick) and *Der letzte Mann,* German cameras rarely tracked, which may explain in part why the camera mobility in these films evoked such a strong critical response.

In other countries, in contrast, cameras had regularly left their stationary position since the mid-1910s. Noteworthy examples include the lateral, diagonal, or axial tracking shots in *Traffic in Souls* (IMP, 1913, dir. George Loane Tucker) and Giovanni Pastrone's *Cabiria* (Itala, 1914), the three-minute receding tracking shot at the princess's soirée in Yevgeni Bauer's *Posle smerti* (Khanzhonkov, 1915), the crane shot of the Babylon set in Griffith's *Intolerance* (Triangle, 1916), or the pull back through a window in Benjamin Christensen's *Hævnens Nat* (Dansk Biograf, 1916). As Barry Salt has observed, in American cinema tracking shots all but disappeared again toward the end of the war.[19] In contrast, post-war French films like Henri Pouctal's *Au travail* (Le Film d'Art, 1920), Marcel L'Herbier's *El Dorado* (Gaumont, 1921), Abel Gance's *La Roue* (Films Abel Gance, 1922), Jean Epstein's *L'Auberge rouge* (Pathé Consortium, 1923), or Raymond Bernard's *Le Miracle des loups* (Société des Films Historiques, 1924) continued to boldly expand the limits of camera mobility. As this overview suggests, German films were neither devoid of notable camera movements prior to *Sylvester* nor did expressive and spectacular moving camera feats originate in Germany. Instead, the central contribution of German cinematography, which captured the attention of observers worldwide, was the (techno-romantically informed) effort to employ devices such as handheld or aerial photography as daring new means to convey ideas and emotions. French camerawork had already made major strides in this direction, but the film that was widely perceived as the catalyst of a cinematographic revolution was *Der letzte Mann*.

18 It is unknown whether this shot was realized. Lotte H. Eisner, *Murnau* [1964], trans. Martin Secker (London: Secker & Warburg, 1973), 31.

19 Barry Salt, *Film Style and Technology: History and Analysis* (London: Starwood, 1983), 138.

Ufa production chief Erich Pommer played a key role in facilitating this development. Convinced that German cinema's path to worldwide recognition was through aesthetic-technological innovation, he promoted the production of artistic yet accessible films and fostered teamwork and a dedication to cinema's artistic development among production crews. As art director Robert Herlth recalled, Pommer set the creators of *Der letzte Mann* the task "to invent something mad."[20] They consequently began to experiment with unconventional camerawork to express their ideas. According to Herlth, the breakthrough came in the context of a sequence that is not part of the surviving film:

> The attendant was supposed to sniff the cigar smoke after the millionaire had already departed up the stairs. At this point Murnau said: "No, that doesn't work." "Why not?" Karl Freund and I asked. "Because you don't see anything – it doesn't have any effect," answered Murnau emphatically. "What are we to do then?" "We need something more intense, if only we could fly with the smoke." "What?" ... "The stairs ..." said Murnau. "With the camera?" asked Freund. "Of course – what else?" "We'd need a fireman's ladder," I said timidly. [...] Someone was sent for a ladder, the camera was fixed at the top, and the not insubstantial Freund took up his position. We removed half of the set and moved the ladder slowly towards the stairs; the camera followed the smoke, rising with it up the stairs as the ladder was wound upwards. "We've got it!" cried Murnau. After this, there was no stopping us. [...] Each day on the set was a fresh opportunity for us to surpass ourselves in inventiveness and ingenuity. At each shot we were anxious for the rest of the team's reactions; we were enthusiasts, although none of us could help feeling a twinge of jealousy when it was someone else who thought of the solution to some knotty problem. In short, all of us felt we had invented new visual processes. And we had not "unchained" the camera for merely technical reasons. On the contrary, we had found a new and more exact way of isolating the image and of intensifying dramatic incident.[21]

Herlth's recollections illustrate the collective quest for emotionally effective visualizations of thought that characterized German prestige filmmaking. The most virtuoso camera feats in European silent cinema, including the

20 Erich Pommer quoted in Robert Herlth, "With Murnau on the Set," in Lotte H. Eisner, *Murnau* [1964], trans. Martin Secker, 59-70 (London: Secker & Warburg, 1973), 62.
21 Herlth, "With Murnau on the Set," in Eisner, *Murnau*, 62-65.

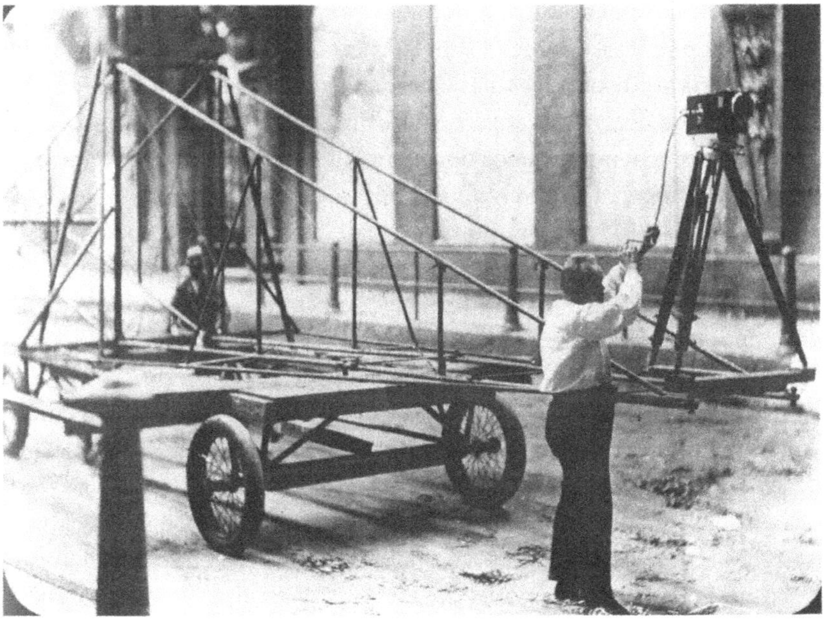

Fig. 6.1. Dolly with a pivoting horizontal boom used for *Sylvester* (1924). Guido Seeber operates his Debrie Parvo remotely.
Source: Deutsche Kinemathek.

"flying tone," the depiction of the sound of a horn in the courtyard reaching the protagonist's ear on the third floor, which Karl Freund created for *Der letzte Mann,* the swinging camera in the deluge sequence in *Metropolis*, Carl Hoffmann's depiction of the flight on Mephisto's cloak or the advancing Horsemen of the Apocalypse in *Faust* (Ufa, 1926, dir. F. W. Murnau), Hoffmann's and Freund's trapeze shots in *Varieté* or the "convention wave" that Jules Kruger and Simon Feldman realized for *Napoléon vu par Abel Gance* (Ciné France et al, 1927, dir. Abel Gance), reflect the techno-romantic impetus to employ the camera as a tool to convey ideas and emotions and facilitate a more immediate, visceral access to the drama on screen.

Given the longstanding close collaborations between leading German filmmakers, it is problematic to credit certain innovations to specific individuals. However, F. W. Murnau and screenwriter Carl Mayer were among the first to publically assert the expressivity of cinematography in motion. Carl Mayer prescribed moving camera effects in *Scherben* (Rex-Film, 1921, dir. Lupu Pick), where Friedrich Weinmann's camera tracks behind the protagonist as he walks along the lonely, wintery railway tracks. For *Sylvester*, Mayer proposed ambulatory camera movements in three dimensions to portray the environment ("*Umwelt*") surrounding the main psychological drama. Guido

Seeber realized this vision using an inventive dolly with a long, pivoting boom, which facilitated horizontal camera motions in all directions (Figure 6.1.). With regard to *Der letzte Mann*, Mayer's contribution was also substantial. Karl Freund exclaimed: "Mayer's imagination convinced us that we could do anything."[22] F. W. Murnau, for his part, described in January 1924 the filmmaker's ultimate tool as a "camera that can be freely moved in space. [...] The camera that surmounts film technology, thereby fulfilling film technology's ultimate artistic purpose."[23] What Murnau formulates here is a techno-romantic vision of technology's triumph over its material limitations in the service of art. Consistent with this, his films featured moving camera effects even before they actually employed mobile cinematography. In *Phantom* (Decla-Bioscop, 1922), which was scripted by Thea von Harbou and photographed by Axel Graatkjær, mobile camera effects appear at key junctures to portray the abnormal perceptions of the troubled protagonist Lorenz Lubota. For instance, a circular canted pan of a dance floor evokes hand-held camerawork in order to create a dizzying and claustrophobic feeling. Later, to convey his state of confusion and loss of control, Lupota's table recedes from the camera. Although the shot looks similar to a crane shot, in actuality the set was moved instead of the camera: Art director Hermann Warm designed an elevator that allowed table and characters to descend into a dark abyss (Figure 6.2.).[24] Mayer's and Murnau's advocacy notwithstanding, without likeminded collaborators—the encouragement of producer Erich Pommer and the audacity and creativity of cinematographers like Karl Freund, Guido Seeber, and Carl Hoffmann and art directors like Robert Herlth, Walter Röhrig, Otto Hunte, and Erich Kettelhut—their innovative capacity would have been limited.[25] Thomas Brandlmeier consequently emphasized that "the unchained camera was a collective achievement."[26]

22 Karl Freund, "Ein Filmkünstler [1947]," in *Caligari und Caligarismus*, ed. Walter Kaul, 79-81 (Berlin: Deutsche Kinemathek, 1970), 80. Camera feats such as the "flying tone" were already laid out in Mayer's script, which only survives in fragments, for instance in Ewald André Dupont and Fritz Podehl, *Wie ein Film geschrieben wird und wie man ihn verwertet*, 90-96 (Berlin: Kühn, 1926), 91.

23 F. W. Murnau, "...der frei im Raum zu bewegende Aufnahmeapparat [1924]," in Fred Gehler and Ulrich Kasten, *Friedrich Wilhelm Murnau* (Berlin: Henschelverlag Kunst und Gesellschaft, 1990), 141.

24 See Warm's recollections in Eisner, *Murnau*, 99-100.

25 See Herlth's recollections in Eisner, *Murnau*, 62-67.

26 Thomas Brandlmeier, "'Entfesseltes Sehen:' Von der *Ausfahrt der Chinakrieger* bis *Varieté*," in *Die Perfektionierung des Scheins: Das Kino der Weimarer Republik im Kontext der Künste*, ed. Harro Segeberg, 143-157 (Munich: Wilhelm Fink Verlag, 2000), 147. Karl Prümm argues along similar lines in "Das schwebende Auge: Zur Genese der bewegten Kamera," in *Die Medien und ihre Technik: Theorien, Modelle, Geschichte*, ed. Harro Segeberg, 235-256 (Marburg: Schüren-Verlag, 2004), 244.

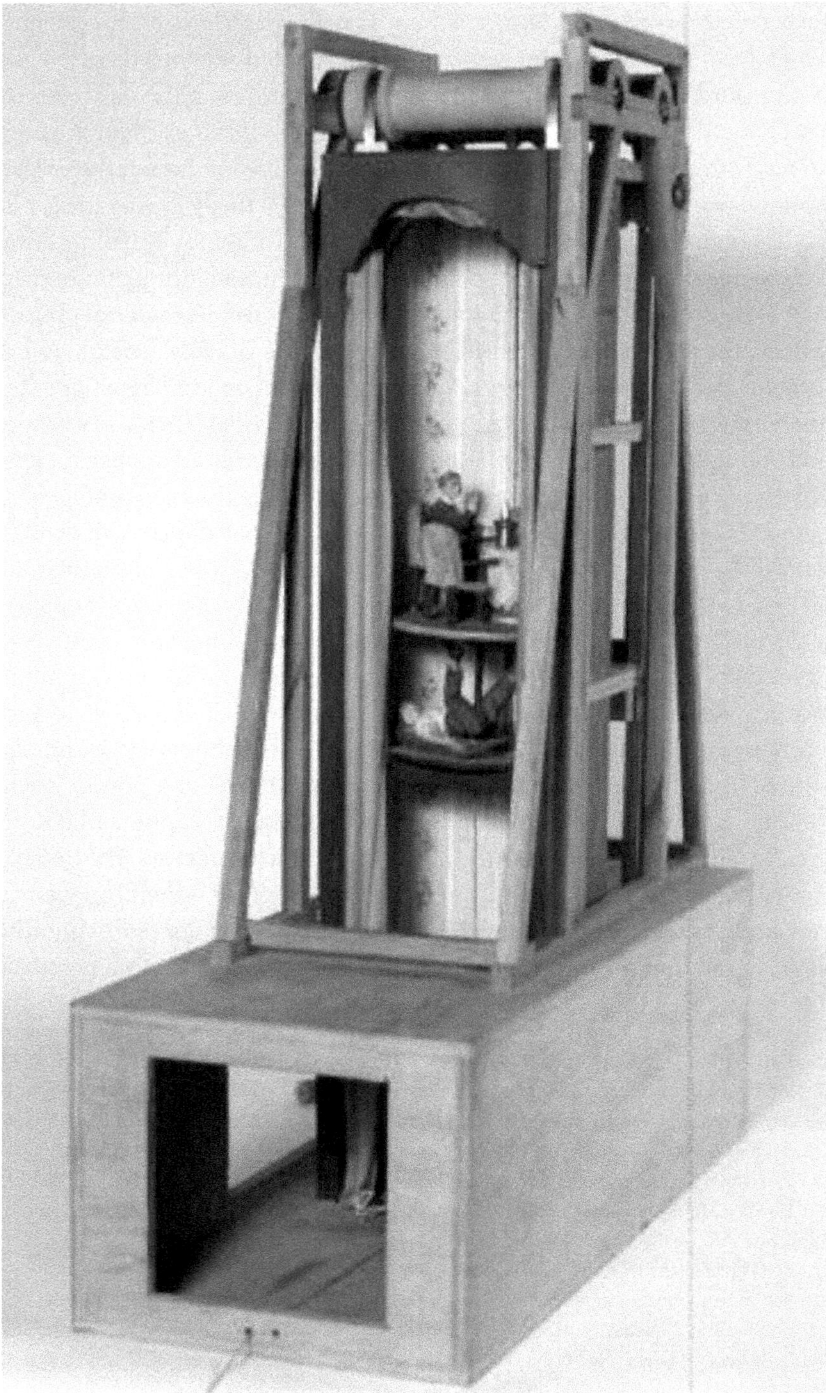

Fig. 6.2. The elevator Herrmann Warm designed for *Phantom* (1922) to simulate a crane shot. Source: Deutsche Kinemathek.

Although, from a historical perspective, the concept "unchained camera" is deeply flawed, it is important to recognize why the label came to play such a prominent role in the critical discourse at the time: Camera movement embodied medium-specific expressivity in accordance with techno-romantic sentiments. For critic Georg Otto Stindt, the mobile camera set cinema apart from photography, thus realizing medium specificity and artistic autonomy.[27] Commentator Lotar Holland pointed to the links between kinematics and cinema's ontology. According to Holland, movement found expression in cinema in three ways, namely the "objective movements of the things and people being filmed, the subjective movement of the camera itself and the formal movement of the filmstrip during projection. [...] Camera movement, or what goes today by the name of 'unchaining' the camera, belongs therefore to the second category, and thus plays the central role in the artistic creation of a film in general."[28] In contrast to the automatic recording of movement and the mechanical transportation of the film through the projector, Holland's second category affords the possibility of human intervention in the apparatus and thus levels of meaning beyond the reproduction of material reality.

Many German critics welcomed the fact that mobile shots—like trick effects—tend to call attention to the human agency that has shaped them. It was precisely this aspect, however, that engendered criticism in the United States, where "camera-consciousness"—be it on the part of actors or spectators—was widely seen as undesirable. American director Edward Sedgwick, for instance, told *Motion Picture News* that "the real basic difference between the European 'regisseur' and his American confrere is in the attitude toward the camera as a medium of story telling. 'The foreign director [...] never for a moment allows his audience to lose consciousness of the camera: the outstanding comment on any foreign picture is for its camera effects. The American director, on the other hand, strives to make his audience forget the camera. He tries to make his audience feel that it is in the midst of life in what is going on on the screen.'"[29] Sedgwick characterized a "neutral" stationary camera as imperceptible and therefore capable of conveying an impression of immediacy. For European commentators, in contrast, the opposite was the case. It was the mobile camera and its human

27 Georg Otto Stindt, "Die entfesselte Kamera," *Filmtechnik* 25 (1926): 491-494.

28 Lotar Holland, "Subjective Movement [1927]," in *The Promise of Cinema: German Film Theory, 1907-1933*, eds. Anton Kaes, Nicholas Baer, and Michael Cowan, trans. Michael Cowan, 512-515 (Berkeley, CA: University of California Press, 2016), 514.

29 "What's a Camera for?" *Motion Picture News* 7 (14 August 1926): 565.

touch that afforded a previously unattainable feeling of proximity to the events on screen. For Béla Balázs, the moving camera had "eliminated the spectator's position of fixed distance: a distance that hitherto has been an essential feature of the visual arts. [...] The camera takes my eye along with it. Into the very heart of the image. I see the world from within the filmic space. I am surrounded by the figures within the film and involved in the action, which I see from all sides."[30] As Balázs points out, the mobile camera facilitates an immersive experience and also promises to overcome the inherent rift between artwork and recipient, which, in cinema, the interpolation of a technical apparatus had further exacerbated. In the same vein, Murnau valued the illusion of immediate experience and therefore considered it "important that the mechanics of the cinema should not be interposed between the spectator and the picture."[31] From a European perspective, the purpose of the mobile camera was thus to minimize the affective distance between audience and onscreen events. The recourse to state-of-the-art technology to make the viewers forget the presence of the apparatus is of course a quintessential techno-romantic endeavour. The "unchained camera" thus can be said to epitomize Walter Benjamin's famous dictum: "The equipment-free aspect of reality has here become the height of artifice, and the vision of immediate reality the Blue Flower in the land of technology."[32] According to the German discourse, this penetration of reality by technology was mitigated by the humanization of technology in the cinema.

The more the camera moved, the more it began to resemble a living being. Karl Freund for instance pointed out: "Up till then the camera registered action with impartial exactness, perched on its tripod, fixed, and immovable as a god. With its descent it became human and we see the result in angles, rolling-shots, eyeline close-ups, etc., which all introduced the human element into mechanics and started to identify the audience with the feelings of the actors."[33] Lotar Holland took the notion of the

30 Béla Balázs, *Early Film Theory: Visible Man and The Spirit of Film*, ed. Erica Carter, trans. Rodney Livingstone (New York, NY/Oxford: Berghahn Books, 2010), 99.

31 F. W. Murnau, "The Ideal Picture Needs No Titles: By Its Very Nature the Art of the Screen Should Tell a Complete Story Pictorially (1928)," in *German Essays on Film*, eds. Richard W. McCormick and Alison Guenther-Pal, 66-68 (New York, NY: Continuum, 2004), 68.

32 Walter Benjamin, "The Work of Art in the Age of Its Technological Reproducibility [second version, 1935/1936]," in *Walter Benjamin Selected Writings*, vol. 3*: 1935-1938*, eds. Howard Eiland and Michael W. Jennings, trans. Edmund Jephcott and Harry Zohn, 101-133 (Cambridge, MA: Harvard University Press, 2002): 115.

33 Karl Freund, "Lines, Lights and Lenses," *Bioscope* (December 1928): 247.

humanized camera further. For him, the process of transferring emotions to the audience involved the camera acting as a physical proxy for the viewer: "[T]he motion of camera thus must correspond exactly to the movements that the spectator, if placed in the situation shown in the film, would execute with his own body, and more precisely with his head; in this way the camera lens can record only what the spectator himself would see in reality, be it as a result of conscious or unconscious movements."[34] According to Holland, the camera should simulate physical movements and postures, which in turn generate empathetic responses in the viewer. Holland's embodied camera draws on contemporary theories of aesthetic empathy, which describe the ensoulment of external phenomena through an instinctive objectification of the observer's own experiences.[35] They also relate to French acting theorist François Delsarte, who taught that body movements could generate emotional and psychological responses (rather than the other way around).[36] The techno-romantic concept of the humanized camera allowed contemporaries to imagine a machine as a means for embodying, transmitting, and evoking emotions.

However, likely due to the obvious limitations of the analogy between camera and human body, critics and practitioners including Vsevolod Pudovkin, Willi Haas, and F. W. Murnau conceptualized the apparatus more abstractly as a disembodied eye, an ideal observer who is "endowed with a ghostly ubiquity."[37] Murnau described the camera as "the eye of a person, through whose mind one is watching the events on the screen. [...] It must whirl and peep and move from place to place as swiftly as thought itself, when it is necessary to exaggerate for the audience the idea or emotion that is uppermost in the mind of the character."[38] While David

34 Holland, "Subjective Movement," in *The Promise of Cinema*, ed. Kaes et al., 513.

35 For comprehensive overviews of the concept of aesthetic empathy see *Einfühlung: Zur Geschichte und Gegenwart eines ästhetischen Konzeps*, eds. Robin Curtis and Gertrud Koch (Munich: Wilhelm Fink Verlag, 2009); Scott Curtis, *The Shape of Spectatorship: Art, Scene and Early Cinema in Germany* (New York, NY: Columbia University Press, 2015), 214-230; Karsten R. Stueber, *Rediscovering Empathy. Agency, Folk Psychology and the Human Sciences* (Cambridge, MA: MIT Press, 2006).

36 See Genevieve Stebbins, *Delsarte System of Expression* [1886] (New York, NY: Dance Horizons, 1977) and Ted Shawn, *Every Little Movement. A Book About François Delsarte, the Man and His Philosophy, His Science and Applied Aesthetics, the Application of This Science* (New York, NY: Dance Horizons, 1968).

37 Willi Haas, "Wird hier gearbeitet? Ein Besuch im Ufagelände Neubabelsberg," in *Willy Haas: Der Kritiker als Mitproduzent. Texte zum Film 1920-1933*, eds. Wolfgang Jacobsen, Karl Prümm, and Benno Wenz, 109-113 (Berlin: Edition Hentrich, 1991), 110; Vsevolod Pudovkin, *Film Technique and Film Acting*, trans. Ivor Montagu (London: Vision Press, 1954), 86-87.

38 F. W. Murnau, "Films of the Future," *McCall's Magazine* (September 1928): 27; 90.

Bordwell rightly points to the extent to which the notion of the camera as a disembodied eyewitness is deeply flawed, it is important to acknowledge the fundamental shift that this novel understanding of the role of the camera entailed.[39] In contrast to the hitherto common conception of the camera as detached, technical observer, the new outlook construed the camera techno-romantically as an emotionally invested, artistic consciousness, which conveys interpretations of events and mise-en-scène. The concept of the invisible observer also suggests that the customary bifurcation between "objective" and "subjective" camera stances is insufficient. Jean Mitry specified four distinct ways in which the camera can relate to the profilmic events. First, the "descriptive" or "objective image" does not aim to add meaning to the profilmic events. Second, the "personal image" evaluates and interprets the profilmic events, introducing figurative layers of meaning and revealing the artists' perspective. Third, in the case of the "semisubjective" or "associated image" the camera conveys a character's interiority while they are also present in the frame. Fourth, the "subjective" or "analytic image" relates a character's optical point of view. Finally, a combined category, the "total image," can motivate the enunciation of an artistic, figurative perspective by identifying it with a character's optical point of view.[40]

Mitry's categories are useful to differentiate between the ways in which the camera can convey subjectivity. At the same time, as Mitry suggests himself, the classification of shots is necessarily contingent upon interpretation. The implications of an individual shot are often ambiguous or even misleading. What is more, shots derive much of their signification from their context and the juxtaposition of different types by means of montage can create a vastly different overall effect. Finally, individual shots can adhere to multiple categories at the same time. Mitry's "total image" is such an instance. Semisubjective images often intersect with the creators' symbolic perspective. Indeed, German efforts to depict interiority cinematographically at times test the limits of Mitry's categories. A brief sequence from *Der letzte Mann* may illustrate this:

Consisting of three shots (Figure 6.3.-6.5.), it captures the moment in which the enamoured aunt discovers that the former doorman is now a bathroom attendant. We initially see him in close-up as he hesitantly pushes the

39 David Bordwell, *Narration in the Fiction Film* (Madison, WI: University of Wisconsin Press, 1985), 9-12.

40 Jean Mitry, *The Aesthetics and Psychology of the Cinema* [1963], trans. Christopher King (Bloomington, IN: Indiana University Press, 2000): 218.

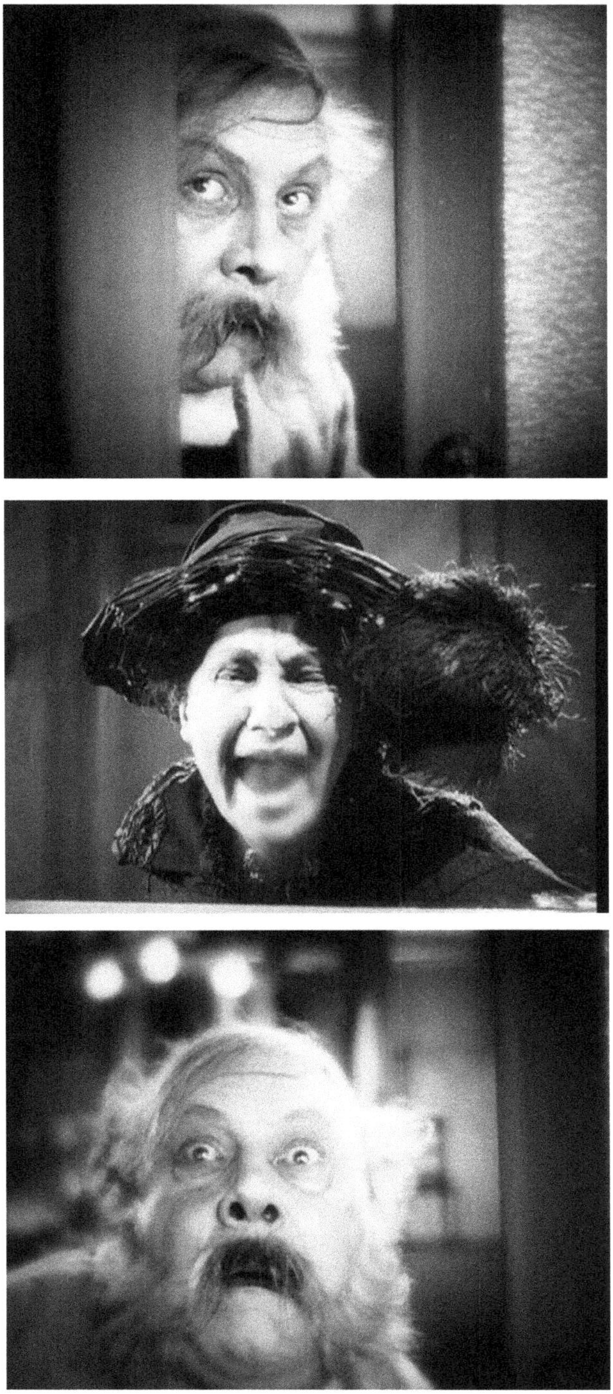

Fig. 6.3-6.5. Moving camera effect synthesizing characters' disparate interiorities and the perspective of a narrating entity in *Der letzte Mann* (1924).
Courtesy of Friedrich Wilhelm Murnau Foundation.

lavatory door ajar, his eyes turned upward towards the left, presumably the top of the stairs where the aunt stands. As if the movement of the bathroom door sets the camera in motion, the following shot swiftly tracks from a slightly low-angle medium close-up of the aunt to an extreme close up of her eyes. She stands directly behind the glass doors at the top of the stairs, looking downward towards the right, which establishes a double eyeline match with the former doorman. As the camera moves towards her, she begins to scream, her nose pressed against the glass pane of the door, which steams up from her breath. When the camera seems only a few inches from her face, Murnau cuts back to an out-of-focus extreme close-up of the former doorman. As he recoils into the bathroom, his body inverses the direction of the camera movement, until he is again in focus and framed in a slightly high-angle medium close-up. It is as if the impact of the camera on her face simultaneously also struck him, causing him to tumble backwards.[41]

Patrick Keating has highlighted the radical paradigm shift embodied in German filmmakers' choice to represent characters' subjective experiences from the outside, which he describes with Mitry as "semisubjective."[42] In a similar vein, Daniel Morgan has called attention to the fact that camera movements in Fritz Lang's work tend to be "object-defined," a concept that likewise corresponds to Mitry's "semisubjective" or "associated image." As Morgan describes, such shots are expressive of the object of the shot rather than the apparent catalyst of the motion, i.e., the person whose view seems to be represented.[43] This is certainly the case here, given that the tracking shot, which originates approximately at *his* optical point of view, captures *her* dismay at the discovery. At the same time, the shot's expressive scope goes beyond the aunt. It is equally shaped by the former doorman: The camera movement is initiated by *his* opening of the bathroom door, is a response to *his* secret getting out, and is mirrored in *his* recoiling. It is as if the moving camera embodies the awful truth escaping through the bathroom door as the former doorman opens it. It jumps in the aunt's face and her horrified reaction seems to hit the former doorman over the head as well. It realizes

41 In his recollections of the genesis of this sequence Karl Freund mentions a downward movement towards Jannings's face, realized by means of a telescoping aerial ladder. This shot does not appear in the surviving version of the film. See Freund, "Ein Filmkünstler," in *Caligari und Caligarismus*, 79-81.

42 Patrick Keating, *The Dynamic Frame: Camera Movement in Classical Hollywood* (New York NY: Columbia University Press, 2019), 24-26.

43 Daniel Morgan, "Beyond Destiny and Design: Camera Movement in Fritz Lang's German Films," in *A Companion to Fritz Lang*, ed. Joe McElhaney, 259-278 (Hoboken, NJ: Wiley-Blackwell, 2015), 260-269.

the shock and anguish that both characters experience in view of that truth. Thus, this tracking shot is not only object- *and* subject-defined *and* speaks to the psychology of both characters; it also reveals something about the relationship between them. Returning to Mitry's categories, the shot can be read as a "total image," which fuses symbolic commentary with a character's point of view. However, it also implies another, the pictured character's perception, adding a "semisubjective" dimension to the "subjective" and "personal" ones already embodied in the "total image." The sequence synthesizes the interiorities of two opposing characters and the enunciation of the perspective of a narrating entity, straining Mitry's categories. The unusual camera effect at the core of this sequence externalizes an affective nexus that would likely be impossible to communicate in another medium. It exemplifies how the techno-romantic paradigm of a "humanized" camera, i.e., a machine that establishes an affective relation to the profilmic events, gave rise to new forms of cinematic expressivity.

In contrast to German experiments that aimed at expanding the camera's ability to convey meaning, American cinematography appeared quite humdrum. European commentators were taken aback by the fact that, despite technologically advanced equipment and impeccable execution, American cinematography was outmoded in practices and stale in conceptions.[44] Comparing *Ben Hur: A Tale of the Christ* (MGM, 1925, dir. Fred Niblo) and *Faust,* for example, *Kinotechnik* pointed out: "One could just as well shoot a modern social drama, even starring Harold Lloyd, in the style of *Ben Hur.* The style of *Faust,* however, is only appropriate for an ancient tale and its style is almost the only one possible here."[45] The notion of "style" played a key role in German discourses about film during the 1920s.[46] "Having style" meant that a film had developed a distinctive manner of rendering abstract thought. When pursued by film-technological means, the quest for style must thus be seen as an essentially techno-romantic undertaking.

While many American observers attributed it to their technological superiority, the emotional potency of German films had less to do with sheer technical prowess than with filmmakers' capacity to employ technology for expressive purposes. Fritz Lang, when recounting an exchange with an American colleague, pointed out:

44 See for instance Eugen Schüfftan, "Mein Verfahren," *Film-Kurier* 24 (19 November 1926).

45 R.T., "Technische Filmkritiken," *Die Kinotechnik* 27 (1926): 637-638.

46 For contemporary discourses about style, see Kristina Köhler, "Nicht der *Stilfilm* also, sondern der *Filmstil* ist wichtig!" in *Filmstil: Perspektivierungen eines Begriffs*, eds. Julian Blunk, Tina Kaiser, Dietmar Kammerer, and Chris Wahl, 91-117 (Munich: edition text + kritik, 2016).

> I recently had the opportunity of showing an American film professional a few scenes from *Metropolis*, in which the beam of an electric flashlight illuminated the pursuit of a young girl through the catacombs of Metropolis. This beam of light pierces the hunted creature like an animal on its sharp pin, does not release her from its clutches, drives her relentlessly forward to the point of utter panic. It brought the amiable American to the confession: 'We can't do that!' Of course they can. But it does not occur to them. For them, the thing is characterless, inanimate, soulless."[47]

The impact of this scene, as Fritz Lang explained, is not actually the result of an extraordinary technical accomplishment, but must be ascribed to the fact that an idea is being visualized. The beam of light embodies an act of aggression against the girl. The example also calls attention to the fact that the ideas and emotions German filmmakers sought to portray were often quite abstract and by no means limited to illustrations of an individual character's thoughts and feelings. The German penchant for rendering abstracta, by definition devoid of physical referents, arises from a techno-romantic impulse to loosen cinema's ties to the immediate and concrete. The discrepancy between levels of abstraction in European and American films may explain in part why European critics and audiences were dissatisfied with what they perceived as the profuse literalness and banality of American films.

Techno-romantic filmmaking comes to Hollywood

There is ample evidence that, following the American releases of *Der letzte Mann* in January 1925 and *Varieté* in June 1926, German cinematography became an important source of inspiration for Hollywood. Erich Pommer's production strategy, which had emphasized export-oriented, artistic, and technically innovative filmmaking, had endowed the Ufa brand with extraordinary cachet in Hollywood. "Ufa" became a stand-in not only for German cinema per se, but also for "technical and artistic excellence, and [...] original and imaginative conceptions."[48] In 1926, Hollywood studios recruited close to a dozen directors from Berlin. They were expected to raise

47 Lang, "The Future of the Feature Film in Germany [1926]," in *The Weimar Republic Sourcebook*, eds. Kaes et al., 622-623. Translation modified.

48 Martha Gruening, "European Revolt against our Films," *New York Times* (31 October 1926): X6.

the artistic reputation of American films and model "European" methods. Cinematographer Hal Moor recalled: "we would try to emulate Mr. Murnau's *Sunrise*, and things like that; trying to do things that no one had ever done before; beyond what anybody had ever—we were trying to outdo Dupont on his *Variety*. And that was the purpose, the idea of the thing."[49]

One of Hollywood's first emphatically "European" productions was *Hotel Imperial*, which premiered in January 1927. Under the supervision of Erich Pommer, Mauritz Stiller directed Pola Negri in an adaptation of Hungarian author Lajos Biró's novel *Hotel Stadt Lemberg*. *Motion Picture News* reported: "That Paramount has recognized the special conditions and needs for the foreign film market is attested by the appointment of Somlyo and also in the acquisition of foreign stars, writers and directors, says Schulberg's statement. He pointed out the production 'Hotel Imperial' which is being produced with and by an almost exclusively foreign staff."[50] The filmmakers were not only expected to deliver a film that would be popular in Europe, but also showcase European-style camera effects, a task that Pommer and Stiller carried out to everybody's satisfaction. According to *Motion Picture Magazine* the film "accomplishes almost to perfection those photographic effects which directors have been striving for. And so simply and directly that one is unconscious of the freakishness of the camera work in one's absorption in the dramatic unfolding of the plot, with its rapid suspense-building."[51] In other words, Pommer and Stiller succeeded in demonstrating the application of state of the art camera effects not as an end in itself but for the purpose of heightening the audience's emotional involvement.

Much of Bert Glennon's and Akeley cinematographer E. Burton Steene's highly fluid cinematography in *Hotel Imperial* was achieved by means of specially constructed camera cranes.[52] For outdoor shots, a fixed-jib crane with a pulley-operated elevator hoist for the camera was mounted on a shooting platform on tracks, which allowed for simultaneous horizontal and vertical movements (Figure 6.6.). For shots of the massive composite indoor set, which consisted of eight connected, four-walled rooms, the camera was located on an electric platform hoist inside a wooden truss structure suspended from rails on the ceiling (Figure 6.7.).[53] Like the fixed-jib crane, this overhead

49 Hal Moor quoted in Richard Koszarski, "Hal Mohr's Cinematography," *Film Comment* (September 1974): 48-53; 52.

50 "Somlyo Joins Famous Players," *Motion Picture News* (4 September 1926): 830.

51 "The Picture Parade," *Motion Picture Magazine* 6 (January 1927): 61.

52 For Steene's involvement see "In Camerafornia," *American Cinematographer* 7 (October 1926): 10.

53 L'Estrange Fawcett, *Films: Facts and Forecasts* (London: Geoffrey Bles, 1927), 183-184.

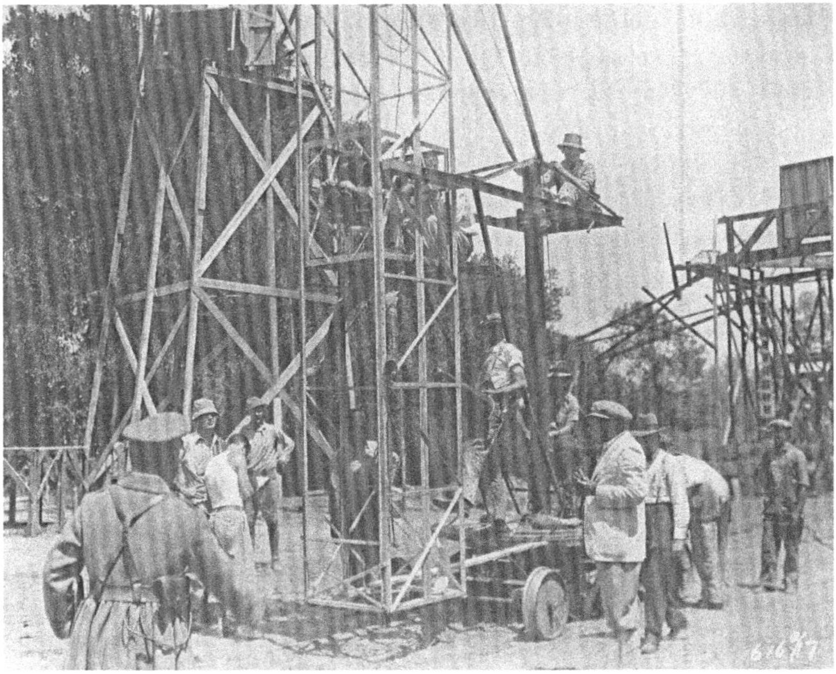

Fig. 6.6. Outdoor fixed-jib crane with pulley-operated elevator hoist for the camera on the set of *Hotel Imperial* (1927). Erich Pommer wearing a newsboy cap in the centre right.
Source: Deutsche Kinemathek.

crane permitted simultaneous movement on two axes, vertical craning and horizontal tracking without floor contact. Combining both types of motion, the overhead crane facilitated dramatic aerial effect shots like the moment when Almasy, from the gallery of the hotel lobby, spots the spy Petroff returning amidst the Russian festivities, and the camera swoops down to a medium shot of Petroff. It not only accounts for visual spectacle, but effectively establishes spatial and affective relationships between characters. It gives expression to Almasy's agitation in view of the spy's ill-boding arrival and underscores the man's danger to the Austrian war effort. Due to the distance and height the camera has to overcome to reach Petroff, the shot's dramatic impact is far greater than what a cut or rapid push-in might have been able to achieve.

The swooping camera in *Hotel Imperial* was evidently modelled on the "flying tone" in *Der letzte Mann*, where the camera ascends quickly to depict the sound of a horn rising through the courtyard.[54] Here, a camera

54 The same approach was presumably also adopted in another scene, which shows the neighbour women exchanging gossip across the courtyard.

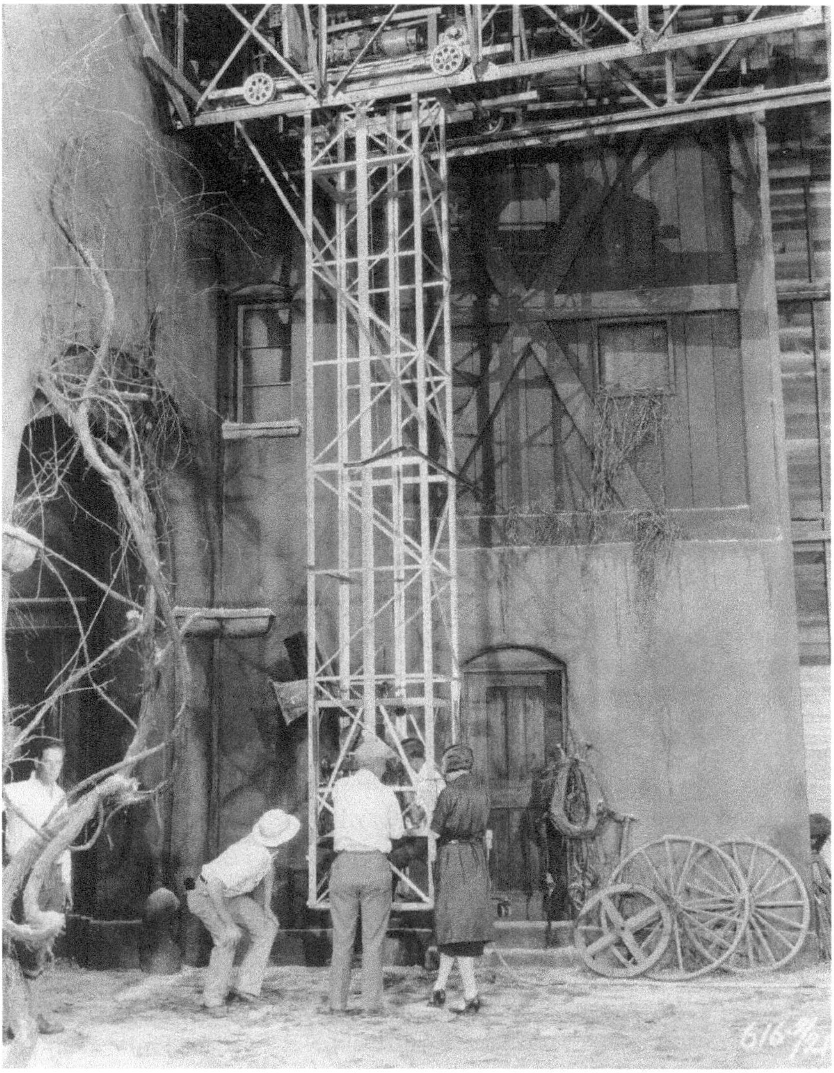

Fig. 6.7. Overhead crane on the indoor set of *Hotel Imperial* (1927). The camera is located on an electric platform hoist inside a wooden truss structure.
Source: Deutsche Kinemathek.

platform was suspended from wooden trusses or steel cables in the tenement courtyard, which allowed for the diagonal aerial movement.[55] This rigging method resembles the cable cam systems that since the 1980s have become ubiquitous in filmmaking and event photography. Art director Robert Herlth recalled, "we [...] fitted Jannings' house with a sort of hoist, with the camera

55 Production photographs document the use of cables and wooden trusses respectively.

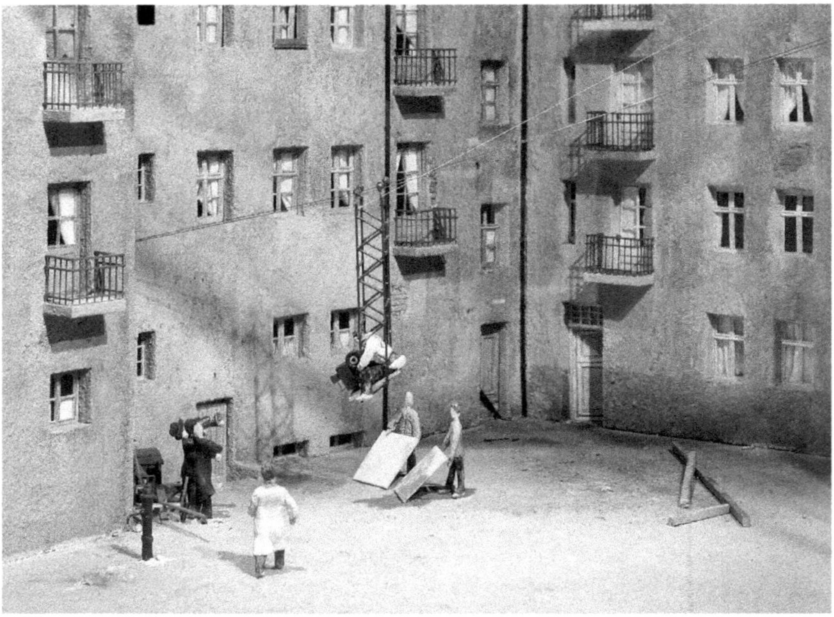

Fig. 6.8. Model depicting the creation of a "flying tone" in *Der letzte Mann* (1924). The camera platform is suspended from steel cables to create diagonal aerial movement.
Courtesy of Filmmuseum Düsseldorf.

in a basket on rails, so that it could slide downwards for about 20 metres."[56] In order to suggest the sound moving upward, the camera was turned upside down to reverse the motion.[57] The "flying tone" became an instant sensation. Because this effect relied on special production processes and resulted in special impressions, it epitomizes a cinematographic special effect. Such spectacular visual feats typically emerged from techno-romantic efforts to visualize the intangible by means of technology by filmmakers under Pommer's aegis, which were subsequently replicated, perhaps per order, by German filmmakers in Hollywood, and then reiterated countless times by their American colleagues.

As many scholars have argued, *Sunrise* marked a watershed moment in Hollywood history. In the mid-1920s, all major Hollywood studios avidly imported well-known European actors and directors. William Fox scored a

56 Herlth, "With Murnau on the Set," in Eisner, *Murnau*, 65.

57 See Klaus Kreimeier, *The Ufa Story: A History of Germany's Greatest Film Company, 1918-1945*, trans. Robert and Rita Kimber (Berkeley, CA: University of California Press, 1999), 107.

coup with the appointment of the "German genius."[58] F. W. Murnau embodied precisely what many Hollywood studios were looking for: cultural prestige, technical ingenuity, and artistic sensibility. Murnau was given far-reaching control over *Sunrise*, which was intended as a showcase production. According to Richard Koszarski, "directors, designers, and cameramen from all over the industry trooped through Murnau's sets and stages. *Sunrise* became a demonstration project of German film techniques applied with Hollywood budgetary resources."[59] In addition, Fox encouraged the studio's leading directors, Frank Borzage, John Ford, and Raoul Walsh, to study Murnau's methods. As a result, Murnau's film-technological experimentation shaped the studio style for years to come.[60] In 1929, the *New York Times* reported: "The Fox West Coast Studios have become known for their facility in handling unusual problems in lighting and photography. Largely through the influence of F. W. Murnau, all manner of camera tracks, trolleys, hoists, cranes, platforms, swinging booms, aerial camera cars and other technical apparatus has been installed to film the scene from the spectator's or an effective viewpoint. The space under the roof of every stage on the Fox lot is a network of cordage and planking."[61]

Among the mechanisms popularized by European filmmakers was the type of overhead crane that Pommer and Stiller used on *Hotel Imperial*. Cinematographer Charles Rosher first encountered it in Berlin on the set of *Faust* and employed it repeatedly in *Sunrise*.[62] According to Rosher, Murnau was "very fond of having the characters followed by the camera. [...] And whenever possible we used a camera suspended somewhere above the actors."[63] Although the pensile, continuously moving camera had already been on display in *Hotel Imperial*, its use in *Sunrise* drew particular attention to cinematography's role in conveying the intangible.[64] Critics on both sides

58 See for instance "Murnau Achievements Praised by Brisbane," *Exhibitors Herald* (17 July 1926): 30. See the title of Janet Bergstrom's essay, "The German Genius," in *Friedrich Wilhelm Murnau: Ein Melancholiker des Films*, eds. Hans Helmut Prinzler, Karin Messlinger, and Vera Thomas, 79-94 (Berlin: Bertz, 2003).

59 Richard Koszarski, *An Evening's Entertainment: The Age of the Silent Feature Picture*, 1915-1928 (Berkeley, CA: University of California Press, 1990), 254-255.

60 For a discussion of the German influence on the camerawork in John Ford's *Four Sons* see Keating, *The Dynamic Frame*, 50-51.

61 "Camera Work for Screen," *New York Times* (2 June 1929): X5.

62 Charles Rosher quoted in Kevin Brownlow, *The Parade's Gone By...* (Berkeley, CA: University of California Press, 1968), 232.

63 Charles Rosher quoted in Lotte H. Eisner, *Murnau* (London: Secker & Warburg, 164), 79.

64 *Hotel Imperial* concluded principal photography before *Sunrise* went into production. See "In Camerafornia," *American Cinematographer* 6 (September 1926): 9; "Murnau Starts Production

of the Atlantic attributed *Sunrise*'s lyrical qualities specifically to its tracking shots. American critic Louise Bogan called the camera in *Sunrise* "a retina reflecting an intricately flowing world," while her German colleague Fritz Rosenfeld observed that "people and objects glide past like melodies."[65] Further, one of Murnau's stated goals for his engagement in Hollywood was to demonstrate the visualization of thought. Months before Murnau's arrival, his assistant Hermann Bing announced to the American trade press: "To put this subjective thought on the screen, to open up the mind, the heart, the soul—that is Murnau's aim."[66] In this context it is important to underscore, however, that the notion of "subjective thought" is not restricted to the perspective of a specific character. Instead, it extends to abstract realms of figurative commentary.

Sunrise's sophisticated approach to subjectivity is on full display in its most frequently discussed scene, where a moving camera accompanies Ansass through the marshes to his rendezvous with the City Woman. As Struss recalled, "the camera was suspended on a small platform from a curved track on the roof of the stage as there was no way to manipulate the camera even with an exceptionally long boom arm to reach the end of the scene, through some willows overlooking the marshes below. [...] I couldn't even sit down to manipulate the camera and it was very difficult to see the small ¾ x one inch image which was inverted and reversed for a full three minutes; the total length of the scene, or one hundred and eighty feet of film."[67] Patrick Keating has called attention to the interweaving of external and internal representation within this sequence.[68] At the outset of the shot, the camera follows Ansass, thereby assuming, as Keating astutely points out, a semisubjective mode in Mitry's sense. The camera adopts Ansass's slow but relentless pace, which makes palpable both his hesitation and compulsion. When the camera begins to take alternate routes, however, it asserts its position as an independent observer and at the same time symbolically comments on the fact that he is going astray. In contrast to Ansass, who has to duck under branches and climb over a fence, the disembodied eye of

on First American Film," *Fox Folks* 10 (October 1926): 22.

65 Louise Bogan, "True to the Medium," *New Republic* (26 October 1927): 263; Fritz Rosenfeld, "Sonnenaufgang," *Arbeiter-Zeitung* 36 (5 February 1928): 19.

66 "Murnau to Produce for Fox Film Corp," *Motion Picture World* (2 January 1926): 69. See also: "Plans Efficiency Plus Art," *Motion Picture News* (2 January 1926): 27-28.

67 Karl Struss, "Recollections of Karl Struss, Interviewed by John Dorr," in *An Oral History of the Motion Picture in America*, Appendix II: *Sunrise* (Los Angeles, CA: University of California, 1969), 1-2.

68 Keating, *The Dynamic Frame*, 30-31.

the camera is not impeded by physical obstacles. At one point, seemingly unaware of the camera's presence, Ansass walks directly towards the lens. As he passes in close proximity, the camera starts panning and moves through a grove, and branches obscure the lens. For a while, it is unclear whether the camera has now taken over Ansass's optical point of view. The camera cuts through a curtain of branches and halts in the clearing, where a picturesque tableau is on display: the vamp under the full moon in front of a body of water. She slowly turns as if sensing a presence, but her manner remains blasé: What we see is in fact not Ansass's viewpoint. The shot's sudden stasis accentuates Ansass's inexplicable delay and attunes us to the vamp's impatience. Throughout the sequence, in the same vein as in *Der letzte Mann*, the camera takes on the position of a compassionate onlooker who has access to and at times embodies the characters' affective states, while also providing symbolic commentary. For Robin Wood, the marsh sequence represents "a perfect balancing of objective and subjective. [...] It is a perfect example of direct emotional communication."[69] While Wood's point is well taken, we must recognize that in this example "subjective" refers to different subjects and perspectives and that "objective" is anything but neutral. Informed by the techno-romantic paradigm, the German tradition of artistic filmmaking conceived of the camera as an omnipresent, empathetic consciousness that materializes the meaning and emotional character of the drama. It externalizes the characters' inner life while simultaneously conveying the drama's affective attitude towards them.

Moving camera feats

The most visible consequence of the American reception of films like *Der letzte Mann* and *Varieté*, and the presence of European filmmakers in Hollywood, was a sudden increase in striking camera effects in American films, in particular moving camera feats and extreme camera angles. As camera effects corroborated a film's pedigree as a "photographic picture," i.e., a film that valued style,[70] they appeared notably in high-budget productions that projected cultural sophistication and were oriented towards international markets. Hardly surprisingly, spectacular camera effects were most prevalent in the work of European filmmakers like E. A. Dupont, Ludwig Berger, Paul Leni, Michael Curtiz, and F. W. Murnau. Joseph von Sternberg, Erich von

69 Robin Wood, "Murnau's Midnight and Sunrise," *Film Comment* 3 (May/June 1976): 4-19; 13.
70 Foster Goss, "The Editors' Lens," *American Cinematographer* (August 1924): 10.

Stroheim, and Paul Fejos, who built careers in America in part based on a "European" public image, also cultivated such devices. Striking camera effects were likewise employed early on by John Ford, William Wellman, Frank Borzage, Raoul Walsh, Clarence Brown, Fred Niblo, and King Vidor at Paramount, Fox, and MGM, studios that pursued ambitious, international production strategies.

When examining instances of spectacular moving camera effects in late 1920s American cinema, it becomes clear that many bear close resemblance to prototypes in *Der letzte Mann* and *Varieté*. Further, it is evident that particular effects became quickly associated with stereotypical applications. For example, a banquet scene regularly involved a suspended camera track over the table. This can for instance be seen in *The Eagle* (Art Finance Corporation, 1925, dir. Clarence Brown), *The Temptress* (MGM, 1926, dir. Fred Niblo), *The Last Command* (Paramount, 1928, dir. Josef von Sternberg), and *The Last Performance* (Universal, 1929, dir. Paul Fejos).[71] Many of the era's moving camera feats fall into two categories: On the one hand, camera plunges, swings, and spins that frequently visualized a character's (imagined) body movement in addition to mental shock or, interestingly, romantic bliss. On the other hand, horizontal, vertical, and diagonal aerial movements, which often acquired symbolic meaning.

To begin with the former group, fast vertical camera drops were likely modelled on Boss's fantasy about dropping his rival Artinelli during their performance in *Varieté* and appeared in similar contexts in *You Never Know Women* (Paramount, 1926, dir. William A. Wellman), *The Man Who Laughs* (Universal, 1928, dir. Paul Leni) or *Polly of the Circus* (MGM, 1932, dir. Alfred Santell). In *The Bat Whispers* (Joseph M. Schenck Productions/United Artists, 1930, dir. Roland West), the camera, imitating a bat, dives from a (miniature) skyscraper. *Varieté* also features a distinctive swinging-camera effect for which the camera moved in unison with Boss on a trapeze, illustrating his disorientation in view of intrusive, violent thoughts. American films used this effect for spectacular stunts like Toberchik's rocking-barrel act in *You Never Know Women* on the one hand and to depict couples' love-struck bliss on outdoor swings, as evident in *Wings* (Paramount, 1927, dir. William Wellman, *The Enemy* (MGM 1927, dir. Fred Niblo), *Four Sons* (Fox, 1928, dir. John Ford), and *The Florodora Girl* (MGM, 1930, dir. Harry Beaumont) on the other. Spinning-camera effects were likely inspired by *Der letzte Mann*, where a gyrating platform carrying

71 In *The Eagle*, the camera was not actually suspended. Clarence Brown describes how the effect was achieved in Brownlow, *The Parade's Gone By…*, 146.

both Jannings and the camera visualized his drunken dizziness. A slower replica is found in *The Garden of Eden* (Feature Productions/United Artists, 1928, dir. Lewis Milestone), where seated lovers shift against the background, again depicting a moment of enamoured rapture. The effect was also frequently used for dancing scenes as in *Hotel Imperial*, where the camera accompanies a woman rapidly pivoting around a central axis and in *Quality Street* (MGM, 1927, dir. Sidney Franklin), where camera and cameraman were suspended from a rotating jib in the centre of the dancing platform.[72]

In contrast to camera plunges, swings, and spins, which tended to visualize a character's experience, horizontal or vertical aerial tracks characterized a certain milieu. As mentioned above, horizontal aerial push-ins and pull-backs across obstacles were commonly employed in scenes involving banquet- and other large tables.[73] In addition, the technique was often employed in conjunction with miniatures, an application that may trace back to *Faust*, where it served to illustrate the all-pervading force of Gretchen's scream for help and Faust's fabulous flight on Mephisto's cloak over the mountains. The latter likely inspired the impressive opening shot of *Tempest* (Feature Productions/United Artists, 1928, dirs. Sam Taylor and Lewis Milestone), where the camera moves over the rooftops of a miniature town, descends and, covered by an invisible cut, arrives in the midst of a full-scale garrison set.[74] Horizontal aerial tracking over a miniature also rendered possible the famous shot in *The Crowd* (MGM, 1928, dir. King Vidor), where the camera seems to ascend a skyscraper and penetrate one of its windows. Vertical aerial movements, likely popularized by the acclaimed hotel elevator shot opening *Der letzte Mann*, were typically employed to follow characters climbing stairs as in *The Monkey Talks* (Fox, 1927, dir. Raoul Walsh), *The Godless Girl* (C. B. DeMille Productions/Pathé Exchange, 1928, dir. Cecil B. DeMille) and incidentally also Alfred Hitchcock's *Blackmail* (British International Pictures, 1929). It attracted particular interest in *7th Heaven* (Fox, 1927, dir. Frank Borzage), where, according to Frank Borzage, the vertical camera movement symbolized

72 For the creation of the moving camera effect in *Quality Street*, see illustration in *Amateur Movie Makers* 2 (February 1928): 88.

73 Non-banquet examples can be found in *The Red Dance* (Fox, 1928, dir. Raoul Walsh), *Hula* (Paramount, 1927, dir. Victor Fleming), and *Wings*. For an illustration of the overhead crane used in *Wings*, see illustration in Brownlow, *The Parade's Gone By…*, 170. Wellman claimed that this was "the first big boom shot," which is evidently not the case. See Brownlow, *The Parade's Gone By…*, 168.

74 See also the opening of *Broadway* (Universal, 1929, dir. Paul Fejos) and the camera's embodiment of the hero's hypnotic gaze in *Svengali* (Warner Bros., 1931, dir. Archie Mayo).

the protagonists' ascent to the "seventh heaven."[75] Rather than serving as an embellishment, it epitomizes a central motif, As Jakob Isak Nielsen points out, "the theme of verticality is written into the very dramatic structure of the film."[76]

Diagonal camera movements between a closer framing at ground level and a wider one at an elevated position (or, more rarely, vice versa) constitute a virtuoso aerial effect that presumably traces back to the "flying tone" in *Der letzte Mann*. It is capable of establishing symbolic relationships between characters and their environment to great effect. In particular, as *Movie Makers* observed, the technique can convey a strong sense of immediacy: In the climactic courtroom scene in *Night Watch* (First National, 1928, dir. Alexander Korda) Carl Struss and Alvin Knechtel's camera "moves down on a cable and passes closely over the witness on the stand, coming to a close-up of the Justice speaking, thus bringing the audience, as spectator, right into the centre of the events taking place."[77] Similar instances appear in *The Last Performance* (Universal, 1929, dir. Paul Fejos), where a suspended camera moves diagonally from the auditorium to the murder scene on stage, capturing the universal commotion in the theatre following the corpse's discovery.[78] In *The Crowd,* the camera, located on an overhead crane, travels diagonally from a high angle extreme long shot of the office hall through the air towards John Sims's table, which eventually fills out the frame completely.[79] Like the elevator shot in *7th Heaven*, the shot captures the film's core theme: the relationship between individual and homogenous mass, which characterizes modern life. At the end of the film, another diagonal aerial shot reverses the direction of the first. Starting on a medium shot of John and Mary in the theatre, the

75 For the camera elevator used in *The Godless Girl*, see illustrations in *Amateur Movie Makers* 3 (March 1928): 160; and *Screenland* 3 (July 1928): 24. For Borzage's thoughts on the vertical camera movement in *7th Heaven* see Hervé Dumont, *Frank Borzage: The Life and Films of a Hollywood Romantic*, trans. Jonathan Kaplansky (Jefferson, NC: McFarland, 2006), 116.

76 Jakob Isak Nielsen, "Camera Movement in Narrative Cinema – Towards a Taxonomy of Functions," PhD. Dissertation, University of Aarhus, 2008, 132. Vertical camera movements are also used for gags in *The Kid Brother* (Harold Lloyd/Paramount, 1927, dirs. Ted Wilde and Harold Lloyd) and *The Cameraman* (MGM, 1928, dir. Edward Sedgwick).

77 Arthur L. Gale, "Critical Focusing," in *Movie Makers* 3 (March 1929): 156.

78 Hal Mohr describes how the effect in *The Last Performance* was achieved in Koszarski, "Hal Mohr's Cinematography," 50. In another scene, magician Eric hypnotizes audience members, which is illustrated by alternating diagonal camera movements created by means of a steeply inclined ramp.

79 King Vidor describes how the sequence was achieved in King Vidor, *A Tree is a Tree* (Hollywood, CA: S. French, 1989), 151.

camera simultaneously pulls back and cranes up until the protagonists have disappeared amidst a sea of audience members, all of whom are also rocking with laughter.

A look at some of the more widely employed moving camera feats highlights the direct impact of *Der letzte Mann* and *Varieté* on 1920s Hollywood filmmaking. It also shows how these initially unconventional techniques became associated with specific applications and how they moved quite seamlessly into sound film. This, in turn, calls into question conventional views of the transitional era as characterized by static, uninteresting visuals. Most importantly, however, attention to these novel devices reveals the extent to which American filmmakers embraced film technology as a means for articulating abstract ideas. The techno-romantic paradigm thus facilitated the vast expansion of American cinema's expressive capacities.

The approach introduced by European filmmakers necessitated new types of equipment, and "German technique" therefore became a major factor in launching the era of camera cranes in Hollywood. Scholars have often undervalued the role of cranes in studio-era filmmaking, yet jib cranes in particular served as a principal camera mount for decades.[80] By the early 1930s, "crane shots, which afford greater flexibility of movement, have to a considerable extent supplanted 'dolly' work on the stage floor."[81] It has been wrongly claimed, even by experts like William Stull, that the crane that Paul Fejos designed for *Broadway* (Universal, 1929) was "probably the first one."[82] One reason for this misconception is a disregard for overhead crane systems, which became widely used in Hollywood following their debuts in *Hotel Imperial* and *Sunrise*.[83] In addition, the freestanding cranes used on *Hotel Imperial*, *4 Devils* (Fox, 1928, dir. F. W. Murnau), and *Street*

80 For MGM's aluminium crane see illustrations in *Cinematographic Annual*, vol. 2 (1931): 354. For a crane in use at RKO see illustration in *Photoplay* 3 (February 1932): 110; For cranes in use at Paramount see illustrations in *Photoplay* 4 (March 1932): 69; *Silver Screen* 5 (March 1932) 69; *Photoplay* 5 (April 1932): 72-73; *Photoplay* 3 (August 1932): 54; "New Apparatus," *Journal of the Society of Motion Picture Engineers* (January 1933): 85-86.

81 Society of Motion Picture Engineers Progress Committee, "Progress in the Motion Picture Industry," *Journal of the Society of Motion Picture Engineers* 2 (August 1932): 123.

82 William Stull, "Development of Mobile Camera Carriages and Cranes," *American Cinematographer* (19 May 1933): 12-13; 36-37.

83 They were for instance used on *The Girl in the Pullman* (DeMille, 1927, dir. Earle C. Kenton), *Wings* (Paramount, 1927, dir. William Wellman), *The Crowd* (MGM, 1928, dir. King Vidor), *The Last Command* (Paramount, 1928, dir. Josef von Sternberg), *The Letter* (Paramount, 1929, dir. Jean de Limur), *Sunnyside Up* (Fox, 1929, dir. David Butler), *Her Man* (Pathé Exchange, 1930, dir. Tay Garnett), or *Kiki* (Feature Productions/United Artists, 1931, dir. Sam Taylor). See for instance "De Mille Studios Get Traveling Camera Crane," *Exhibitors Herald* 13 (10 September 1927): 7.

Angel (Fox, 1928, dir. Frank Borzage) were not widely publicized.[84] Among the few sources of information about the crane Murnau had custom-built for *4 Devils* is a description by Margaret Reed in *Picture Play*:

> It is a great iron mechanism, secured by joists in the ground underneath the floor of the stage, and reaches to the ceiling. At the top, a crane—similar to those seen on derricks—extends downward at an angle. On a small platform at the bottom of the crane, the camera is secured—with place for the camera man and Mr. Murnau. During the action, the main ironwork slowly forming the pivot, swings the crane around and around, encompassing the entire circle as often as is required. It moves by means of a steam engine at the base. On occasion, for some particular shot, the camera platform is slid upward along the crane by a pulley, or the crane itself moves suddenly at a right angle—conveying the camera to a close-up of some one in the audience. Description of this machine is difficult for such as I—who can't tell a valve from a differential—but anyway it is a colossal, demon thing christened "the go-devil," and will be the means of many new and breathtaking camera tricks.[85]

Compared to Murnau's "go-devil," Paul Fejos's "trick camera crane" was faster, more mobile, and had a greater range.[86] The 28-ton, electric mechanism consisted of a ten-metre long steel girder, which could swing 360 degrees horizontally in sixteen seconds and 180 degrees vertically and raise the camera over fifteen metres above ground.[87] It was mounted on a turret on a six-wheeled truck chassis, which could be moved up to 25 miles per hour. The device cost at least $35,000 and necessitated the construction of a new stage at Universal with an eighteen-metre ceiling and a concrete floor that could carry its weight.[88] As cinematographer Hal Mohr recalled,

84 A somewhat inconclusive photograph was published in *Photoplay Magazine* 4 (September 1928): 104. For the production history of *4 Devils* see Janet Bergstrom, "Murnau in America: Chronicle of Lost Films, (*4 Devils, City Girl*)," in *Looking Past the Screen: Case Studies in American Film History and Method*, eds. Jon Lewis and Eric Smoodin, 303-352 (Durham NC: Duke University Press, 2007).

85 Margaret Reid, "After Sunrise Comes What?" *Picture Play* (October 1928): 22-24, 116; 116.

86 Neville Reay, "The Stroller," *Picture Play* 3 (May 1929): 55; "Crane Mechanical Marvel," *American Cinematographer* (May 1929): 14-15; 14.

87 Stull, "Development of Mobile Camera Carriages and Cranes," 12. See also "Gigantic Camera Crane Built for Broadway," *Universal Weekly* 25 (26 January 1929): 10.

88 For the $35,000 figure see Stull, "Development of Mobile Camera Carriages and Cranes," 12. Mohr claims that the price was $50,000 and according to *Universal Weekly* it was $75,000. See Leonard Maltin, "Interview with Hal Mohr," in *The Art of the Cinematographer*, 75-94 (New York

flaunting the prowess of this crane was a prime concern for the filmmakers during the production of *Broadway*, an approach that resulted, as *American Cinematographer* marvelled, in "indescribable" photographic effects.[89] For Paul Fejos, however, state-of-the-art technical equipment was not an end in itself. Like many Europeans, Fejos approached filmmaking from a decidedly techno-romantic perspective. His first American film *The Last Moment* (Samuel Freedman-Edward M. Spitz, 1928), which is considered lost, featured a plethora of visual effects and attempted, as Fejos wrote, "to project visually the vital images of life experiences-incidents that crowd through the brain before death."[90] In *Broadway*, he reportedly intended to capture the "soul" of the nightclub and the city at the centre of the film. As the *New York Times* explained, the camera work was supposed to "be symbolic not only of one of Broadway's nocturnal resorts, not only of Broadway itself, but a compendium of New York."[91]

During the second half of the 1920s, mobile camerawork became pervasive in Hollywood, but initially did not draw much criticism. It was not until the early sound period, specifically following the release of Rouben Mamoulian's *Applause* (Paramount, 1929), that daring moving camera feats were publicly called into question. During the 1920s, the public discourse focused on another device: unusual camera angles.

"Trick angles" and cinematic immersion

Following the U.S. premiere of *Varieté* in June 1926, American cinema witnessed an unprecedented proliferation of unusual camera angles. Harry Potamkin recalled: "When Dupont's *Variety* came to America, it hurtled all the Hollywood shopmen into angles."[92] The device and its perceived sudden ubiquity triggered a heated public debate in the American film industry, which still resonated in the 1940s. "Angles" became a catchword epitomizing

NY: Dover Publications, 1978), 86; and "Gigantic Camera Crane Built for Broadway," *Universal Weekly* 25 (26 January 1929): 10. For Universal's stage 12, see "The American Film Institute and American Society of Cinematographers Seminar with Hal Mohr, ASC, Held March 24, 1973" (Beverly Hills, CA: Center for Advanced Film Studies, 1978): 6-7. Some of these widely differing figures may include the cost of the stage.

89 Maltin, "Interview with Hal Mohr," in *The Art of the Cinematographer*, 86; "Crane Mechanical Marvel," *American Cinematographer* (May 1929): 14.

90 Fejos quoted in Mordaunt Hall, "Prize Play Pictured," *New York Times* (18 March 1928): 125.

91 "Broadway on Screen," *New York Times* (19 May 1929): X4.

92 Harry Alan Potamkin, "Phases of Cinema Unity," *Close-Up* 5 (May 1929): 27-38; 28.

alien influences, which, depending on the perspective, were construed as enriching or harmful, pretentious or innovative.

As I have indicated above, the conception of extreme camera angles as trick effects was not unusual at the time. For instance, Guido Seeber included them in his book on trick effects and Hollywood spoke of "trick angles."[93] Extreme camera angles constitute a case in point for a device that may produce striking visual impressions, involve unusual set-ups, and could therefore be perceived as a special effect. Barry Salt has shown that high and low angle shots already appear in American and Danish films in the early 1910s, sometimes as a result of the topography on set or to represent a character's optical point of view. By the second half of the 1910s, high and low angles were also employed to impart figurative meaning. For instance, a high angle accentuates the wretchedness of the captured Black "renegade" Union soldier in *The Birth of a Nation* (David W. Griffith Corp./Epoch Producing Corporation, 1915, dir. D. W. Griffith) and a low angle the megalomania of Homunculus in *Die Rache des Homunculus* (Deutsche Bioscope, 1916, dir. Otto Rippert). However, cameras were still fairly distant from their objects, which limited their affective impact. Since around 1920, increasingly closer and more extreme low and high angle shots gained currency in continental European cinema. Simultaneously, artists like László Moholy-Nagy, Albert Renger-Patzsch, and Alexander Rodchenko explored their expressive and defamiliarizing potency in still photography, and in the second half of the 1920s, Soviet filmmakers carried the device to unprecedented extremes. In Hollywood, unusual angles only appeared occasionally during the early 1920s most nobly in Erich von Stroheim's *Greed* (MGM, 1924), where unusual angles not only represent characters' optical points of view and portray their interiority from "the outside," they also reflect an artistic, figurative perspective, a narrator's point of view.

Patrick Keating alleges that the ubiquitous term "angles" in 1920s Hollywood also denoted "camera movement." Although imprecise language was certainly widespread in industry discourse, critics *did* differentiate between camera angles and camera movement and even the article Keating references in support of his claim states: "The most eminent of the Europeans in the contribution of camera engineering to the cinema is Karl Freund. Freund established the mobile camera in *The Last Laugh*;

93 Guido Seeber, *Der praktische Kameramann*, vol. 2, *Der Trickfilm in seinen grundsätzlichen Möglichkeiten* [1927] (Frankfurt am Main: Deutsches Filmmuseum, 1979), 159-162; See for instance Carroll Graham, "Ah, Those Were The Days," *Picture Play* 6 (August 1929): 24-25.

the camera angle in *Variety*."[94] The American preoccupation with unusual camera angles amounts to more than simply terminological imprecision, and the fact that public attention was initially almost exclusively focused on perspective instead of movement deserves to be taken seriously. To a far greater extent than most other types of camerawork, extreme camera angles constitute an overtly self-conscious cinematic device. Their sudden proliferation questioned the role of form in filmmaking. Specifically, the controversy about unusual camera angles revolved around the relationship between cinematography and narrative, or, as *Motion Picture News* casually formulated, "what's a camera for?"[95]

Tom Gunning provides a helpful framework for conceptualizing the narrative function of cinematic devices. Following Gérard Genette, he differentiates between four aspects of a narrative. First, there is the story, i.e., the content of the narrative, independent of the medium in which it is told, second, the narrative discourse, which refers to the signifiers that communicate the content, third, the act of narrating, which describes the ways in which an implied audience is addressed and finally narrativization, which signifies the transformation of photographed reality into a narrative.[96] As Gunning points out, because films consist of a series of photographic representational signs, mimesis (showing) necessarily dominates over diegesis (telling). The act of narrating involves the creation of a "hierarchy of narratively important elements within a mass of contingent details."[97] The entity that organizes the narrative discourse, the narrator, must not be mistaken for an actual person. Instead, the narrator is "constructed by the reader or spectator [and] addresses and affects him or her through specific devices."[98]

From an American perspective, German-made films exhibited unusual forms of audience address and blatantly foregrounded their narrative discourse. Almost bewildered, a reviewer of *Varieté* for instance observed in the *Los Angeles Times*: "Dupont places a secondary import on the story in this picture. Treatment, the manner in which the story is told, is what he has aimed at especially."[99] Until the second half of the 1920s, the narrative

94 Harry Alan Potamkin, "Knights of the Camera," *International Photographer* 8 (September 1930): 14-16; 14.

95 "What's a Camera for?" *Motion Picture News* (14 August 1926): 565.

96 Tom Gunning, *D. W. Griffith and the Origins of American Narrative Film* (Urbana, IL: University of Illinois Press, 1991), 17-18.

97 Gunning, *D. W. Griffith and the Origins of American Narrative Film*, 17.

98 Gunning, *D. W. Griffith and the Origins of American Narrative Film*, 24.

99 "Bizarre Effects Attained in German Cinema," *Los Angeles Times* (13 July 1926): C25.

discourse of American cinema largely relied on mise-en-scène and editing, while cinematography played a comparatively minor role. Calling attention to the camera disrupted the illusion of direct, mimetic presentation, which, according to Harry Potamkin, American films rarely relinquished: "The mind of the American film, regarding both content and approach is literal; and that is why the American film is still rudimentary and why no one here has extended or even equalled the compositions of Griffith or logically developed the innovations of Billy Bitzer. [...] Nor has there yet been very much learned by director or cinematographer of Hollywood about the angle as a principle, rather than as an effect device."[100] For many American observers, in contrast, mimetic presentation constituted the prerequisite for audience absorption.

As we have seen, commentators such as director Edward Sedgwick characterized a stationary camera at eyelevel as imperceptible and therefore capable of suggesting immediacy. For European filmmakers and critics, on the other hand, precisely the opposite was the case. Paul Fejos for instance declared: "A picture can be made in such a way that the camera, instead of merely witnessing the action, enters into it."[101] In other words, Europeans proposed a different conception of cinematic immersion. Immersion, as narratologist Marie-Laure Ryan puts it, is the feeling of being "integrated in an art world."[102] Both American and European filmmakers aimed for immersive experiences, yet their views on how to achieve them diverged considerably. For many Hollywood commentators, conspicuous cinematic devices like unusual camera angles had disturbingly alienating, anti-illusionary effects. At the same time, it was precisely those techniques that rendered German films so emotionally effective, as for instance William Cameron Menzies noted: "In these German productions, American directors and cameramen learned that a few feet of film shot from an abnormal angle would heighten the dramatic effect of certain types of sequences more than a whole reel of normally fore-shortened pictures."[103]

Paradoxically, as the example of unusual camera angles indicates, conspicuous cinematic devices can have immersive *and* distancing

100 Harry Alan Potamkin, "Knights of the Camera," *International Photographer* 8 (September 1930): 14-16; 14.

101 Paul Fejos, "Illusion on the Screen," *New York Times* (26 May 1929): X3.

102 Marie-Laure Ryan, *Narrative as Virtual Reality 2: Revisiting Immersion and Interactivity in Literature and Electronic Media* (Baltimore, MD: Johns Hopkins University Press, 2015), 13.

103 William Cameron Menzies, "The Layout of Bulldog Drummond," *Creative Art* (October 1929): 729-734; 729.

effects. They disrupt the illusion of direct presentation, yet by introducing additional semantic levels, they can facilitate more complex modes of audience engagement. According to Rudolf Arnheim, the defamiliarizing quality of an extreme camera angle extends viewers' interest beyond the object itself and encourages them to deliberate how formal qualities reflect on the object's nature.[104] Arnheim's view is aligned with Viktor Shklovsky, for whom making objects unfamiliar and forms difficult was similarly geared towards intensifying recipients' feelings.[105] Thus, although conspicuous cinematic devices undermine the semblance of unmediated presence, they do not necessarily eradicate the feeling of immersion. Moniek M. Kuipers et al. have noted that narrative aesthetic absorption is not limited to content-based "absorption in the story world," but can expand to formal and stylistic elements, a response that they call "artifact absorption."[106] However, in contrast to experimental art, which is routinely geared towards artifact absorption, mainstream films typically proffer absorption simultaneously on multiple levels. Indeed, spectators' awareness of the filmic image *as image* can deepen their attachment to the fictional world onscreen. In German mainstream films of the 1920s, conspicuous formal devices aimed at communicating—be it symbolically or viscerally—the narrator's perspective or characters' mental life, while the viewers' resulting emotional involvement curbed the distancing effects of these techniques. European filmmakers were well aware that this type of emotional immersion was unfamiliar to American observers and saw it as a major competitive advantage. Murnau explained: "So much of the subjective is left out of motion pictures; there is so much violent action or situation, that I thought it would be an innovation, worth emulation in America, to try to make audiences FEEL WITH the main character, rather than at what is happening to him [sic]."[107]

German filmmakers' techno-romantic commitment to evoking affective responses through film technology also relates to contemporary theories of aesthetic empathy (*Einfühlung*), which, as Robin Curtis and Christiane

104 Rudolf Arnheim, *Film as Art* (Berkeley, CA: University of California Press, 1957), 44-45.

105 See Victor Shklovsky, "Art as Technique [1916]," in *Russian Formalist Criticism: Four Essays*, eds. Lee T. Lemon and Marion J. Reiss, 3-24 (Lincoln, NE: University of Nebraska Press, 1965), 12.

106 Moniek M. Kuipers, Frank Hakemulder, Katalin Bálint, Miruna Doicaru, and Ed Tan, "Towards a New Understanding of Absorbing Reading Experiences," in *Narrative Absorption*, eds. Frank Hakemulder, Moniek M. Kuipers, Ed S. Tan, Katalin Bálint, and Miruna M. Doicaru, 29-48 (Amsterdam: John Benjamins Publishing Company, 2017).

107 F. W. Murnau, "Real 'Motion' Pictures," *Film Daily* (7 June 1925): 21.

Voss have argued, corresponds to the more recent concept of immersion.[108] According to this discourse, empathy describes a process of psychological resonance, during which sensually encountered objects are imbued with spiritual content. When for instance perceiving a storm as "wrathful," the observer constructs an external phenomenon as the symbolic representation of human interiority.[109] Theories of aesthetic empathy share with the techno-romantic perspective an urge to substantiate the preponderance of the human spirit over the material. Their focus on the relationship between symbolic form and emotional response had a formative influence on German film culture, shaping both the process of filmmaking and that of reception. Filmmakers sought to create art worlds that embodied mental experiences. Empathic encounters with the filmic universe thus not only involved a film's characters, but also its spaces, atmospheres, ideas, and forms. Art director Walter Reimann explained, "moods and expressions are constructed. If a scene requires an exhaustingly endless country road, the task is not to build a country road, but its exhausting and endless quality."[110]

Of course, in cinema, the empathic immediacy between object and observer is mediated by a technological apparatus. In his third version of "The Work of Art in the Age of its Technological Reproducibility," Walter Benjamin thus claimes that the audience does not actually empathize with the objects onscreen, but rather with the camera. He ascribes this to the fact that the spectators adopt the camera's position, its point of view. Benjamin's understanding has affinities to Lotar Holland's embodied camera discussed above, which conveys mental attitudes by simulating physical postures.[111] In the unusual camera angle the camera's viewpoint itself acquires symbolic meaning. Indeed, as Béla Balázs suggests in *The Spirit of Film*, perspective is never neutral and necessarily involves an

108 Robin Curtis, "Immersion und Einfühlung: Zwischen Repräsentationalität und Materialität bewegter Bilder," *montage A/V* (February 2008): 89-107; 97, 99; Christiane Voss, "Fiktionale Immersion," *montage A/V* (February 2008): 69-86; 72.

109 Joachim Paech, "Disposition der Einfühlung: Anmerkungen zum Einfluss der Einfühlungs-Ästhetik des 19. Jahrhunders auf die Theorie des Kinofilms," in *Der Film in der Geschichte*, eds. Knut Hickethier, Eggo Müller and Rainer Rother, 106-121 (Baden-Baden: Edition Sigma, 1997), 109.

110 Walter Reimann, "Filmarchitektur – Filmarchitekt?! [1925]," in *Werkstatt Film: Selbstverständnis und Visionen von Filmleuten der zwanziger Jahre*, eds. Rolf Aurich and Wolfgang Jacobsen, 111-115 (Munich: edition text + kritik, 1998), 115.

111 Walter Benjamin, "The Work of Art in the Age of its Technological Reproducibility [third version, 1936/1939]," in *Walter Benjamin: Selected Writings*, vol. 4: *1938-1940*, eds. Howard Eiland and Michael W. Jennings, trans. Harry Zohn and Edmund Jephcott, 251-284 (Cambridge, MA: Harvard University Press, 2002), 260.

emotional connection: "Every image implies a camera point of view, every point of view implies a relation. And that relation is more than merely spatial. Every view of the world contains a world view. Similarly, every camera set-up points to an inner human attitude. For nothing is more subjective than objectivity. Once captured in an image, every impression becomes an expression, whether by design or not."[112] Like Benjamin, Balázs implies that the viewers inevitably adopt the "inner human attitude" presented by the camera's perspective. In contrast to Benjamin, however, he does not dismiss the possibility of empathy with onscreen events. Rather, it is the camera's perspective that facilitates medium-specific, subjective expression. Rudolf Arnheim shared this view. As he expounds in *Film as Art*, unusual camera angles constitute a quintessentially medium-specific formal device precisely because they do not tamper with the basic functionality of the cinematic apparatus. Instead, the camera records the profilmic events without interference. When rendered from an unusual perspective, objects appear as they do in real life and yet they become interpretable.[113] For both Balázs and Arnheim, unusual camera angles demonstrated the possibility of conveying ideas and emotions in a photographic medium. For Balázs, they exemplified "the camera's subjective gaze" and for Arnheim an instance in which "the conditions of representation [...] mould the object."[114] Unusual camera angles testify to the human ascendancy over the mechanically produced image, thus reconciling, in line with techno-romantic tenets, the material and immaterial, apparatus and artist.

In the German context, filmmakers and critics rarely felt obliged to comment specifically on unusual camera angles, presumably because they were perceived as an innate aspect of expressive cinematography. In American cinema, in contrast, they were controversial, in part because adequate applications first had to be negotiated. For instance, unusual camera angles became a common technique to make inconsequential sequences visually more appealing. As *Picture Play* argued, "unexpected camera angles are an important factor in keeping the spectator interested."[115] However, many commentators disapproved of formal ornamentation without narrative motivation. *Motion Picture World* for instance asserted: "While the benefit to American motion pictures from the general

112 Balázs, *Early Film Theory*, 112-113.
113 Arnheim, *Film as Art*, 35-58.
114 Balázs, *Early Film Theory*, 112-113; Arnheim, *Film as Art*, 57.
115 Norbert Lusk, "The Screen in Review," *Picture Play* 5 (January 1929): 70-73; 96, 72.

'loosening up' of camera technique and greater attention to placement as a result of a few German pictures is undoubtedly great, there is a grave danger in the tendency to regard 'camera angles' as an end in themselves. They are not. When they enhance the telling of the story, they belong. When they do not, they are an inexcusable evil."[116] The discussion about unusual camera angles thus served in part to strengthen the consensus about the primacy of the narrative and the mandate to conceal narrational activity. At the same time, it prompted Hollywood observers to reflect on the role of narrative discourse and audience address. Strong narrative motivation could serve to mitigate the distancing effects of formal devices and make them acceptable. Screenwriter Lillian Case Russell wrote: "Study the pictures that employ these unusual effects and you will find they are intelligently used to illustrate a mood to add value to a situation. But in the mad rush to imitate many producers and directors are losing sight of their story in planning new gymnastics for their cameraman."[117]

That angles should never be used ornamentally became somewhat of a mantra in Hollywood. European directors, and particularly F. W. Murnau and Paul Fejos, were repeatedly asked about angles. The *New York Times* reported that Fejos was "annoyed at the phrase, 'European trick shots.' He has been accused of using the method and insists that 'camera angles' are put in for a reason, not per se."[118] In the same vein, Murnau felt obliged to emphasize: "There should not be any such thing as an 'interesting camera angle,' he explained. 'As an angle in itself, it does not mean a thing. If it does not intensify the dramatic action of the scene, it is even a detriment.'"[119] More so than any other émigré director, Murnau set the pattern for how camera techniques could be employed to communicate ideas and emotions, and thus heighten the film's affective impact. *The Los Angeles Times* marvelled: "Murnau has made a life study of camera angles. His study has introduced a new spirit into the film industry. [...] If Murnau's approach is analyzed, it will be discovered that there is a certain psychological element permeating his work. It is not so much camera angles as some particular idea that each scene embodies, making for harmony, rhythm of theme and pictorial beauty."[120]

116 "Angles," *Motion Picture News* (25 December 1926): 2407.
117 L. Case Russell, "Baa, Baa, Black Sheep," *Hollywood Topics* 12 (19 February 1927): 18.
118 "Broadway Screen Tips," *New York Times* (14 October 1928): X4.
119 "Director Tendered Luncheon," *Los Angeles Times* (4 March 1927): A9.
120 "Murnau's Study of Camera Noted," *Los Angeles Times* (19 June 1929): A8.

The fervent discussion about extreme camera angles in 1920s Hollywood raised major conceptual questions about the role of the camera in filmic narration. Possibly for the first time, the public discourse paid heed to cinematography as an expressive tool. The German films that startled American observers were characterized by a much stronger emphasis on "telling" and little interest in simulating direct presentation. Particularly impressive was their "psychological" nature, their ability to convey mental aspects and their emotional effectiveness. This was the result of the camera, rather than serving as an imperceptible chronicler of profilmic events, acting as the principal narrator and communicating figurative and implicit levels of meaning. Frequently, the camera's perspective symbolized human interiority without it being necessarily tied to individual characters.

A sizable segment of Hollywood practitioners and critics did not welcome this approach. They either objected to conspicuous camera work because they saw mimetic presentation as necessary for audience absorption or they interpreted camera effects as pretentious and gratuitous. Not without reason, it seems, given that camera effects were widely employed for purely decorative purposes. As late as 1943, Karl Freund lamented: "Too often the director seems to ignore the camera entirely, except as a machine for recording his scenes in the most literal way. Or else he may go to the other extreme and become too exaggeratedly conscious of the camera that he overloads his picture with 'arty' tricks of focus and angles—copied, perhaps, from something he saw in somebody else's picture—which play no real part in advancing the story or in building up dramatic moods."[121] As Freund's comments suggest, the same attitudes toward camerawork persisted for decades. Simultaneously, however, cinematography became a vital tool for cinematic narration, for conveying mental and emotional meaning.

Indeed, the influx of German films and European directors changed American cinema. Scholars have deliberated extensively on the connections between German cinema of the 1920s and American film noir of the 1940s, a historical lineage that, as Thomas Elsaesser has rightly pointed out, "does not ultimately make sense."[122] In the second half of the 1920s, however, German filmmaking practices had a palpable impact on Hollywood. Devices that

121 Karl Freund, "An Open Letter to Arthur Edeson, ASC and Michael Curtiz [March 1943]," *American Cinematographer* (August 1994): 75-77.

122 Thomas Elsaesser, *Weimar Cinema and After: Germany's Historical Imaginary* (London: Routledge, 2000), 420.

were widely associated with German-style filmmaking such as moving camera feats and extreme camera angles quickly gained currency. However the ultimate significance of "German technique" was not the popularization of any specific method. Instead, what affected American cinema the most was the German understanding of the camera as cinema's core creative tool. German-style films demonstrated how cinematography and special effects could be used for dramatic ends. Adhering to techno-romantic tenets, German filmmakers sought to harness the camera, as the embodiment of cinema's machine nature, for medium-specific forms of creativity and expressivity and enunciated a film's intentionality as a mark of its artistic nature. Neither the quest for film art nor techno-romantic zeal was as pervasive in Hollywood, yet under the impression of German films the camera's role in shaping narrative discourse and audience address expanded noticeably. American filmmakers closely studied German efforts to represent ideas and feelings and to immerse viewers emotionally. German filmmakers, in turn, were well aware of the significance their approach to cinematography had had on the evolution of film art. Cinematographer Carl Hoffmann stated: "Despite the vast development which the film industry has experienced in America, the fact remains, that the development of the artistic film-photography originated in Germany, just as Germany deserves a lion's share of the credit for raising the artistic level of film production in general."[123] While the core concerns of American cinema—entertainment taking priority over art, invisible style over ostentatious narrative discourse, and action over interiority—continued to dominate, German cinema's focus on immaterial and abstract content and its emphasis on creativity and experimentation left clear marks. Approaches to filmmaking based in techno-romantic thought prompted Hollywood to pay increased attention to the expressive potential of technical tools—with lasting impact on American cinema.

Bibliography

"Angles," *Motion Picture News* (25 December 1926): 2407.
Rudolf Arnheim, *Film as Art* (Berkeley, CA: University of California Press, 1957).
Béla Balázs, *Early Film Theory: Visible Man and The Spirit of Film*, ed. Erica Carter, trans. Rodney Livingstone (New York, NY/Oxford: Berghahn Books, 2010).

123 Carl Hoffmann, "Camera Problems," *Close Up* 1 (July 1929): 29-31.

Randolph Bartlett, "German Film Revision Upheld as Needed Here," *New York Times* (13 March 1927): X3.
Walter Benjamin, "The Work of Art in the Age of Its Technological Reproducibility [second version, 1935/1936]," in *Walter Benjamin Selected Writings, vol. 3, 1935-1938*, eds. Howard Eiland and Michael W. Jennings, trans. Edmund Jephcott and Harry Zohn, 101-133 (Cambridge, MA: Harvard University Press, 2002).
Walter Benjamin, "The Work of Art in the Age of its Technological Reproducibility [third version, 1936/1939]," in *Walter Benjamin: Selected Writings*, vol. 4: *1938-1940*, eds. Howard Eiland and Michael W. Jennings, trans. Harry Zohn and Edmund Jephcott, 251-284 (Cambridge, MA: Harvard University Press, 2002).
Janet Bergstrom, "Murnau in America: Chronicle of Lost Films (*4 Devils, City Girl*)," in *Looking Past the Screen: Case Studies in American Film History and Method*, eds. Jon Lewis and Eric Smoodin, 303-352 (Durham, NC: Duke University Press, 2007).
Janet Bergstrom, "The German Genius," in *Friedrich Wilhelm Murnau: Ein Melancholiker des Films*, eds. Hans Helmut Prinzler, Karin Messlinger, and Vera Thomas, 79-94 (Berlin: Bertz, 2003).
"Bizarre Effects Attained in German Cinema," *Los Angeles Times* (13 July 1926): C25.
Louise Bogan, "True to the Medium," *New Republic* (26 October 1927): 263.
David Bordwell, *Narration in the Fiction Film* (Madison, WI: University of Wisconsin Press, 1985).
Thomas Brandlmeier, "'Entfesseltes Sehen:' Von der *Ausfahrt der Chinakrieger* bis *Varieté*," in *Die Perfektionierung des Scheins: Das Kino der Weimarer Republik im Kontext der Künste*, ed. Harro Segeberg, 143-157 (Munich: Wilhelm Fink Verlag, 2000).
"Broadway on Screen," *New York Times* (19 May 1929): X4.
"Broadway Screen Tips," *New York Times* (14 October 1928): X4.
Kevin Brownlow, *The Parade's Gone By...* (Berkeley, CA: University of California Press, 1968).
"Bell & Howell Pioneer Professional Standard," *American Cinematographer* 1 (January 1927): 12.
"Camera Work for Screen," *New York Times* (2 June 1929): X5.
"Case of Jonathan Drew," *Variety* (13 June 1928): 12.
"Crane Mechanical Marvel," *American Cinematographer* (May 1929): 14-15.
Robin Curtis, "Immersion und Einfühlung: Zwischen Repräsentationalität und Materialität bewegter Bilder," *montage A/V* (February 2008): 89-107.
Robin Curtis and Gertrud Koch, eds., *Einfühlung: Zur Geschichte und Gegenwart eines ästhetischen Konzeps* (Munich: Wilhelm Fink Verlag, 2009).
Scott Curtis, *The Shape of Spectatorship: Art, Scene and Early Cinema in Germany* (New York, NY: Columbia University Press, 2015).

"De Mille Studios Get Traveling Camera Crane," *Exhibitors Herald* 13 (10 September 1927): 7.
"Die entfesselte Kamera," *Ufa-Magazin* (1927): 216-217.
"Director Tendered Luncheon," *Los Angeles Times* (4 March 1927): A9.
Hervé Dumont, *Frank Borzage: The Life and Films of a Hollywood Romantic*, trans. Jonathan Kaplansky (Jefferson, NC: McFarland, 2006).
Ewald André Dupont and Fritz Podehl, *Wie ein Film geschrieben wird und wie man ihn verwertet* (Berlin: Kühn, 1926).
Lotte Eisner, *Murnau* [1964], trans. Martin Secker (London: Secker & Warburg, 1973).
Thomas Elsaesser, *Weimar Cinema and After: Germany's Historical Imaginary* (London: Routledge, 2000).
Paul Fejos, "Illusion on the Screen," *New York Times* (26 May 1929): X3.
Karl Freund, "An Open Letter to Arthur Edeson, ASC and Michael Curtiz [March 1943]," *American Cinematographer* (August 1994): 75-77.
Karl Freund, "Ein Filmkünstler [1947]," in *Caligari und Caligarismus*, ed. Walter Kaul, 79-81 (Berlin: Deutsche Kinemathek, 1970).
Karl Freund, "Lines, Lights and Lenses," *Bioscope* (December 1928): 247.
Arthur L. Gale, "Critical Focusing," *Movie Makers* 3 (March 1929): 156.
"Gigantic Camera Crane Built for Broadway," *Universal Weekly* 25 (26 January 1929): 10.
Foster Goss, "The Editors' Lens," *American Cinematographer* (August 1924): 10.
Carroll Graham, "Ah, Those Were The Days," *Picture Play* 6 (August 1929): 24-25.
Carl Louis Gregory, "An Optical Printer for Trick Work," *Transactions of SMPE* 34 (1928): 419-426.
Carl Louis Gregory, *Motion Picture Photography* (New York, NY: Falk Publishing, 1927).
Martha Gruening, "European Revolt against our Films," *New York Times* (31 October 1926): X6.
Tom Gunning, *D. W. Griffith and the Origins of American Narrative Film* (Urbana, IL: University of Illinois Press, 1991).
Willi Haas, "Wird hier gearbeitet? Ein Besuch im Ufagelände Neubabelsberg," in *Willy Haas: Der Kritiker als Mitproduzent. Texte zum Film 1920-1933*, eds. Wolfgang Jacobsen, Karl Prümm, and Benno Wenz, 109-113 (Berlin: Edition Hentrich, 1991).
Hal Hall, "Cinematographers and Directors Meet Discuss Camera Trucking Problems," *American Cinematographer* (August 1932): 10.
Mordaunt Hall, "Prize Play Pictured," *New York Times* (18 March 1928): 125.
Robert Herlth, "With Murnau on the Set," in Lotte Eisner, *Murnau* [1964], trans. Martin Secker, 59-70 (London: Secker & Warburg, 1973).
Carl Hoffmann, "Camera Problems," *Close Up* 1 (July 1929): 29-31.

Lotar Holland, "Subjective Movement [1927]," in *The Promise of Cinema: German Film Theory, 1907-1933*, eds. Anton Kaes, Nicholas Baer, and Michael Cowan, trans. Michael Cowan, 512-515 (Berkeley, CA: University of California Press, 2016).

"In Camerafornia," *American Cinematographer* 6 (September 1926): 9.

"In Camerafornia," *American Cinematographer* 7 (October 1926): 10.

Patrick Keating, *Hollywood Lighting from the Silent Era to Film Noir* (New York, NY: Columbia University Press, 2009).

Patrick Keating, *The Dynamic Frame: Camera Movement in Classical Hollywood* (New York, NY: Columbia University Press, 2019).

Kristina Köhler, "Nicht der *Stilfilm* also, sondern der *Filmstil* ist wichtig!" in *Filmstil: Perspektivierungen eines Begriffs*, eds. Julian Blunk, Tina Kaiser, Dietmar Kammerer, and Chris Wahl, 91-117 (Munich: edition text + kritik, 2016).

Richard Koszarski, *An Evening's Entertainment: The Age of the Silent Feature Picture*, 1915-1928 (Berkeley, CA: University of California Press, 1990).

Richard Koszarski, "Hal Mohr's Cinematography," *Film Comment* (September 1974): 48-53.

Klaus Kreimeier, *The Ufa Story: A History of Germany's Greatest Film Company, 1918-1945*, trans. Robert and Rita Kimber (Berkeley, CA: University of California Press, 1999).

Moniek M. Kuipers, Frank Hakemulder, Katalin Bálint, Miruna Doicaru, and Ed Tan, "Towards a New Understanding of Absorbing Reading Experiences," in *Narrative Absorption*, eds. Frank Hakemulder, Moniek M. Kuipers, Ed S. Tan, Katalin Bálint, and Miruna M. Doicaru, 29-48 (Amsterdam: John Benjamins Publishing Company, 2017).

Fritz Lang, "The Future of the Feature Film in Germany [1926]," in *The Weimar Republic Sourcebook*, eds. Anton Kaes, Martin Jay, and Edward Dimendberg, trans. Don Reneau, 622-623 (Berkeley, CA: University of California Press, 1994).

James Mitchell Leisen, "Some Problems of the Art Director," *Transactions of S.M.P.E.* 33 (1928): 71-80.

Norbert Lusk, "The Screen in Review," *Picture Play* 5 (January 1929): 70-73; 96.

Leonard Maltin, "Interview with Hal Mohr," in *The Art of the Cinematographer*, 75-94 (New York, NY: Dover Publications, 1978).

William Cameron Menzies, "The Layout of Bulldog Drummond," *Creative Art* (October 1929): 729-734.

G. A. Mincenty, "Echo aus Amerika," *Filmtechnik* 12 (1925): 254.

Jean Mitry, *The Aesthetics and Psychology of the Cinema* [1963], trans. Christopher King (Bloomington, IN: Indiana University Press, 2000).

Dan Morgan, "Beyond Destiny and Design: Camera Movement in Fritz Lang's German Films," in *A Companion to Fritz Lang*, ed. Joe McElhaney, 259-278 (Hoboken, NJ: Wiley-Blackwell, 2015).

F. W. Murnau, "...der frei im Raum zu bewegende Aufnahmeapparat [1924]," in Fred Gehler and Ulrich Kasten, *Friedrich Wilhelm Murnau* (Berlin: Henschelverlag Kunst und Gesellschaft, 1990).
F. W. Murnau, "Films of the Future," *McCall's Magazine* (September 1928): 27; 90.
F. W. Murnau, "Real 'Motion' Pictures," *Film Daily* (7 June 1925): 21.
F. W. Murnau, "The Ideal Picture Needs No Titles: By Its Very Nature the Art of the Screen Should Tell a Complete Story Pictorially (1928)," in *German Essays on Film*, eds. Richard W. McCormick and Alison Guenther-Pal, 66-68 (New York, NY: Continuum, 2004).
"Murnau Achievements Praised by Brisbane," *Exhibitors Herald* (17 July 1926): 30.
"Murnau's Study of Camera Noted," *Los Angeles Times* (19 June 1929): A8.
"Murnau Starts Production on First American Film," *Fox Folks* 10 (October 1926): 22.
"Murnau to Produce for Fox Film Corp," *Motion Picture World* (2 January 1926): 69.
Jakob Isak Nielsen, "Camera Movement in Narrative Cinema – Towards a Taxonomy of Functions," PhD. Dissertation, University of Aarhus, 2008.
"New Apparatus," *Journal of the Society of Motion Picture Engineers* (January 1933): 85-86.
"Now They'll Get 'Acting Continuity,'" *Photoplay* (June 1925): 53.
Joachim Paech, "Disposition der Einfühlung: Anmerkungen zum Einfluss der Einfühlungs-Ästhetik des 19. Jahrhunders auf die Theorie des Kinofilms," in *Der Film in der Geschichte*, eds. Knut Hickethier, Eggo Müller, and Rainer Rother, 106-121 (Baden-Baden: Edition Sigma, 1997).
"Paul Fejos Praised For Broadway Direction," *Universal Weekly* (15 June 1929): 25.
"Plans Efficiency Plus Art," *Motion Picture News* (2 January 1926): 27-28.
Harry Alan Potamkin, "Knights of the Camera," *International Photographer* 8 (September 1930): 14-16.
Harry Alan Potamkin, "Phases of Cinema Unity," *Close-Up* 5 (May 1929): 27-38.
Karl Prümm, "Das schwebende Auge: Zur Genese der bewegten Kamera," in *Die Medien und ihre Technik: Theorien, Modelle, Geschichte*, ed. Harro Segeberg, 235-256 (Marburg: Schüren-Verlag, 2004).
Vsevolod Pudovkin, *Film Technique and Film Acting*, trans. Ivor Montagu (London: Vision Press, 1954).
Neville Reay, "The Stroller," *Picture Play* 3 (May 1929): 55.
Margaret Reid, "After Sunrise Comes What?" *Picture Play* (October 1928): 22-24, 116.
Walter Reimann, "Filmarchitektur – Filmarchitekt?! [1925]," in *Werkstatt Film: Selbstverständnis und Visionen von Filmleuten der zwanziger Jahre*, eds. Rolf Aurich and Wolfgang Jacobsen, 111-115 (Munich: edition text + kritik, 1998).
Fritz Rosenfeld, "Sonnenaufgang," *Arbeiter-Zeitung* 36 (5 February 1928): 19.
R. T., "Technische Filmkritiken," *Die Kinotechnik* 27 (1926): 637-638.
L. Case Russell, "Baa, Baa, Black Sheep," *Hollywood Topics* 12 (19 February 1927): 18.

Marie-Laure Ryan, *Narrative as Virtual Reality 2: Revisiting Immersion and Interactivity in Literature and Electronic Media* (Baltimore, MD: Johns Hopkins University Press, 2015).
Barry Salt, *Film Style and Technology: History and Analysis* (London: Starwood, 1983).
Eugen Schüfftan, "Mein Verfahren," *Film-Kurier* 24 (19 November 1926).
Guido Seeber, *Der praktische Kameramann*, vol. 2, *Der Trickfilm in seinen grundsätzlichen Möglichkeiten* [1927] (Frankfurt am Main: Deutsches Filmmuseum, 1979).
Ted Shawn, *Every Little Movement. A Book About François Delsarte, the Man and His Philosophy, His Science and Applied Aesthetics, the Application of This Science* (New York, NY: Dance Horizons, 1968).
Victor Shklovsky, "Art as Technique [1916]," in *Russian Formalist Criticism: Four Essays*, eds. Lee T. Lemon and Marion J. Reiss, 3-24 (Lincoln, NE: University of Nebraska Press, 1965).
Society of Motion Picture Engineers Progress Committee, "Progress in the Motion Picture Industry," *Journal of the Society of Motion Picture Engineers* 2 (August 1932): 123.
"Somlyo Joins Famous Players," *Motion Picture News* (4 September 1926): 830.
Genevieve Stebbins, *Delsarte System of Expression* [1886] (New York, NY: Dance Horizons, 1977).
Georg Otto Stindt, "Die entfesselte Kamera," *Filmtechnik* 25 (1926): 491-494.
Karl Struss, "Recollections of Karl Struss, Interviewed by John Dorr," in *An Oral History of the Motion Picture in America*, Appendix II: *Sunrise* (Los Angeles, CA: University of California, 1969).
Karsten R. Stueber, *Rediscovering Empathy. Agency, Folk Psychology and the Human Sciences* (Cambridge, MA: MIT Press, 2006).
William Stull, "Development of Mobile Camera Carriages and Cranes," *American Cinematographer* (19 May 1933): 12-13; 36-37.
"The American Film Institute and American Society of Cinematographers Seminar with Hal Mohr, ASC, Held March 24, 1973" (Beverly Hills, CA: Center for Advanced Film Studies, 1978).
"The Picture Parade," *Motion Picture Magazine* 6 (January 1927): 61.
Kristin Thompson, *Herr Lubitsch Goes to Hollywood: German and American Film after World War I* (Amsterdam: Amsterdam University Press, 2005).
Julie Turnock, *Plastic Reality: Special Effects, Technology and the Emergence of 1970s Blockbuster Aesthetics* (New York, NY: Columbia University Press, 2015).
King Vidor, *A Tree is a Tree* (Hollywood, CA: S. French, 1989).
Christiane Voss, "Fiktionale Immersion," *montage A/V* (February 2008): 69-86.
Richard Watts, "Some Additional Remarks on that New German Photoplay," *New York Herald Tribune* (11 July 1926): E3.
"What's a Camera for?" *Motion Picture News* 7 (14 August 1926): 565.

Robin Wood, "Murnau's Midnight and Sunrise," *Film Comment* 3 (May/June 1976): 4-19.

Conclusion: Techno-Romantic Cinema from the Silent to the Digital Era

Abstract
Techno-romantic thought, which construes machine technology as a means to reach beyond material reality, is still with us today. It is reflected in the vogue of speculative fiction in contemporary moving image media, which has been made possible by radical advances in digital visual effects. Computer-generated imagery has brought into reach the fully malleable photograph, a dream that epitomizes a major triumph of the human mind over outside reality and thus an essentially techno-romantic fantasy. The same ambition already animated German silent filmmakers, who saw special effects as a way to shape mechanically produced images. Their use of trick technology for conveying thoughts and emotions gives rise to a new research area: special/visual effects as artistic tools.

Keywords: CGI, digital cinema, visual effects, expressive special effects

Techno-romantic thought has been with us for at least two hundred fifty years. Every wave of technological innovation during the industrial, technological, and most recently the digital revolution has engendered new iterations of the same paradoxical response: Technological progress calls forth hubristic fantasies of unlimited, quasi-magical powers, while also triggering deep-seated anxieties about subjugation, dehumanization, and annihilation. This tension manifests, for instance, in Mary Shelley's *Frankenstein* (1818), where the eponymous hero's command of fantastical technology allows him to overcome death and assume the godlike status of a "modern Prometheus." At the same time, he renders his creature a victim to cruel oppression and thus turns it into a lethal danger. Techno-romantic perspectives help articulate and mitigate fears about modernity. Rendering it possible to savour the fascinating aspects of technology while grappling with its threats, techno-romantic thought is neither inherently technophile

Loew, Katharina, *Special Effects and German Silent Film: Techno-Romantic Cinema*.
Taylor & Francis Group, 2021
DOI 10.5117/9789463725231_CONCL

nor technophobic, but can be found in the context of technological utopias like Ian M. Banks's *The Culture* series (1987-2012) as well as dystopias like the Wachowskis' *The Matrix* franchise (1999-).

The techno-romantic paradigm construes modern technology as a force that can reach beyond the limits of physical reality. This, on the one hand, magnifies technology's powers and thus its perils, but also envisions it as a means to safeguard the human soul against modern reification. Associated with emotional, imaginary, or spiritual qualities, technology then corroborates the primacy of human consciousness and facilitates retreats to an immaterial realm. As Mark Coeckelbergh has suggested, "As children of twentieth-century romantic counterculture, we seamlessly fuse technology and romanticism. Engaging with our many screens and smart gadgets and shielded from the inner, machine-like workings of our devices (developed by science), we try to satisfy our romantic desires and are more like Rousseau, Novalis, or Wordsworth than we think."[1]

Offering access to remote, mysterious, adventurous, or idealized spheres, techno-romantic views and ventures aim at satisfying essential human needs for reprieves from the here and now. There is little doubt that the massive socio-economic transformation processes in the wake of the digital revolution have further aggravated feelings of depersonalization, estrangement, and lack of agency that have accompanied the introduction of major disruptive technologies such as the steam engine, the assembly line, or nuclear energy, ever since the first industrial revolution. Simultaneously, mediated means of escape from the everyday have become more readily available than ever. By diverting attention from present realities, techno-romantic tendencies not only make modern existence more tolerable, they also empower its most oppressive forces, including militaristic, totalitarian, and capitalist ideologies. Indeed, the staggering box office successes of recent transmedia franchises like *Star Wars* (1977-), *Jurassic Park* (1990-), *Harry Potter* (2001-2011), or *Avengers* (2011-) bespeak both a pervasive desire for techno-romantic transcendence and its exploitability for monetary gain.

Many of the dazzling renderings of speculative fiction that dominate the output of contemporary media industries embrace an idealist outlook and feature remote worlds that, by the power of magic and fanciful technology, transcend the everyday. Simultaneously, they address visions and fears that ultimately pertain to the relationship between humanity and technology.

1 Mark Coeckelbergh, *New Romantic Cyborgs: Romanticism, Information Technology and the End of the Machine* (Cambridge, MA: MIT Press, 2017), 4.

The vogue of fantasy, dystopian, science fiction, and superhero media does not solely spring from cognitive and affective needs of mass audiences in the information age, however. It has been made possible by radical advances in the digital creation and manipulation of moving images. Since roughly the 1980s, computer-generated imagery (CGI) has facilitated ever more convincing renderings of synthetic environments, creatures, and forces. Today, their ubiquity has all but erased, as Stephen Prince has argued, the boundaries between normative live-action photography and visual effects.[2] The dream of creating fully malleable photographs, which traces as far back as the 1840s, constitutes a major triumph of the human mind over outside reality and must thus be recognized as an essentially techno-romantic ambition. It has finally come into reach.

Special/visual effects are usually analyzed in terms of two distinct functions: spectacle and narrative support. This conceptual bifurcation likely traces back to Tom Gunning's and André Gaudreault's notion of "attraction," a mode of audience address that is characteristic for early cinema in that it differs "from later narrative cinema through its fascination in the thrill of display rather than its construction of a story."[3] The dichotomy between attraction and narration subsequently became, as Thomas Elsaesser has described, a ubiquitous, if problematic, typological tool in film studies.[4] In line with this, special/visual effects have been regarded as either ostentatious or surreptitious. The first category is understood as constituting spectacles of excess, as is the case with fantastical creatures and awe-inspiring events, which call the viewers' attention to the trick itself and are therefore at odds with the goal of seamless narration. The second category, in contrast, is imperceptible as effects and fully subordinated to a narrative, as is the case for instance with set extensions, colour grading, or fluid simulation.

However, as silent filmmaking practice shows, a third domain must be taken into account: The use of special effects as expressive devices. German filmmakers certainly utilized tricks for set extensions and magical incidents, but they were particularly interested in employing unconventional and elaborate film technological tools to convey thoughts and emotions. A primary purpose of image manipulations therefore was, as cinematographer Carl Hoffmann put

2 Stephen Prince, *Digital Visual Effects in Cinema: The Seduction of Reality* (New Brunswick, NJ: Rutgers University Press, 2011), 4.

3 Tom Gunning, "'Primitive' Cinema: A Frame-up? Or the Trick's on Us," *Cinema Journal* 2 (Winter 1989): 3-12; 9.

4 Thomas Elsaesser, "The New Film History as Media Archaeology," *Cinémas: Journal of Film Studies* 2-3 (2004): 75-117; 84.

it, "to render pictorially all of the script's thought content."[5] Special effects like the white forest in *Nosferatu* (Prana, 1922, dir. F. W. Murnau), the lama's gaze controlling the ocean liner in *Lebende Buddhas* (Paul-Wegener-Film, 1923/1925, dir. Paul Wegener), or, as seen in Figure 7.1., the montage shot of the audience's eyes devouring the dancing false Maria in *Metropolis* (Ufa, 1927, dir. Fritz Lang) emphatically call attention to themselves as effects. They seem to correspond to what Christian Metz has called "visible trucage," namely effects that do not represent an objective reality within the fiction and are "meant to be discernible (accelerated motion, slow motion, etc.)."[6] Metz did not elaborate on the characteristics of this effects category, but instead referred to the examples of *Ballada o soldate* (Mosfilm, 1959, dir. Grigori Chukhrai) and *Staroye i novoye* (Sovkino, 1929, dir. Sergei Eisenstein). In the former, a young soldier travels on a train and mentally evokes (by means of superimposition) a girl he has encountered shortly before. In the latter, a peasant girl and a worker literally speed up (by means of time-lapse photography) inert bureaucrats. As these instances demonstrate, special/visual effects can act as formal devices and convey figurative meaning. This emblematic function is evocative of Kristen Whissel's notion of the visual-effects emblem, which refers to a dazzling digital effect that appears at a key turning point and serves as the "spectacular elaboration of the concepts and themes central to the film in which it appears."[7] The examples cited above fit this description in so far as they bespeak their films' broader concerns. At the same time, however, it is evident that their primary function is to visualize abstract ideas or feelings and not objective events within the fiction. Visual-effects emblems, in contrast, ostensibly present a diegetic reality, even if they ultimately reference larger concerns.

Special/visual effects are notoriously difficult to define, in part due to changing industrial contexts, pervasive experimentation, a vast body of heterogeneous applications, and because creators and viewers conceive of them in different ways. As the title of the leading visual effects periodical *Cinefex: The Journal of Cinematic Illusions* suggests, special/visual effects are widely regarded as essentially deceptive. However, expressive uses show that they are not limited to simulating objective phenomena. In fact, when taking on expressive functions, special/visual effects work in much the same vein as devices like lighting or camera techniques, which can equally

5 Hermann Treuner, "Carl Hoffmann," in *Filmkünstler: Wir über uns selbst* (Berlin: Sibyllen-Verlag, 1928), n.p.

6 Christian Metz, "Trucage and the Film," trans. Françoise Meltzer, *Critical Inquiry* 4 (Summer 1977): 657-675, 663.

7 Kristen Whissel, *Spectacular Digital Effects: CGI and Contemporary Cinema* (Durham, NC: Duke University Press, 2014), 7.

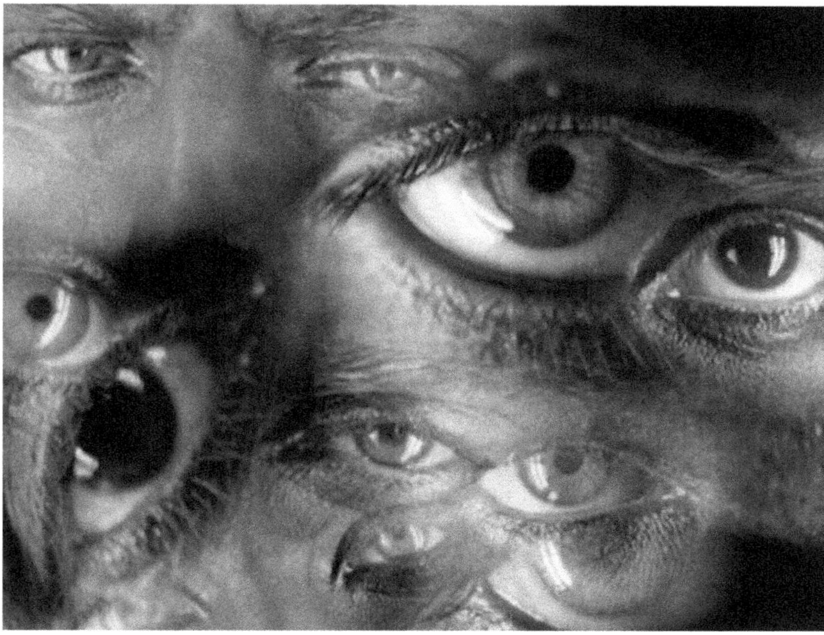

Fig. 7.1. Montage shot in *Metropolis* (1927).
Source: Deutsche Kinemathek.

solicit the viewers' attention by means of startling impressions ("effects") and warrant elaborate production schemes. As CGI technologies such as virtual cinematography and digital lighting become increasingly widespread, categorical differences blur even further. German silent filmmakers' approaches to trick technology give occasion to re-examine special/visual effects specifically as artistic tools, a project that holds out the prospect of productive new avenues for analytical scrutiny.

Special effects came to play such a prominent role in the development of cinematic expressivity because their value as artistic tools was recognized early on. The filmmakers and critics who defined German silent film culture strove to reconcile the technological nature of the cinematic medium with their idealist conceptions of art. Proceeding from techno-romantic assumptions, they identified trick technology as the most feasible way that would allow human consciousness to shape mechanically produced images. Special effects afforded imaginative experiments with the cinematic apparatus and inspired filmmakers to develop film's expressive abilities. Most importantly, they established technology as a creative means. Special effects initiated the emergence of German film art from the spirit of technology.

Bibliography

Mark Coeckelbergh, *New Romantic Cyborgs: Romanticism, Information Technology and the End of the Machine* (Cambridge, MA: MIT Press, 2017).

Thomas Elsaesser, "The New Film History as Media Archaeology," *Cinémas: Journal of Film Studies* 2-3 (2004): 75-117.

Tom Gunning, "'Primitive' Cinema: A Frame-up? Or the Trick's on Us," *Cinema Journal* 2 (Winter 1989): 3-12.

Christian Metz, "Trucage and the Film," trans. Françoise Meltzer, *Critical Inquiry* 4 (Summer 1977): 657-675.

Hermann Treuner, "Carl Hoffmann," in *Filmkünstler: Wir über uns selbst* (Berlin: Sibyllen-Verlag, 1928). n.p.

Kristen Whissel, *Spectacular Digital Effects: CGI and Contemporary Cinema* (Durham, NC: Duke University Press, 2014).

Bibliography

Richard Abel, "Photogénie and Company," in *French Film Theory and Criticism: A History/Anthology, 1907-1939*, ed. and trans. Richard Abel, 95-124 (Princeton, NJ: Princeton University Press, 1988).

Theodor W. Adorno, *Minima Moralia: Reflections from Damaged Life* [1951], trans. E. F. N. Jephcott, 238-244 (London: Verso, 1974).

Henri Alekan, *Des lumières et des ombres* (Paris: Librairie du Collectionneur, 1991).

Henri Alekan, *Le vécu et l'imaginaire: Chroniques d'un homme d'images* (Geneva: Publishing La Sirène, 1999).

Henri Alekan, *Procédé et dispositif de prise de vues combinées*. République Française Brevet d'invention 1,098,128, filed 7 January 1954, and issued 2 March 1955.

Emilie Altenloh, *Zur Soziologie des Kino: Die Kino-Unternehmung und ihre Besucher* (Leipzig: Spamerische Buchdruckerei, 1913).

J. Dudley Andrew, *The Major Film Theories: An Introduction* (London: Oxford University Press, 1976).

Stefan Andriopoulos, "The Terror of Reproduction: Early Cinema's Ghostly Doubles and the Right to One's Own Image," *New German Critique* 99 (Fall 2006): 151-170.

"Angebot der Universum-Film AG an die Deutsche Spiegeltechnik GmbH &Co, Berlin W9, Köthenerstrasee 1/4 vom 15. September 1926." *Files of the Deutsche Spiegeltechnik GmbH & Co. K.-G. Berlin* (R109/I2070). Bundesarchiv Berlin.

"Angles," *Motion Picture News* (25 December 1926): 2407.

Rudolf Arnheim, *Film as Art* (Berkeley, CA: University of California Press, 1957).

Rudolf Arnheim, "Zum Geleit [für Hugo Münsterberg: Das Lichtspiel]," *montage A/V* 2 (2000): 55-57.

Helmut G. Asper, *Filmexil in Hollywood: 'Etwas Besseres als den Tod.' Portraits, Filme, Dokumente* (Marburg: Schüren-Verlag, 2002).

Helmut G. Asper, "'Man muss eben struggeln, um oben zu bleiben.' Eugen Schüfftan und Siegfried Kracauer im amerikanischen Exil," in *Nachrichten aus Hollywood, New York und anderswo: Der Briefwechsel Eugen und Marlise Schüfftans mit Siegfried und Lili Kracauer*, ed. Helmut G. Asper, 3-30 (Trier: Wissenschaflicher Verlag, 2003).

"Auflösungsvertrag vom 28. Februar 1928." *Files of the Deutsche Spiegeltechnik GmbH & Co. K.-G. Berlin in the Bundesarchiv Berlin* (R109/I2070). Bundesarchiv Berlin.

Jacques Aumont, Alain Bergala, Michel Marie, and Marc Vernet, *Aesthetics of Film* [1983], trans. Richard Neupert (Austin, TX: University of Texas Press, 2004).

Alfred A. Baeumler, "The Effects of the Film Theater: An Attempt at an Apologia for the Cinematographic Theater [1912]," in *The Promise of Cinema: German*

Film Theory, 1907-1933, eds. Anton Kaes, Nicholas Baer, and Michael Cowan, trans. Alex H. Bush, 381-384 (Berkeley, CA: University of California Press, 2016).

Béla Balázs, *Early Film Theory: Visible Man and The Spirit of Film*, ed. Erica Carter, trans. Rodney Livingstone (New York, NY/Oxford: Berghahn Books, 2010).

Klaus Bartels, "Über das Technisch-Erhabene," in *Das Erhabene: Zwischen Grenzerfahrung und Größenwahn*, ed. Christiane Pries, 295-316 (Berlin: Verlag Walter de Gruyter, 1989).

Klaus Bartels, "Vom Erhabenen zur Simulation. Eine Technikgeschichte der Seele: Optische Medien bis 1900 (Guckkasten, Camera Obscura, Panorama, Fotografie) und der menschliche Innenraum," in *Armaturen der Sinne: Literarische und technische Medien 1870 bis 1920*, eds. Jochen Hörisch and Michael Wetzel, 17-42 (Munich: Wilhelm Fink Verlag, 1990).

Randolph Bartlett, "German Film Revision Upheld as Needed Here," *New York Times* (13 March 1927): X3.

Tessel M. Bauduin, "Abstract Art as By-Product of Astral Manifestation: The Influence of Theosophy on Modern Art in Europe," in *Handbook of the Theosophical Current*, eds. Olav Hammer and Mikael Rothstein, 429-451 (Leiden: Brill, 2013).

Tessel M. Bauduin and Henrik Johnsson, "Introduction: Conceptualizing Occult Modernism," in *The Occult in Modernist Art, Literature, and Cinema*, eds. Tessel M. Bauduin and Henrik Johnsson, 1-30 (Cham: Palgrave Macmillan, 2019).

André Bazin, "The Life and Death of Superimposition [1946]," in *Bazin at Work: Major Essays and Reviews from the Forties and Fifties*, ed. Bert Cardullo, trans. Alain Piette and Bert Cardullo, 73-77 (Oxford: Routledge, 1997).

"Bell & Howell Pioneer Professional Standard," *American Cinematographer* 1 (January 1927): 12.

Dagmar Bellmann, *Von Höllengefährten zu schwimmenden Palästen: Die Passagierschifffahrt auf dem Atlantik (1840-1930)* (Frankfurt am Main: Campus Verlag, 2015).

Raymond Bellour, "Ideal Hadaly," trans. Stanley E. Grey, *Camera Obscura* 15 (Fall 1986): 111-136.

John Belton, *Cinema Stylists* (Metuchen, NJ: Scarecrow Press, 1983).

Walter Benjamin, "On Some Motifs in Baudelaire [1939]," in *Selected Writings*, vol. 4: *1938-1940*, eds. Howard Eiland and Michael W. Jennings, trans. Harry Zohn, 313-355 (Cambridge, MA: Harvard University Press, 2003).

Walter Benjamin, "Reply to Oskar A. H. Schmitz [1927]," in *Selected Writings*, vol. 2: 1927-1934, eds. Howard Eiland and Michael W. Jennings, trans. Rodney Livingstone, 16-19 (Cambridge, MA: Harvard University Press, 1999).

Walter Benjamin, "The Work of Art in the Age of Its Technological Reproducibility [second version, 1935/1936]," in *Walter Benjamin Selected Writing*, vol. 3, *1935-1938*, eds. Howard Eiland and Michael W. Jennings, trans. Edmund Jephcott and Harry Zohn, 101-133 (Cambridge, MA: Harvard University Press, 2002).

Walter Benjamin, "The Work of Art in the Age of its Technological Reproducibility [third version, 1936/1939]," in *Walter Benjamin: Selected Writings*, vol. 4: *1938-1940*, eds. Howard Eiland and Michael W. Jennings, trans. Harry Zohn and Edmund Jephcott, 251-284 (Cambridge, MA: Harvard University Press, 2002).

A. Berein, "Kinematographisches," *Der Komet* 1109 (23 June 1906): 6.

Gretchen Berg, "The Viennese Night: A Fritz Lang Confession, Parts One and Two (1965)," in *Fritz Lang Interviews*, ed. Barry Keith Grant, 50-76 (Jackson, MS: University Press of Mississippi, 2003).

Ludwig Berger, *Wir sind vom gleichen Stoff aus dem die Träume sind: Summe eines Lebens* (Tübingen: R. Wunderlich, 1953).

Janet Bergstrom, "Murnau in America: Chronicle of Lost Films (*4 Devils*, *City Girl*)," in *Looking Past the Screen: Case Studies in American Film History and Method*, eds. Jon Lewis and Eric Smoodin, 303-352 (Durham NC: Duke University Press, 2007).

Janet Bergstrom, "The German Genius," in *Friedrich Wilhelm Murnau: Ein Melancholiker des Films*, eds. Hans Helmut Prinzler, Karin Messlinger, and Vera Thomas, 79-94 (Berlin: Bertz, 2003).

"Bericht über Revision der Geschäftsbücher und der Bilanz für das am 30. September 1927 abgelaufene Geschäftsjahr." *Files of the Deutsche Spiegeltechnik GmbH & Co. K.-G. Berlin* (R109/I2456). Bundesarchiv Berlin.

Bericht über Zwischenrevision der Geschäftsbücher und der Bilanz per 28. Februar 1927." *Files of the Deutsche Spiegeltechnik GmbH & Co. K.-G. Berlin* (R109/I2455). Bundesarchiv Berlin.

Luciano Berriatúa, *Nosferatu: Un film Erótico – Ocultista – Espiritista – Metafísico* (Valladolid: Divisa Red, 2009).

"Bizarre Effects Attained in German Cinema," *Los Angeles Times* (13 July 1926): C25.

J. Stuart Blackton and Albert E. Smith, "Some Tricks of the Moving Picture Maker," *Scientific American* (26 June 1909): 476.

Helena Blavatsky, "Are Chelas Mediums?" *The Theosophist* 9 (June 1884): 210-211.

Helena P. Blavatsky, *Isis Unveiled: A Master-Key to the Mysteries of Ancient and Modern Science and Theology*, vol. 1 (New York, NY: J.W. Bouton, 1877).

Helena Petrovna Blavatsky, *The Key to Theosophy* (London: The Theosophical Publishing Company, 1889).

Helena P. Blavatsky, *The Secret Doctrine* [1888], ed. Boris de Zirkoff, vol. 1 (Wheaton, IL: The Theosophical Publishing House, 1893).

Helena Blavatsky, *The Theosophical Glossary* (London: The Theosophical Publishing Company, 1892).

Karl Bleibtreu, "Filmkritik" [1914]," in *Kein Tag ohne Kino: Schriftsteller über den Stummfilm*, ed. Fritz Güttinger, 272 (Frankfurt am Main: Deutsches Filmmuseum, 1984).

Ernst Bloch, *The Heritage of Our Times* [1935], trans. Neville and Stephen Plaice (Berkeley, CA: University of California Press, 1990).

Hans-Michael Bock and Tim Bergfelder, eds. *The Concise Cinegraph: Encyclopaedia of German Cinema* (New York, NY: Berghahn, 2009).

Lisa Bode, *Making Believe: Screen Performance and Special Effects in Popular Cinema* (New Brunswick, NJ: Rutgers University Press, 2017).

Louise Bogan, "True to the Medium," *New Republic* (26 October 1927): 263.

Peter Bogdanovich, *Fritz Lang in America* (London: Studio Vista, 1967).

Peter Bogdanovich, *Who the Devil Made It* (New York, NY: Alfred A. Knopf, 1997).

Wilhelm Bölsche, "Die Poesie der Großstadt," *Das Magazin für Literatur* 40 (4 October 1890): 622-625.

David Bordwell, *Narration in the Fiction Film* (Madison, WI: University of Wisconsin Press, 1985).

David Bordwell, "The Classical Hollywood Style, 1917-1960," in David Bordwell, Janet Staiger, and Kristin Thompson, *The Classical Hollywood Cinema: Film Style and Mode of Production to 1960*, 1-84 (New York, NY: Columbia University Press, 1985).

Michel Bouvier and Jean Louis Leutrat, *Nosferatu* (Paris: Gallimard, 1981).

Thomas Brandlmeier, "'Entfesseltes Sehen:' Von der *Ausfahrt der Chinakrieger* bis *Varieté*," in *Die Perfektionierung des Scheins: Das Kino der Weimarer Republik im Kontext der Künste*, ed. Harro Segeberg, 143-157 (Munich: Wilhelm Fink Verlag, 2000).

Thomas Brandlmeier, *Kameraautoren: Technik und Ästhetik* (Marburg: Schüren Verlag, 2008).

Bertolt Brecht, "Singing Steyr Motor Cars [1927]," in *The Collected Poems of Bertolt Brecht*, trans. and eds. Tom Kuhn and David Constantine, 361-362 (New York, NY: Liveright Publishing, 2018).

André Breton, "Surrealist Situation of the Object: Situation of the Surrealist Object [1935]," in *Manifestoes of Surrealism*, trans. Richard Seaver and Helen Lane, 255 (Ann Arbor, MI: University of Michigan Press, 1972).

"British National Strengthened," *Kinematograph Weekly* (13 January 1927): 56.

"Broadway on Screen," *New York Times* (19 May 1929): X4.

"Broadway Screen Tips," *New York Times* (14 October 1928): X4.

Max Brod, "Höhere Welten," *Pan* 16 (16 June 1911): 538-545.

Kevin Brownlow, *The Parade's Gone By...* (Berkeley, CA: University of California Press, 1968).

Reinhard Bruck, "Kinematograph und Theater," *Der Kinematograph* 247 (20 September 1911).

Scott Bukatman, "Disobedient Machines: Animation and Autonomy," in *Beyond the Finite: The Sublime in Art and Science*, eds. Ronald Hoffmann and Iain Boyd Whyte, 128-149 (Oxford: Oxford University Press, 2011).

Scott Bukatman, *Matters of Gravity: Special Effects and Supermen in the 20th Century* (Durham, NC: Duke University Press, 2003).
Luis Buñuel, "*Metropolis* [1927]," in *An Unspeakable Betrayal: Selected Writings of Luis Buñuel*, trans. Garrett White, 99-102 (Berkeley, CA: University of California Press, 2000).
Edmund Burke, *A Philosophical Enquiry into the Sublime and Beautiful* [1757], ed. James T. Boulton (London/New York, NY: Routledge, 2008).
W. Stephen Bush, "'The Life of Richard Wagner' Made by Messter & Co. in Four Reels and Imported by Klaw and Erlanger," *Moving Picture World* 8-13 (1913): 995.
Ken Calhoon, "F. W. Murnau, C. D. Friedrich, and the Conceit of the Absent Spectator," *MLN* 3 (April 2005): 633-653.
"Camera Work for Screen," *New York Times* (2 June 1929): X5.
Edward Carrick, *Designing for Films* [1941] (London: The Studio Publications, 1949).
Noël Carroll, "Film, Emotion, and Genre," in *Passionate Views*, eds. Carl Plantinga and Greg M. Smith, 21-47 (Baltimore, MD: Johns Hopkins University Press, 1999).
"Case of Jonathan Drew," *Variety* (13 June 1928): 12.
Carl Casinghino, *Moving Images: Making Movies, Understanding Media* (Clifton Park, NY: Delmar, 2010).
C. B., "Opium," *Der Film* 7 (15 February 1919).
Blaise Cendrars, "The Modern: A New Art, the Cinema [1919]," in *French Film Theory and Criticism: A History/Anthology, 1907-1939*, ed. and trans. Richard Abel, 182-183 (Princeton, NJ: Princeton University Press, 1988).
Arthur Brunel Chatwood, *The New Photography* (London: Downey, 1896).
Mark Coeckelbergh, *New Romantic Cyborgs: Romanticism, Information Technology and the End of the Machine* (Cambridge, MA: MIT Press, 2017).
Richard Nikolaus Coudenhove-Kalergi, *Apologie der Technik* (Leipzig: Verlag der neue Geist Dr. Peter Reinhold, 1922).
Michael Cowan, "Advertising, Rhythm and the Filmic Avant-Garde in Weimar Guido Seeber and Julius Pinschewer's Kipho Film," *October* 131 (Winter 2010): 23-50.
Michael Cowan, "Moving Picture Puzzles: Training Urban Perception in the Weimar 'Rebus Films,'" *Screen* 51:3 (Autumn 2010): 197-218.
Michael Cowan, "The Heart Machine Rhythm and Body in Weimar Film and Fritz Lang's *Metropolis*," in *Modernism/modernity* 2 (April 2007): 225-248.
Michael Cowan, *Walter Ruttmann and the Cinema of Multiplicity: Avant-Garde – Advertising –Modernity* (Amsterdam: Amsterdam University Press, 2014).
"Crane Mechanical Marvel," *American Cinematographer* (May 1929): 14-15.
Istvan Csicsery-Ronay, Jr, *The Seven Beauties of Science Fiction* (Middleton, CT: Wesleyan University Press, 2008).
Robin Curtis, "Immersion und Einfühlung: Zwischen Repräsentationalität und Materialität bewegter Bilder," *montage A/V* (February 2008): 89-107.

Robin Curtis and Gertrud Koch, eds., *Einfühlung: Zur Geschichte und Gegenwart eines ästhetischen Konzeps* (Munich: Wilhelm Fink Verlag, 2009).

Scott Curtis, *The Shape of Spectatorship: Art, Scene and Early Cinema in Germany* (New York, NY: Columbia University Press, 2015).

Angela Dalle Vacche, *Cinema and Painting: How Art Is Used in Film* (Austin, TX: University of Texas Press, 1996).

Linda Dalrymple Henderson, "Abstraction, the Ether, and the Fourth Dimension: Kandinsky, Mondrian, and Malevich in Context," in *Kandinsky, Malevich, Mondrian: The Infinite White Abyss*, eds. Marian Ackermann and Isabelle Malz, 233-44 (Düsseldorf: Kunstsammlung Nordrhein-Westfalen, 2014).

Linda Dalrymple Henderson, "Rethinking Modern Art, Science, and Occultism in Light of the Ether of Space: Wassily Kandinsky, Umberto Boccioni, and Kazimir Malevich," in *The History of Art and 'Rejected Knowledge': From the Hermetic Tradition to the 21st Century*, ed. Anna Korndorf, 218-237 (Moscow: The State Institute of Art Studies, 2018).

Karlheinz Daniels, "Expressionismus und Technik," in *Technik in der Literatur*, ed. Harro Segeberg, 351-386 (Frankfurt am Main: Suhrkamp-Verlag, 1987).

"Das Ende der Dekoration," *Kinematograph* 950 (3 May 1925).

"Das Schüfftan'sche Kombinationsverfahren!" *Lichtbild-Bühne* 11 (13 January 1927).

Carol Margaret Davison, *Anti-Semitism and British Gothic Literature* (New York, NY: Palgrave, 2004).

James Searle Dawley, *Art of Making Motion Pictures*, U.S. patent 1,278,117, filed 17 August 1914, and issued 10 September 1918.

"De Mille Studios Get Traveling Camera Crane," *Exhibitors Herald* 13 (10 September 1927): 7.

"Der Autorenfilm und seine Bewertung," *Der Kinematograph* 326 (26 March 1913).

"Der Student von Prag," *Das kleine Journal* (25 August 1913).

Der Student von Prag, dir. Hanns-Heinz Ewers (1913, Edition Filmmuseum 80, Munich: film&kunst, 2016), DVD.

Deutsche Bioscop Gesellschaft, ed., *Unsere Künstler* (Promotion Booklet, 1913).

Deutsche Bioscop Ges.m.b.H. Berlin, ed., *Der Student von Prag* [1913], in the Scripts, Gray Literature, and Audio Documents Archive of the Deutsche Kinemathek Berlin.

Deutsche Bioscop, "Unsere Prophezeihung über *Der Student von Prag*," *Der Kinematograph* 349 (3 September 1913).

"Die entfesselte Kamera," *Ufa-Magazin* (1927): 216-217.

Helmut H. Diederichs, "Frühgeschichte deutscher Filmtheorie: Ihre Entstehung und Entwicklung bis zum Ersten Weltkrieg." Postdoctoral Thesis, J. W. Goethe-Universität Frankfurt am Main, 1996.

"Director Tendered Luncheon," *Los Angeles Times* (4 March 1927): A9.

Royal Dixon, *Americanization* (New York, NY: Macmillan, 1916).

Emil du Bois-Reymond, *Reden: Erste Folge. Literatur – Philosophie – Zeitgeschichte* (Leipzig: Veit u. Comp., 1886).

Paul Dubro, "Frau im Mond, Fritz Lang und ein Interview," *Ufa Feuilleton* 42 (16 October 1929): 4.

Hermann Duenschmann, "Cinematograph and Crowd Psychology: A Sociopolitical Study [1912]," in *The Promise of Cinema: German Film Theory, 1907-1933*, eds. Anton Kaes, Nicholas Baer, and Michael Cowan, trans. Eric Ames, 256-258 (Berkeley, CA: University of California Press, 2016).

Hervé Dumont, *Frank Borzage: The Life and Films of a Hollywood Romantic*, trans. Jonathan Kaplansky (Jefferson, NC: McFarland, 2006).

Ewald André Dupont and Fritz Podehl, *Wie ein Film geschrieben wird und wie man ihn verwertet* (Berlin: Kühn, 1926).

Carl du Prel, *Der Spiritismus* (Leipzig: Reclam-Verlag, 1893).

Matthias Eberle, *Der Weltkrieg und die Künstler der Weimarer Republik. Dix, Grosz, Beckmann, Schlemmer* (Stuttgart: Belser, 1989).

Thomas Edison, "The Realms Beyond [1920]," in *The Diary and Sundry Observations of Thomas Alva Edison*, ed. Dagobert D. Runes, 238-241 (New York, NY: Philosophical Library, 1948).

Axel Eggebrecht, "*Metropolis*," *Weltbühne* 3 (18 January 1927): 115-116.

eht, "Theater und Musik," *Vossische Zeitung* (23 August 1913).

Sergei Eisenstein, "A Dialectic Approach to Film Form [1929]," in *Film Form: Essays in Film Theory*, ed. and trans. Jay Leyda, 45-63 (New York, NY: Harcourt, 1949).

Sergei Eisenstein, "Dickens, Griffith and the Film Today [1944]," in *Film Form: Essays in Film Theory*, ed. and trans. Jay Leyda, 195-255 (New York, NY: Harcourt, 1949).

Sergei Eisenstein, "The Cinematographic Principle and the Ideogram [1929]," in Sergei Eisenstein, *Film Form: Essays in Film Theory*, ed. and trans. Jay Leyda, 28-44 (New York, NY: Harcourt, 1949).

Lotte Eisner, "Der produktive Kameramann wird zum Mitschöpfer," *Film-Kurier* 272 (26 November 1931).

Lotte H. Eisner, *Murnau* [1964], trans. Martin Secker (London: Secker & Warburg, 1973).

Lotte H. Eisner, "Stile und Gattungen des Films," in *Das Fischer Lexikon. Film, Rundfunk, Fernsehen*, eds. Lotte H. Eisner and Heinz Friedrich, 259-283 (Frankfurt am Main: Fischer, 1958).

Lotte H. Eisner, *The Haunted Screen: Expressionism in the German Cinema and the Influence of Max Reinhardt* [1952], trans. Roger Greaves (Berkeley, CA: University of California Press, 2008).

R. Bruce Elder, *Harmony and Dissent: Film and Avant-garde Art Movements in the Early Twentieth Century* (Waterloo, ON: Wilfrid Laurier University Press, 2008).

Thomas Elsaesser, "No End to *Nosferatu* (1922)," in *Weimar Cinema*, ed. Noah Isenberg, 79-94 (New York, NY: Columbia University Press, 2008).

Thomas Elsaesser, "The New Film History as Media Archaeology," *Cinémas: Journal of Film Studies* 2-3 (2004): 75-117.

Thomas Elsaesser, *Weimar Cinema and After: Germany's Historical Imaginary* (London: Routledge, 2000).

Horst Emscher (= Josef Coböken), "Große Kunst im Film," *Zeit im Bild*, vol. 3 (1913): 2524-2525.

Horst Emscher (= Josef Coböken), "Neue Wege in der Filmkunst," *Der Kinematograph* 363 (10 December 1913).

Marguerite Engberg, "*Studenten fra Prag* og den gådefulde Stellan Rye," in *Sekvens. Filmvidenskabelig årbog*, 161-185 (Copenhagen: Reitzel, 1982).

Paul Ernst, "Möglichkeiten einer Kinokunst [1913]," in *Kein Tag ohne Kino: Schriftsteller über den Stummfilm*, ed. Fritz Güttinger, 69-73 (Frankfurt am Main: Deutsches Filmmuseum, 1984).

Hanns Heinz Ewers, "Der Kientopp," *Der Morgen* (Berlin) 18 (11 October 1907): 578.

Hanns Heinz Ewers, "*Der Student von Prag*," in Helmut H. Diederichs, *Der Student von Prag: Einführung und Protokoll* (Stuttgart: Focus Verlagsgemeinschaft, 1985).

Hanns Heinz Ewers, *Edgar Allan Poe*, trans. Adèle Lewisohn (New York, NY: B.W. Huebsch, 1917).

Hanns Heinz Ewers, "Schlangenfang auf Jawa [1910]," in *Kein Tag ohne Kino: Schriftsteller über den Stummfilm*, ed. Fritz Güttinger, 14-20 (Frankfurt am Main: Deutsches Filmmuseum, 1984).

Hanns Heinz Ewers, "Vom Kinema [1910]," in *Kein Tag ohne Kino: Schriftsteller über den Stummfilm*, ed. Fritz Güttinger, 20-23 (Frankfurt am Main: Deutsches Filmmuseum, 1984).

Sylvain Exertier, "La Lettre oubliée de *Nosferatu*," *Positif* (March 1980): 47-51.

Max Feige, "Der Film *Metropolis*," *Der Film* (Berlin) 1 (15 January 1927).

Paul Fejos, "Illusion on the Screen," *New York Times* (26 May 1929): X3.

Christine Ferguson, "Dracula and the Occult," in *The Cambridge Companion to* Dracula, ed. Roger Luckhurst, 57-65 (Cambridge, UK: Cambridge University Press, 2018).

"Film – Extrablatt," *Illustriertes Wiener Extrablatt* (4 July 1924).

Friedhelm Wilhelm Fischer, "Geheimlehren und moderne Kunst," in *Fin de Siècle: Zur Literatur und Kunst der Jahrhundertwende*, ed. Roger Bauer, 344-377 (Frankfurt am Main: Klostermann, 1977).

Barbara Flückinger, *Visual Effects: Filmbilder aus dem Computer* (Marburg: Schüren-Verlag, 2008).

Theodor Fontane, *Ein Sommer in London* (Dessau: Gebrüder Katz, 1854).

Carl Forch, "Thrills in Film Drama and Elsewhere [1913]," in *The Promise of Cinema: German Film Theory, 1907-1933*, eds. Anton Kaes, Nicholas Baer, and Michael Cowan, trans. Sara Hall, 35-38 (Berkeley, CA: University of California Press, 2016).

E. M. Forster, "The Machine Stops [1909]," in *Selected Stories*, 91-124 (New York, NY: Penguin Classics, 2001).

Friedrich Freksa, "Theater, Pantomime and Cinema [1916]," in *The Promise of Cinema: German Film Theory, 1907-1933*, eds. Anton Kaes, Nicholas Baer, and Michael Cowan, trans. Alex H. Bush, 111-114 (Berkeley, CA: University of California Press, 2016).

Sigmund Freud, "The Uncanny [1919]," in *The Standard Edition of the Complete Works of Sigmund Freud*, vol. 17, eds. James Strachey et al., 219-256 (London: The Hogarth Press and the Institute of Psychoanalysis, 1953-1974).

Karl Freund, "An Open Letter to Arthur Edeson, ASC and Michael Curtiz [March 1943]," *American Cinematographer* (August 1994): 75-77.

Karl Freund, "Ein Filmkünstler [1947]," in *Caligari und Caligarismus*, ed. Walter Kaul, 79-81 (Berlin: Deutsche Kinemathek, 1970).

Karl Freund, "Just What Is 'Montage'?" *American Cinematographer* 5 (September 1934): 204; 210.

Karl Freund, "Lines, Lights and Lenses," *Bioscope* (December 1928): 247.

Egon Friedell, "Prologue before the Film [1913]," in *The Promise of Cinema: German Film Theory, 1907-1933*, eds. Anton Kaes, Nicholas Baer, and Michael Cowan, trans. Don Reneau, 168-169 (Berkeley, CA: University of California Press, 2016).

Ludwig Fulda, "Theater und Kinematograph," *Die Woche* (Berlin) 16 (20 April 1912): 639-642.

Arthur L. Gale, "Critical Focusing," in *Movie Makers* 3 (March 1929): 156.

Joseph Garncarz, *Maßlose Unterhaltung: Zur Etablierung des Films in Deutschland 1896-1914* (Frankfurt am Main: Stroemfeld, 2010).

Evelyn Gerstein, "*Metropolis*," *The Nation* (23 March 1927): 323-324.

Fritz Giese, *Girlkultur: Vergleiche zwischen amerikanischem und europäischem Rhythmus und Lebensgefühl* (Munich: Delphin-Verlag, 1925).

Rolf Giesen, "Eugen Schüfftan," in *Cinegraph: Lexikon zum deutschsprachigen Film*, ed. Hans-Michael Bock, vol. 7 (Munich: edition text + kritik, 1984-).

Rolf Giesen, "Spezialeffekte Made in Germany," *Künstliche Welten: Tricks, Special Effects und Computeranimation im Film von den Anfängen bis heute*, eds. Rolf Giesen and Claudia Meglin, 69-111 (Hamburg/Vienna: Europa Verlag, 2000).

Rolf Giesen, *The Nosferatu Story: The Seminal Horror Film, Its Predecessor and Its Enduring Legacy* (Jefferson, NC: McFarland, 2019).

"Gigantic Camera Crane Built for Broadway," *Universal Weekly* 25 (26 January 1929): 10.

Jeremy Gilbert-Rolfe, *Beauty and the Contemporary Sublime* (New York, NY: Allworth Press, 1999).

Sander Gilman, *The Jew's Body* (Milton Park: Taylor & Francis, 1991).

G.K., "Künstlerfilms," *Der Kinematograph* 347 (20 August 1913).

Johann Wolfgang Goethe, "Kunst und Handwerk [1797]," in *Gedenkausgabe der Werke, Briefe und Gespräche*, ed. Ernst Beutler, vol. 13, (Zurich: Artemis, 1971-1977).

Ludwig Max Goldberger, *Das Land der unbegrenzten Möglichkeiten: Beobachtungen über das Wirtschaftsleben der Vereinigten Staaten von Amerika* (Berlin/Leipzig: F. Fontane, 1905).

Nicholas Goodrick-Clarke, *The Occult Roots of Nazism: The Ariosophists of Austria and Germany, 1890-1935* (Wellingborough: The Aquarian Press, 1985).

Maxim Gorky, "A Review of the Lumière Programme at the Nizhni-Novgorod Fair [1896]," in Jay Leyda, *Kino: A History of the Russian and Soviet Film*, trans. Leda Swan, 407-410 (London: George Allen and Unwin, 1960).

Foster Goss, "The Editors' Lens," *American Cinematographer* (August 1924): 10.

Rémy de Gourmont, "Epilogues: Cinematograph [1907]," in *French Film Theory and Criticism: A History/Anthology, 1907-1939*, ed. and trans. Richard Abel, 47-50 (Princeton, NJ: Princeton University Press, 1988).

Carroll Graham, "Ah, Those Were The Days," *Picture Play* 6 (August 1929): 24-25.

Albin Grau, "Vampire," *Bild und Film* 21 (1921).

Carl Louis Gregory, "An Optical Printer for Trick Work," *Transactions of SMPE* 34 (1928): 419-426.

Carl Louis Gregory, *Motion Picture Photography* (New York, NY: Falk Publishing, 1927).

Carl Louis Gregory, "Trick Photography Methods Summarized," *American Cinematographer* (June 1926): 21-22.

Ludwig Greve, Margot Pehle, and Heidi Westhoff, eds. *Hätte ich das Kino! Die Schriftsteller und der Stummfilm* (Stuttgart: Klett-Verlag, 1976).

Martha Gruening, "European Revolt against our Films," *New York Times* (31 October 1926): X6.

Uta Grund, *Zwischen den Künsten: Edward Gordon Craig und das Bildertheater um 1900* (Berlin: Akademie Verlag, 2002).

Rosemary Ellen Guiley, *The Encyclopedia of Vampires, Werewolves, and Other Monsters* (New York, NY: Visionary Living, 2005).

Tom Gunning, *D. W. Griffith and the Origins of American Narrative Film* (Urbana, IL: University of Illinois Press, 1991).

Tom Gunning, "Like Unto A Leopard: Figurative Discourse in *Cat People* (1942) and Todorov's *The Fantastic*," *Wide Angle* 3 (1988): 30-39.

Tom Gunning, "'Primitive' Cinema: A Frame-up? Or the Trick's on Us," *Cinema Journal* 2 (Winter 1989): 3-12.

Tom Gunning, *The Films of Fritz Lang: Allegories of Vision and Modernity* (London: British Film Institute, 2000).
Tom Gunning, "To Scan a Ghost: The Ontology of Mediated Vision," *Grey Room* 26 (Winter 2007): 94-127.
Tom Gunning, "Uncanny Reflections, Modern Illusions: Sighting the Modern Optical Uncanny," in *Uncanny Modernity: Cultural Theories, Modern Anxieties.* eds. Jo Collins and John Jervis, 68-90 (New York, NY: Palgrave Macmillan, 2008).
Tom Gunning, "What's the Point of an Index? Or, Faking Photographs," *Nordicom Review* 1/2 (September 2004).
A. Günsberg, "Künstlerische Regie bei kinematographischen Aufnahmen und Vorführungen [1907]," in *Geschichte der Filmtheorie: Kunsttheoretische Texte von Méliès bis Arnheim,* ed. Helmut H. Diederichs, 44-47 (Frankfurt am Main: Suhrkamp-Verlag, 2004).
Willy Haas, *"Metropolis," Film-Kurier* (11 January 1927).
Willy Haas, "November-Filme [1923]," in *Hätte ich das Kino! Die Schriftsteller und der Stummfilm* eds. Ludwig Greve, Margot Pehle, and Heidi Westhoff, 172-173 (Stuttgart: Klett-Verlag, 1976).
Willi Haas, "Wird hier gearbeitet? Ein Besuch im Ufagelände Neubabelsberg," in *Willy Haas: Der Kritiker als Mitproduzent. Texte zum Film 1920-1933*, eds. Wolfgang Jacobsen, Karl Prümm, and Benno Wenz, 109-113 (Berlin: Edition Hentrich, 1991).
Hermann Häfker, "Atlantis [1914]," in *Prolog vor dem Film: Nachdenken über ein neues Medium, 1909-1914*, ed. Jörg Schweinitz, 405-406 (Leipzig: Reclam-Verlag, 1992).
Hermann Häfker, "Hauptmanns Atlantis," *Bild & Film* III:6 (1913/1914): 139-140.
Hermann Häfker, *Kino und Kunst* (M. Gladbach: Volksvereinsverlag, 1913).
Malte Hagener, *Moving Forward, Looking Back: The European Avant-Garde and the Invention of Film Culture, 1919-1939* (Amsterdam: Amsterdam University Press, 2007).
Sabine Hake, "Das Kino, die Werbung und die Avantgarde," in *Die Spur durch den Spiegel: Der Film in der Kultur der* Moderne, eds. Malte Hagener, Johann N. Schmidt, and Michael Wedel, 193-206 (Berlin: Bertz und Fischer, 2004).
Hal Hall, "Cinematographers and Directors Meet Discuss Camera Trucking Problems," *American Cinematographer* (August 1932): 10.
Mordaunt Hall, "Prize Play Pictured," *New York Times* (18 March 1928): 125.
Réjanne Hamus Vallée and Caroline Renouard, *Les Effets spéciaux au cinéma 120 ans de créations en France et dans le monde* (Vanves: Armand Colin, 2018).
Miriam Bratu Hansen, *Cinema and Experience: Siegfried Kracauer, Walter Benjamin, and Theodor W. Adorno* (Berkeley, CA: University of California Press, 2011).
Miriam Hansen, "Decentric Perspectives: Kracauer's Early Writings on Film and Mass Culture," *New German Critique* 54 (Autumn 1991): 47-76.

Miriam Bratu Hansen, "Introduction," in Siegfried Kracauer, *Theory of Film: The Redemption of Physical Reality*, vii-xlv (Princeton, NJ: Princeton University Press, 1997).

Miriam Hansen, "Mass Culture as Hieroglyphic Writing: Adorno, Derrida, Kracauer," *New German Critique* 56 (Spring/Summer 1992): 42-73.

Thea von Harbou, *Metropolis* [screenplay, 1924], Gottfried Huppertz Collection, Filmmuseum Berlin-Deutsche Kinemathek.

Thea von Harbou, *Metropolis* [novel, 1926], anonymous trans. (Mineola, NY: Dover Publications, 2015).

Robert Hare, *Experimental Investigation of the Spirit Manifestations* (New York, NY: Partridge & Brittan, 1855).

Franz Hartmann, *Magic, White and Black* [1886] (New York, NY: The Path, 1895).

Franz Hartmann, *The Life and Doctrines of Philippus Theophrastus, Bombast of Hohenheim Known by the Name of Paracelsus* [1891] (New York, NY: Theosophical Publishing, 1902).

Siegfried Hartmann, "*Metropolis* Technik," *Deutsche Allgemeine Zeitung* (20 January 1927): 2.

Gustav Hartung, "Kinematographie und Theater," *Der Kinematograph* 170 (30 March 1910).

Gerhart Hauptmann, "Flagman Thiel [1887]," in *German Novellas of Realism II*, ed. Jeffrey L. Sammons, trans. Adele S. Seltzer, 304-333 (New York, NY: Continuum, 1989).

Raoul Hausmann, "Photomontage [1931]," in *The Weimar Republic Sourcebook*, eds. Anton Kaes, Martin Jay, and Edward Dimendberg, 651-653 (Berkeley, CA: University of California Press, 1994).

H. H. P., "Das Zerrbild der Technik im Film: Ein Nachwort zu *Metropolis*," *Tages-Post* (Linz, Austria) 52 (6 March 1927): 3-4.

Jörg Heiniger, "Erhaben," *Ästhetische Grundbegriffe: Historisches Wörterbuch in 7 Bänden*, vol. 2, 275-310 (Stuttgart: Metzler Verlag, 2001).

Karl Henckell, "Viadukt," in *Gesammelte Werke*, vol. 2, *Buch des Kampfes* (Munich: J. Michael Müller Verlag, 1921).

Jeffrey Herf, *Reactionary Modernism: Technology, Culture, and Politics in Weimar and the Third Reich* (Cambridge, UK: Cambridge University Press, 1984).

Robert Herlth, "Erinnerungen (1958)," in *Filmarchitektur Robert Herlth*, 48-54 (Munich: Deutsches Institut für Film und Fernsehen, 1965).

Robert Herlth, "With Murnau on the Set," in Lotte Eisner, *Murnau* [1964], trans. Martin Secker, 59-71 (London: Secker & Warburg, 1973).

Benedikt Hjartarson, "Ghosts Before Breakfast: The Appetite for the Beyond in Early Avant-Garde Film," in *The Occult in Modernist Art, Literature, and Cinema*, eds. Tessel M. Bauduin and Henrik Johnsson, 137-162 (Cham: Palgrave Macmillan, 2019).

Carl Hoffmann, "Camera Problems," *Close Up* 1 (July 1929): 29-31.
E. T. A. Hoffmann, *Phantastische Geschichten*, ed. Hanns Heinz Ewers (Munich: Georg Müller Verlag, n.d.).
E. T. A. Hoffmann, *Sämtliche Werke: Historisch-kritische Ausgabe*, ed. Carl Georg von Maassen, 10 vols. (Munich: Georg Müller Verlag, 1908-1928).
Patrick Colm Hogan, "Narrative Universals, Nationalism and Sacrificial Terror from *Nosferatu* to Nazism," *Film Studies* 8 (Summer 2006): 93-105.
Arthur Holitscher, *Amerika heute und morgen* [1912] (Berlin: S. Fischer, 1923).
Lotar Holland, "Subjective Movement [1927]," in *The Promise of Cinema: German Film Theory, 1907-1933*, eds. Anton Kaes, Nicholas Baer, and Michael Cowan, trans. Michael Cowan, 512-515 (Berkeley, CA: University of California Press, 2016).
F. M. Hübner, "Der Dichter und das Kino," *Allgemeine Zeitung* (Munich) 39 (27 September 1913): 629-630.
Otto Hunte, "Der Baumeister von *Metropolis*," *Presse- und Propagandaheft Metropolis* (Berlin: Parufamet Presse- und Propagandabteilung, 1927).
Andreas Huyssen, "The Vamp and the Machine: Technology and Sexuality," *New German Critique* 24/25 (Autum 1981/Winter 1982): 221-237.
"In Camerafornia," *American Cinematographer* 6 (September 1926): 9.
"In Camerafornia," *American Cinematographer* 7 (October 1926): 10.
"Interview avec Henri Alekan," *Cinématographe* 68 (June 1981): 23-28.
"Inventory List." *Norman O. Dawn Collection of Special Effects Cinematography.* Harry Ransom Center at the University of Texas at Austin.
Noah Isenberg, ed., *Weimar Cinema: An Essential Guide to Classic Films of the Era* (New York, NY: Columbia University Press, 2008).
Gertrud Isolani, "Gespräch mit Eugen Shuftan," *Basler Nachrichten* (19 October 1965): 9.
Leopold Jacoby, *Die Idee der Entwicklung* [1874/1876] (Zurich: Schabelitz, 1887).
Alexander Jason, *Handbuch der Filmwirtschaft: Filmstatistiken und Verzeichnisse der Filmschaffenden, Filmfirmen, der Filme und der Tonfilmkinos*, vol. 2, Film-Europa (Berlin: Verlag für Presse, Wirtschaft und Politik, 1931).
Will F. Jenkins, *Apparatus for Production of Light Effects in Composite Photography.* United States Patent 2,727,427, filed 3 March 1952, and issued 20 December 1955.
Jay Johnston, "Vampirism, Lycanthropy, and Otherkin," in *The Occult World*, ed. Christopher Partridge, 412-423 (New York, NY: Routledge, 2015).
Jason A. Josephson-Storm, *The Myth of Disenchantment: Magic, Modernity, and the Birth of the Human Sciences* (Chicago, IL: University of Chicago Press, 2017).
William Q. Judge, *Echoes from the Orient: A Broad Outline of Theosophical Doctrines* (New York, NY: The Path, 1890).
Anton Kaes, *Shell Shock Cinema: Weimar Culture and the Wounds of War* (Princeton, NJ: Princeton University Press, 2009).

Wassily Kandinsky, *Concerning the Spiritual in Art* [1912], trans. Michael T. H. Sadler (Mineola, NY: Dover Publications, 1977).

Wassily Kandinsky, "Reminiscences," in *Modern Artists on Art*, ed. Robert L. Herbert, trans. Eugenia W. Herbert, 19-44 (Mineola, NY: Dover Publications, 1999).

Immanuel Kant, *Observations on the Feeling of the Beautiful and Sublime* [1764], trans. John T. Goldthwait (Berkeley, CA: University of California Press, 2003).

Immanuel Kant, *The Critique of the Power of Judgment* [1790], ed. Paul Guyer, trans. Paul Guyer and Eric Matthews (Cambridge, UK: Cambridge University Press, 2000).

Allan Kardec, *Experimental Spiritism: Book on Mediums*, trans. Emma A. Wood (Boston: Colby and Rich, 1874).

Jürgen Kasten, "Filmstil als Markenartikel. Der expressionistische Film und das Stilexperiment *Von morgens bis mitternachts*," in *Die Perfektionierung des Scheins: Das Kino der Weimarer Republik im Kontext der Künste*, ed. Harro Segeberg, 37-66 (Munich: Wilhelm Fink Verlag, 2000).

Rudolf Kayser, "Amerikanismus," *Vossische Zeitung* 458 (27 September 1925).

Patrick Keating, *Hollywood Lighting from the Silent Era to Film Noir* (New York, NY: Columbia University Press, 2009).

Patrick Keating, *The Dynamic Frame: Camera Movement in Classical Hollywood* (New York, NY: Columbia University Press, 2019).

Charlie Keil and Kristen Whissel, "Introduction," in *Editing and Special/Visual Effects*, eds. Charlie Keil and Kristen Whissel, 1-21 (New Brunswick, NJ: Rutgers University Press, 2016).

Sarah Keller, *Anxious Cinephilia. Pleasure and Peril at the Movies* (New York, NY: Columbia University Press, 2020).

Frank Kessler, Sabine Lenk, and Martin Loiperdinger, eds., *KINtop: Jahrbuch zur Erforschung des frühen Films*, (Frankfurt am Main: Stroemfeld, 1992-2006).

Erich Kettelhut, *Der Schatten des Architekten*, ed. Werner Sudendorf (Munich: Belleville, 2009).

Hermann Kienzl, "Theater und Kinematograph," *Der Strom* (Vienna) 8 (November 1911): 219-221.

Kinematheksverbund, Arbeitsgruppe Deutsche Filmografie, "Deutsche Spielfilme 1925: Jahresproduktion und Filmbestand Bundesarchiv," Bundesarchiv-Filmarchiv, October 2007.

Grace Kingsley, "Flashes," *Los Angeles Times* (17 September 1925): A11.

Friedrich A. Kittler, "Dracula's Legacy," trans. William Stephen Davis, *Stanford Humanities Review* 1 (1989): 143-173.

Ludwig Klages, *Mensch und Erde* [1913] (Stuttgart: A. Kroener Verlag, 1956).

Hermann Koch, "Kino und Theater," *Der Strom* (Vienna) 2 (May 1914): 61-63.

Kristina Köhler, "Nicht der *Stilfilm* also, sondern der *Filmstil* ist wichtig!" in *Filmstil: Perspektivierungen eines Begriffs*, eds. Julian Blunk, Tina Kaiser, Dietmar Kammerer, and Chris Wahl, 91-117 (Munich: edition text + kritik, 2016).

Willy Köhler, Einrichtung für kinematographische Aufnahmen, Deutsches Reichspatent 429,755 (addition to patent 428,589), filed 4 March 1924, and issued 2 June 1926.

Willy Köhler, *Verfahren und Einrichtung für kinematographische Aufnahmen*, Deutsches Reichspatent 428,589, filed 14 September 1922, and re-issued 10 May 1926.

Willy Köhler, *Verfahren und Einrichtung für kinematographische Aufnahmen*, Deutsches Reichspatent 429,753 (addition to patent 428,589), filed 11 February 1923, and issued 2 June 1926.

Willy Köhler, *Verfahren für kinematographische Aufnahmen*, Deutsches Reichspatent 429,754 (addition to patent 428,589), filed 10 May 1923, and issued 2 June 1926.

Alexander Kossowsky, "Die Männer der Kurbel II: Reimar Kuntze," *Film-Kurier* (14 May 1925).

Alexander Kossowsky, "Die Männer der Kurbel III: Werner Brandes," *Film-Kurier* (19 May 1925).

Alexander Kossowsky, "Die Männer der Kurbel IV: Karl Freund," *Film-Kurier* (30 May 1925).

Alexander Kossowsky, "Die Männer der Kurbel VI: Guido Seeber," *Film-Kurier* (22 August 1925).

Richard Koszarski, *An Evening's Entertainment: The Age of the Silent Feature Picture*, 1915-1928 (Berkeley, CA: University of California Press, 1990).

Richard Koszarski, "Hal Mohr's Cinematography," *Film Comment* (September 1974): 48-53.

Siegfried Kracauer, *From Caligari to Hitler: A Psychological History of the German Film* [1947] (Princeton, NJ: Princeton University Press, 2004).

Siegfried Kracauer, "Hochstaplerfilme [1923]," in Siegfried Kracauer, *Kleine Schriften zum Film*, vol. 6.1., *1921-1927*, ed. Inka Mülder-Bach, 37 (Frankfurt am Main: Suhrkamp-Verlag, 2004).

Siegfried Kracauer, *Theory of Film: The Redemption of Physical Reality* [1960] (Princeton, NJ: Princeton University Press, 1997).

Siegfried Kracauer, "Zwischen Flammen und Bestien [1923]," in Siegfried Kracauer, *Kleine Schriften zum Film*, vol. 6.1., *1921-1927*, ed. Inka Mülder-Bach, 35-36 (Frankfurt am Main: Suhrkamp-Verlag, 2004).

Karl Kraus, "Das technoromantische Abenteuer," *Die Fackel* 474-483 (May 1918): 41-45.

Klaus Kreimeier, *The Ufa Story: A History of Germany's Greatest Film Company, 1918-1945*, trans. Robert and Rita Kimber (Berkeley, CA: University of California Press, 1999).

Émile Kress, "Trucs et illusions. Applications de l'optique et de la mécanique au cinématographe (c. 1912)," *1895. Revue d'histoire du cinéma* 27 (September 1999), 7-20.

Wilfried Kugel, *Der Unverantwortliche: Das Leben des Hanns Heinz Ewers* (Düsseldorf: Grupello Verlag, 1992).

Wilfried Kugel, "Die Rekonstruktion des Stummfilms *Der Student von Prag* (Deutsche Bioscop GmbH, Berlin 1913)." Unpublished manuscript, Berlin, 1988.

Wilfried Kugel, "Entstehung, Umfeld und Folgen: *Der Student von Prag*." Unpublished manuscript, Berlin, 1988/2007.

Moniek M. Kuipers, Frank Hakemulder, Katalin Bálint, Miruna Doicaru, and Ed Tan, "Towards a New Understanding of Absorbing Reading Experiences," in *Narrative Absorption*, eds. Frank Hakemulder, Moniek M. Kuipers, Ed S. Tan, Katalin Bálint, and Miruna M. Doicaru, 29-48 (Amsterdam: John Benjamins Publishing Company, 2017).

Harald Lamprecht, *Neue Rosenkreuzer: Ein Handbuch* (Göttingen: Vandenhoeck & Ruprecht, 2004).

Hans Land, "Lichtspiele [1910]," in *Prolog vor dem Film: Nachdenken über ein neues Medium, 1909-1914*, ed. Jörg Schweinitz, 18-19 (Leipzig: Reclam-Verlag, 1992).

Fritz Lang, "Arbeitsgemeinschaft im Film," *Kinematograph* 887 (17 February 1924): 10.

Fritz Lang, "The Future of the Feature Film in Germany [1926]," in *The Weimar Republic Sourcebook*, eds. Anton Kaes, Martin Jay, and Edward Dimendberg, trans. Don Reneau, 622-623 (Berkeley, CA: University of California Press, 1994).

Fritz Lang, "Was ich in Amerika sah: Neuyork – Los Angeles," *Film-Kurier* (11 December 1924): 1.

Konrad Lange, "Bühne und Lichtspiel," *Deutsche Revue* (October 1913): 119-125.

Konrad Lange, "Die 'Kunst' des Lichtspieltheaters [1913]," in *Geschichte der Filmtheorie: Kunsttheoretische Texte von Méliès bis Arnheim*, ed. Helmut H. Diederichs, 75-88 (Frankfurt am Main: Suhrkamp-Verlag, 2004).

Michelle Langford, "Lola and the Vampire: Technologies of Time and Movement in German Cinema," in *Cinema and Technology: Cultures, Theories and Practices*, eds. Bruce Bennett, Marc Furstenau, and Adrian Mackenzie, 187-200 (New York, NY: Palgrave Macmillan, 2008).

Kurd Lasswitz, *Wirklichkeiten* (Berlin: Verlag von Emil Felber, 1900).

Standish D. Lawder, *The Cubist Cinema* (New York, NY: New York University, 1975).

Volker Lechler, *Die ersten Jahre der Fraternitas Saturni* (Stuttgart: Verlag Volker Lechler, 2015).

Laura Lee, *Japanese Cinema Between Frames* (London: Palgrave Macmillan, 2017).

Thierry Lefebvre, "Les Métamorphoses de Nosferatu," *1895. Revue d'histoire du cinéma* 29 (December 1999): 61-77.

James Mitchell Leisen, "Some Problems of the Art Director," *Transactions of S.M.P.E.* 33 (1928): 71-80.

Carl Lemcke, *Populäre Ästhetik* (Leipzig: E.A. Seemann, 1873).

Gotthold Ephraim Lessing, *Laocoön: An Essay on the Limits of Painting and Poetry* [1766], trans. Edward Allen McCormick (Baltimore: Johns Hopkins University Press, 1984).

Éliphas Lévi, *Transcendental Magic: Its Doctrine and Ritual* [1854-1856], trans. Arthur Edward Waite (York Beach, ME: Weiser Books, 2001).

Willy Ley, *Die Möglichkeit der Weltraumfahrt* (Leipzig: Hachmeister & Thal, 1928).

Wilhelm Liebknecht, "Knowledge is Power – Power is Knowledge [1872]," in *Wilhelm Liebknecht and German Social Democracy: A Documentary History*, ed. William A. Pelz, trans. Erich Hahn (Chicago, IL: Haymarket Books, 2016).

Vachel Lindsay, *The Art of the Moving Picture* [1915] (New York, NY: Liveright Publishing, 1970).

A. Lion, "Das Schüfftan-Verfahren: Umwälzung in der Filmaufnahme," *Die Umschau* (1927): 248-251.

V. Loers and P. Witsman, eds., *Okkultismus und Avant-Garde: Von Munch bis Mondrian* (Frankfurt am Main: Schirn-Kunsthalle, 1995).

Ernst Lorenzen, "Hanns Heinz Ewers: Student von Prag," *Bild & Film* III:6 (1913/1914): 141.

Rachel Low, *The History of British Film: 1918-1929* (London: George Allen & Unwin, 1971).

Herbert G. Luft, "Der Mann hinter der Kamera: Karl Freund – schon Filmgeschichte," *Der Filmkreis* (July 1963): 16-19.

Georg Lukács, "Thoughts Toward an Aesthetic of the Cinema [1911/1913]," in *The Promise of Cinema: German Film Theory, 1907-1933*, eds. Anton Kaes, Nicholas Baer, and Michael Cowan, trans. Janelle Blankenship, 377-381 (Berkeley, CA: University of California Press, 2016).

Norbert Lusk, "The Screen in Review," *Picture Play* 5 (January 1929): 70-73; 96.

Joseph August Lux, "Die Muse des Films," *Velhagen & Klasings Monatshefte* 11 (July 1913): 429.

Waldemar Lydor, "Kristallene Filmbauten," *Das Kino-Journal* (Vienna) 848 (30 October 1926): 14-15.

Jean-François Lyotard, *The Postmodern Condition: A Report on Knowledge*, trans. Geoff Bennington and Brian Massumi (Manchester: Manchester University Press, 1984).

Leonard Maltin, "Interview with Hal Mohr," in *The Art of the Cinematographer*, 75-94 (New York, NY: Dover Publications, 1978).

Heinrich Mann, *Macht und Mensch: Essays*, ed. Peter-Paul Schneider (Frankfurt am Main: Fischer, 1989).

Thomas Mann, "Germany and the Germans," in *Thomas Mann's Addresses Delivered at the Library of Congress, 1942-1949*, 45-66 (Cabin John, MD: Wildside Press, 2008).

Odo Marquard, "Über die Unvermeidlichkeit der Geisteswissenschaften. Hermann Lübbe zum 60. Geburtstag," *Sprache und Literatur* 57 (1986): 72-81.

Karl Marx, *Capital: A Critique of Political Economy*, vol. 1, *The Process of Production of Capital* [1867], trans. Ben Fowkes (Harmondsworth: Penguin, 1990).

Leo Marx, *The Machine in the Garden: Technology and the Pastoral Ideal in America* (New York, NY: Oxford University Press, 1965).

Theodor Heinrich Mayer, "Lebende Photographien [1912]," in *Kein Tag ohne Kino: Schriftsteller über den Stummfilm*, ed. Fritz Güttinger, 119-129 (Frankfurt am Main: Deutsches Filmmuseum, 1984).

Judith Mayne, "Dracula in the Twilight: Murnau's *Nosferatu* (1922)," in *German Film & Literature: Adaptations and Transformations*, ed. Eric Rentschler, 25-39 (New York, NY: Methuen, 1986).

Hugh Mcleod, *Secularisation in Western Europe, 1848-1914* (New York, NY: St. Martin's Press, 2000).

Walter Mehring, *Berlin Dada* (Zurich: Verlag der Arche, 1959).

Walter Mehring, *Die verlorene Bibliothek: Autobiographie einer Kultur* (Hamburg: Rowohlt-Verlag, 1952).

"Meinungen von Karl Freund," *Filmtechnik* 1 (5 January 1926): 4-5, 27-28.

Gustav Melcher, "Die künstlerischen Vorzüge der Kinematographie," *Der Kinematograph* 116 (17 March 1909).

Gustav Melcher, "On Living Photography and the Film Drama [1909]," in *The Promise of Cinema: German Film Theory, 1907-1933*, eds. Anton Kaes, Nicholas Baer, and Michael Cowan, 17-20, trans. Alex H. Bush and Jon Cho-Polizzi, 17-20 (Berkeley, CA: University of California Press, 2016).

Georges Méliès, "Cinematographic Views [1907]," in *French Film Theory and Criticism: A History/Anthology, 1907-1939*, ed. and trans. Richard Abel, 35-47 (Princeton, NJ: Princeton University Press, 1988).

Moses Mendelssohn, "Über das Erhabene und Naive in den schönen Wissenschaften [1758]," in *Moses Mendelssohn: Ästhetische Schriften*, ed. Anne Pollok, 216-259 (Hamburg: Felix Meinder Verlag, 2006).

William Cameron Menzies, "The Layout of Bulldog Drummond," *Creative Art* (October 1929): 729-734.

G. Mercator (=Josef Krämer), "Ist die Photographie eine Kunst?" *Deutsche Photographen-Zeitung* 15 (1891).

"Metropolis – A Movie Based on Science," *Science and Invention* 170 (June 1927): 106-107.

"Metropolis Film Seen," *New York Times* (11 January 1927): 36.

Metropolis Ufa Magazin Sondernummer (Berlin: Bukwa, 1927).

Christian Metz, "Trucage and the Film," trans. Françoise Meltzer, *Critical Inquiry* 4 (Summer 1977): 657-675.

Alfred Richard Meyer, "Der Student von Prag," *Die Bücherei Maiandros* (1 October 1913): 11.

Jochen Meyer, "Die Hochhausdiskussion in Deutschland am Anfang der zwanziger Jahre," in *Der Schrei nach dem Turmhaus: der Ideenwettbewerb Hochhaus am Bahnhof Friedrichstrasse, Berlin 1921/22*, eds. Florian Zimmermann and Bruno Bruognolo, 187-214 (Berlin: Argon, 1988).

Heinz Michaelis, "Art and Technology in Film [1923]," in *The Promise of Cinema: German Film Theory, 1907-1933*, eds. Anton Kaes, Nicholas Baer, and Michael Cowan, trans. Alex H. Bush, 581-582 (Berkeley, CA: University of California Press, 2016).

Carlo Mierendorff, "If I Only Had the Cinema! [1920]," in *The Promise of Cinema: German Film Theory, 1907-1933*, eds. Anton Kaes, Nicholas Baer, and Michael Cowan, trans. Jeffrey Timon, 426-433 (Berkeley, CA: University of California Press, 2016).

G. A. Mincenty, "Echo aus Amerika," *Filmtechnik* 12 (1925): 254.

Jean Mitry, *The Aesthetics and Psychology of the Cinema* [1963], trans. Christopher King (Bloomington, IN: Indiana University Press, 2000).

M.-l., "Technische Kritik zu Metropolis," *Lichtbild-Bühne* 10 (12 January 1927).

Daniel Morgan, "Beyond Destiny and Design: Camera Movement in Fritz Lang's German Films," in *A Companion to Fritz Lang*, ed. Joe McElhaney, 259-278 (Hoboken, NJ: Wiley-Blackwell, 2015).

Herbert Moulton, "Magical Effects Brought to Screen by Unique Process," *Los Angeles Times* (18 April 1926): C25-6.

Inka Mülder-Bach, "Der Umschlag der Negativität. Zur Verschränkung von Phänomenologie, Geschichtsphilosophie und Filmästhetik in Siegfried Kracauers Metaphorik der 'Oberfläche,'" *Deutsche Vierteljahrsschrift für Literaturwissenschaft und Geistesgeschichte* 61:2 (June 1987): 359-373.

Corinna Müller, "Das 'andere' Kino? Autorenfilme in der Vorkriegsära," in *Die Modellierung des Kinofilms: Zur Geschichte des Kinoprogramms zwischen Kurzfilm und Langfilm 1905/06-1918, Mediengeschichte des Films*, vol. 2, ed. Harro Segeberg, 153-192 (Munich: Wilhelm Fink Verlag, 1998).

Corinna Müller, *Frühe deutsche Kinematographie: Formale, wirtschaftliche und kulturelle Entwicklungen 1907-1912* (Stuttgart: Metzler-Verlag, 1994).

Richard Müller-Freienfels, *Geheimnisse der Seele* (Munich: Delphin-Verlag, 1927).

Helmut Müller-Sievers, *The Cylinder: Kinematics of the Nineteenth Century* (Berkeley, CA: University of California Press, 2012).

Hugo Münsterberg, *Die Amerikaner* (Berlin: Ernst Siegfried Mittler und Sohn, 1904).

Hugo Münsterberg, *The Photoplay: A Psychological Study* (New York, NY: D. Appleton and Company, 1916).
Margaret Münsterberg, *Hugo Münsterberg: His Life and Work* (New York, NY: D. Appleton and Company, 1922).
F. W. Murnau, "...der frei im Raum zu bewegende Aufnahmeapparat [1924]," in Fred Gehler and Ulrich Kasten, *Friedrich Wilhelm Murnau* (Berlin: Henschelverlag Kunst und Gesellschaft, 1990).
F. W. Murnau, "Films of the Future," *McCall's Magazine* (September 1928): 27; 90.
F. W. Murnau, "Real 'Motion' Pictures," *Film Daily* (7 June 1925): 21.
F. W. Murnau, "The Ideal Picture Needs No Titles: By Its Very Nature the Art of the Screen Should Tell a Complete Story Pictorially (1928)," in *German Essays on Film*, eds. Richard W. McCormick and Alison Guenther-Pal, 66-68 (New York, NY: Continuum, 2004).
"Murnau Achievements Praised by Brisbane," *Exhibitors Herald* (17 July 1926): 30.
"Murnau's Study of Camera Noted," *Los Angeles Times* (19 June 1929): A8.
"Murnau Starts Production on First American Film," *Fox Folks* 10 (October 1926): 22.
"Murnau to Produce for Fox Film Corp," *Motion Picture World* (2 January 1926): 69.
Tobias Nagl, *Die unheimliche Maschine: Rasse und Repräsentation im Weimarer Kino* (Munich: edition text + kritik, 2009).
Oskar Negt, "Erbschaft aus Ungleichzeitigkeit und das Problem der Propaganda," in *Es muss nicht immer Marmor sein. Ernst Bloch zum 90. Geburtstag*, ed. Detlef Horster (Berlin: Wagenbach, 1975).
Dietrich Neumann, "The Urbanistic Vision in Fritz Lang's *Metropolis*," in *Dancing on the Volcano: Essays on the Culture of the Weimar Republic*, 143-154 (New York, NY: Camden House, 1994).
"New Apparatus," *Journal of the Society of Motion Picture Engineers* (January 1933): 85-86.
Isaac Newton, *Opticks: Or, a Treatise of the Reflections, Refractions, Inflections and Colours of Light* [1704] (Mineola, NY: Dover Publications, 1979).
Jakob Isak Nielsen, "Camera Movement in Narrative Cinema – Towards a Taxonomy of Functions," PhD. Dissertation, University of Aarhus, 2008.
Hans Nieter, "The Schüfftan Process of Model Photography," *Photographic Journal*, London (January 1930): 16.
Judith Noble, "Clear Dreaming: Maya Deren, Surrealism and Magic," in *Surrealism, Occultism and Politics: In Search of the Marvellous*, eds. Tessel M. Bauduin, Victoria Ferentinou, and Daniel Zamani, 210-226 (New York, NY: Routledge, 2018).
Kristoffer Noheden, *Surrealism, Cinema, and the Search for a New Myth* (Cham: Palgrave Macmillan, 2017).
Dan North, *Performing Illusions: Cinema, Special Effects and the Virtual Actor* (London: Wallflower Press, 2008).

Dan North, Bob Rehak, and Michael S. Duffy, eds., *Special Effects: New Histories/Theories/Contexts* (London: BFI/Palgrave, 2016).
"Now They'll Get 'Acting Continuity,'" *Photoplay* (June 1925): 53.
Geoffrey Nowell-Smith, *The History of Cinema: A Very Short Introduction* (Oxford: Oxford University Press, 2017).
Franz Anton Nüsslein, *Lehrbuch der Ästhetik als Kunstwissenschaft* (Regensburg: Verlag von G. Joseph Manz, 1837).
David E. Nye, *American Technological Sublime* (Cambridge, MA: MIT University Press, 1994).
Jan Olsson, "Framing Silent Calls: Coming to Cinematographic Terms with Telephony," in *Allegories of Communication: Intermedial Concerns from Cinema to the Digital*, eds. John Fullerton and Jan Olsson, 157-192 (Bloomington, IN: Indiana University Press, 2005).
Max Ophüls, *Spiel im Dasein: Eine Rückblende* [1959], ed. Helmut G. Asper (Berlin: Alexander Verlag, 2015).
Alex Owen, *The Place of Enchantment: British Occultism and the Culture of the Modern* (Chicago, IL: University of Chicago Press, 2004).
Joachim Paech, "Disposition der Einfühlung: Anmerkungen zum Einfluss der Einfühlungs-Ästhetik des 19. Jahrhunders auf die Theorie des Kinofilms," in *Der Film in der Geschichte*, eds. Knut Hickethier, Eggo Müller, and Rainer Rother, 106-121 (Baden-Baden: Edition Sigma, 1997).
Hans Pander, "Intertitles [1923]," in *The Promise of Cinema: German Film Theory, 1907-1933*, eds. Anton Kaes, Nicholas Baer, and Michael Cowan, trans. Alex H. Bush, 489-492 (Berkeley, CA: University of California Press, 2016).
Christopher Partridge, "Occulture is Ordinary," in *Contemporary Esotericism*, eds. Egil Asprem and Kennet Granholm, 113-133 (New York, NY: Routledge, 2014).
Christopher Partridge, "Orientalism and the Occult," in *The Occult World*, ed. Christopher Partridge, 611-625 (New York, NY: Routledge, 2015).
"Paul Fejos Praised For Broadway Direction," *Universal Weekly* (15 June 1929): 25.
Friedrich Paulsen, *Die deutschen Universitäten und das Universitätsstudium* (Berlin: Verlag von Asher & Co, 1902).
Friedrich Paulsen, "Theater und Kino," *Die Grenzboten* (Berlin) 45 (1913): 285-288.
Gilbert Perez, *The Material Ghost: Films and Their Medium* (Baltimore, MD: Johns Hopkins University Press, 2000).
Emil Perlmann, *Das Kino als modernes Volkstheater: Eine Entgegnung auf unberechtigte Angriffe* (Berlin: Agitations-Komitee der kinematographischen Fachpresse zur Förderung der Lichtbildkunst, 1912).
Patrice Petro, *Joyless Streets: Women and Melodramatic Representation in Weimar Germany* (Princeton, NJ: Princeton University Press, 1989).

Kurt Pinthus, "Quo Vadis, Cinema? On the Opening of the Königspavillon-Theater [1913]," in *The Promise of Cinema: German Film Theory, 1907-1933*, eds. Anton Kaes, Nicholas Baer, and Michael Cowan, trans. Don Reneau, 186-188 (Berkeley, CA: University of California Press, 2016).
Kurt Pinthus, "The Photoplay: A Serious Introduction for Those Who Think Ahead and Reflect [1913]," in *The Promise of Cinema: German Film Theory, 1907-1933*, eds. Anton Kaes, Nicholas Baer, and Michael Cowan, trans. Don Reneau, 199-203 (Berkeley, CA: University of California Press, 2016).
"Plans Efficiency Plus Art," *Motion Picture News* (2 January 1926): 27-28.
Helmuth Plessner, "Die Utopie in der Maschine [1924]," in *Gesammelte Schriften in zehn Bänden*, vol. X: *Schriften zur Soziologie und Sozialphilosophie* (Berlin: Suhrkamp-Verlag, 2003).
Gerhard Plumpe, *Der tote Blick: Zum Diskurs der Photographie in der Zeit des Realismus* (Munich: Wilhelm Fink Verlag, 1990).
Edgar Allan Poe, *Edgar Allan Poes Werke in zehn Bänden*, eds. Heda and Arthur Moeller-Bruck, trans. Hedda Moeller-Bruck and Hedwig Lachmann, 10 vols. (Minden: J.C.C. Bruns Verlag, 1901-1904).
Edgar Allan Poe, *Nebelmeer*, ed. Hanns Heinz Ewers, trans. Gisela Etzel (Munich: Georg Müller, 1914).
Alfred Polgar, "Das Drama im Kinematographen [1911]," in *Prolog vor dem Film: Nachdenken über ein neues Medium, 1909-1914*, ed. Jörg Schweinitz, 159-164 (Leipzig: Reclam-Verlag, 1992).
Erich Pommer, "Geschäftsfilm und künstlerischer Film," *Der Film* (10 December 1922): 1.
"Erich Pommer Interview with Radio Frankfurt [1950]," Eric Pommer Collection, Doheny Memorial Library, University of Southern California, Los Angeles, CA.
Friedrich Porges, "Luftschlösser im Film," *Die Bühne* (Vienna) 120 (25 February 1927).
Harry Alan Potamkin, "Knights of the Camera," *International Photographer* 8 (September 1930): 14-16.
Harry Alan Potamkin, "Phases of Cinema Unity," *Close-Up* 5 (May 1929): 27-38.
Arthur Powell, *The Astral Body and Other Astral Phenomena* (Wheaton, IL: The Theosophical Publishing House, 1927).
René Prédal, "Les Grands opérateurs (IV): Henri Alekan," *Cinéma* 173 (February 1973): 88-95.
Presse- und Propagandaheft Metropolis (Berlin: Parufamet Presse- und Propagandabteilung, 1927).
Christiane Pries, "Einleitung," in *Das Erhabene: Zwischen Grenzerfahrung und Größenwahn*, ed. Christiane Pries, 1-30 (Berlin: Verlag Walter de Gruyter, 1989).
Stephen Prince, *Digital Visual Effects in Cinema: The Seduction of Reality* (New Brunswick, NJ: Rutgers University Press, 2011).

Karl Prümm, "Das schwebende Auge: Zur Genese der bewegten Kamera," in *Die Medien und ihre Technik: Theorien, Modelle, Geschichte*, ed. Harro Segeberg, 235-256 (Marburg: Schüren-Verlag, 2004).
Vsevolod Pudovkin, *Film Technique and Film Acting*, trans. Ivor Montagu (London: Vision Press, 1954).
Lisa Purse, *Digital Imagining in Popular Cinema* (Edinburgh: Edinburgh University Press, 2019).
Priska Pytlik, *Okkultismus und Moderne ein kulturhistorisches Phänomen und seine Bedeutung für die Literatur um 1900* (Paderborn: Ferdinand Schöningh, 2005).
Otto Rank, *The Double: A Psychoanalytic Study* [1914], ed. and trans. Harry Tucker, Jr. (Chapel Hill, NC: The University of North Carolina Press, 1971).
Willy Rath, *Kino und Bühne* (M. Gladbach: Volksvereinsverlag, 1913).
Willy Rath, "Künstlerische Möglichkeiten des Lichtspiels [1913]," in *Geschichte der Filmtheorie: Kunsttheoretische Texte von Méliès bis Arnheim*, ed. Helmut H. Diederichs, 121-131 (Frankfurt am Main: Suhrkamp-Verlag, 2004).
Walter Rathenau, *Zur Kritik der Zeit* [1912] (Berlin: Fischer, 1929).
Ulrich Rauscher, "Das Kintop-Epos [1913]," in *Kein Tag ohne Kino: Schriftsteller über den Stummfilm*, ed. Fritz Güttinger, 137-140 (Frankfurt am Main: Deutsches Filmmuseum, 1984).
Neville Reay, "The Stroller," *Picture Play* 3 (May 1929): 55.
Hans-Peter Reichmann, *Hans Poelzig: Bauten für den Film* (Frankfurt am Main: Deutsches Filmmuseum, 1997).
Margaret Reid, "After Sunrise Comes What?" *Picture Play* (October 1928): 22-24, 116.
Walter Reimann, "Filmarbeit," *Gebrauchsgraphik* 6 (1924): 28-32.
Walter Reimann, "Filmarchitektur – Filmarchitekt?! [1925]," in *Werkstatt Film: Selbstverständnis und Visionen von Filmleuten der zwanziger Jahre*, eds. Rolf Aurich and Wolfgang Jacobsen, 111-115 (Munich: edition text + kritik, 1998).
Walter Reimann, "Filmarchitektur – heute und morgen," *Filmtechnik* (20 February 1926).
Malwine Rennert, "Heureka [1913]," in *Geschichte der Filmtheorie: Kunsttheoretische Texte von Méliès bis Arnheim*, ed. Helmut H. Diederichs, 143-146 (Frankfurt am Main: Suhrkamp-Verlag, 2004).
Richard Rickitt, *Special Effects: The History and Technique* (New York, NY: Billboard Books, 2007).
Rainer Maria Rilke, *Sonnets to Orpheus* [1922], trans. Edward Snow (New York, NY: North Point Press, 2004), 65.
Fritz K. Ringer, *The Decline of the German Mandarins: The German Academic Community, 1890-1933* (Cambridge, MA: Harvard University Press, 1969).
Günther Rittau, "Günther Rittau über Filmkunst und Filmtechnik," *Filmtechnik* 3 (5 February 1926): 50.

Günther Rittau, "Special Effects in *Metropolis*, [1927]," in *Fritz Lang's Metropolis: Cinematic Visions of Technology and Fear*, eds. Michael Minden and Holger Bachmann, 78-80 (Rochester, NY: Camden House, 2002).

Ariel Rogers, *On the Screen: Displaying the Moving Image, 1926-1942* (New York, NY: Columbia University Press, 2019).

Christian Rogowsky, ed., *The Many Faces of Weimar Cinema: Rediscovering Germany's Filmic Legacy* (Columbia, MD: Camden House, 2011).

Paul Rohnstein, "Beiträge zur wirtschaftlichen Entwicklung der Deutschen Film-Industrie (unter besonderer Berücksichtigung des Kinomatographentheatergewerbes [sic])," Ph.D. Dissertation, Julius-Maximilians-Universität Würzburg, 1922.

Jules Romains, "The Crowd at the Cinematograph [1911]," in *French Film Theory and Criticism: A History/Anthology, 1907-1939*, ed. and trans. Richard Abel, 53-54 (Princeton, NJ: Princeton University Press, 1988).

Theodore Roosevelt, "True Americanism [1894]," in *American Ideals and Other Essays, Social and Political*, 16-37 (Philadelphia, PA: Gebbie and Company, 1903).

Fritz Rosenfeld, "Sonnenaufgang," *Arbeiter-Zeitung* 36 (5 February 1928): 19.

Joseph Roth, "Der auferstandene Mensch [1923]," in *Joseph Roth Werke*, vol. 1, *Journalistische Arbeiten 1915-1923*, ed. Klaus Westermann, 936-939 (Cologne: Kiepenheuer & Witsch, 1989-1991).

Rudolf Rotheit, "Kasperle und Kinematograph," *Der Kinematograph* 5 (3 February 1907).

R.T., "Technische Filmkritiken," *Die Kinotechnik* 27 (1926): 637-638.

L. Case Russell, "Baa, Baa, Black Sheep," *Hollywood Topics* 12 (19 February 1927): 18.

R. L. Rutsky, "The Mediation of Technology and Gender: *Metropolis*, Nazism, Modernism," *New German Critique* (Autumn 1993): 3-32.

Tom Ryall, "Blackmail," in *British Film Institute Film Classics*, vol. 1, eds. Rob White and Edward Buscombe (London: British Film Institute, 2003).

Marie-Laure Ryan, *Narrative as Virtual Reality 2: Revisiting Immersion and Interactivity in Literature and Electronic Media* (Baltimore, MD: Johns Hopkins University Press, 2015).

–s, "Kritik der Leinwand," *Die Filmwoche* 22 (1925): 519.

Piotr Sadowski, *The Semiotics of Light and Shadows: Modern Visual Arts and Weimar Cinema* (London: Bloomsbury, 2017).

Barry Salt, *Film Style and Technology: History and Analysis* (London: Starwood, 1983).

Roland Schacht, "Nosferatu," *Das Blaue Heft* (15 April 1922).

Will Scheller, "The New Illusion [1914]," in *The Promise of Cinema: German Film Theory, 1907-1933*, eds. Anton Kaes, Nicholas Baer, and Michael Cowan, trans. Alex H. Bush, 196-199 (Berkeley, CA: University of California Press, 2016).

Robert Scheu, "Das Kino der Kommenden," *Vossische Zeitung* 117 5 (March 1914).

Dietrich Scheunemann, "Activating the Differences: Expressionist Film and Early Weimar Cinema," in *Expressionist Film: New Perspectives*, ed. Dietrich Scheunemann, 1-31 (Rochester, NY: Camden House, 2003).
Friedrich Schiller, *On the Aesthetic Education of Man* [1795], trans. Reginald Snell (Mineola, NY: Dover Publications, 1992).
Friedrich Schiller, "On the Pathetic [1793]," in *Friedrich Schiller Essays*, eds. Walter Hinderer and Daniel O. Dahlstrom, trans. Daniel O. Dahlstrom, 45-69 (New York, NY: Continuum, 1993).
Friedrich Schiller, *The Works of Friedrich Schiller: Aesthetical and Philosophical Essays*, ed. Nathan Haskell Dole (Boston, MA: The Wyman-Fogg Company, 1902).
Hans Schliepmann, "Wolkenkratzer," *Tägliche Rundschau* 49 (30 January 1921).
Heide Schlüpmann, "Der Spiegel des Grauens: Murnaus *Nosferatu*," *Frauen und Film* 49 (December 1990): 38-51.
Heide Schlüpmann, "The First German Art Film: Rye's *The Student of Prague*," in *German Film and Literature: Adaptations and Transformations*, ed. Eric Rentschler, 9-24 (New York, NY: Methuen, 1986).
Heide Schlüpmann, "'Je suis la solitude:' Zum Doppelgängermotiv in *Der Student von Prag*," *Frauen und Film* 36 (1984): 10-24.
Heide Schlüpmann, "Phenomenology of Film: On Siegfried Kracauer's Writings of the 1920s," *New German Critique* 40 (Winter 1987): 97-114.
Heide Schönemann, *Paul Wegener: Frühe Moderne im Film* (Stuttgart: Edition Axel Menges, 2003).
Arthur Schopenhauer, *Metaphysik des Schönen* [1820] (Munich: Piper Verlag, 1985).
Arthur Schopenhauer, *The World as Will and Representation* [1819], vol. 1, trans. E. F. J. Payne (New York, NY: Dover Publications, 1969).
Sime, "Metropolis," *Variety* (16 March 1927): 16-17.
Eugen Schüfftan, "Mein Verfahren," *Film-Kurier* 24 (19 November 1926).
Eugen Schüfftan, "Verbesserungen in der Miniatur-Aufnahmetechnik," in Friedrich Porges, *Mein Film-Buch*, 35-37 (Vienna: Mein Film Verlag, 1929).
"Schufftan Process Acquired," *Kinematograph Weekly* (13 January 1927): 56.
Jörg Schweinitz, ed., *Prolog vor dem Film: Nachdenken über ein neues Medium, 1909-1914* (Leipzig: Reclam-Verlag, 1992).
Guido Seeber, *Der praktische Kameramann* [1927] (Frankfurt am Main: Deutsches Filmmuseum, 1980).
Guido Seeber, *Der praktische Kameramann*, vol. 2, *Der Trickfilm in seinen grundsätzlichen Möglichkeiten* [1927] (Frankfurt am Main: Deutsches Filmmuseum, 1979).
Guido Seeber, "Der Student von Prag," *Lichtbild-Bühne* (15 November 1926).
Guido Seeber, "Doppelgängerbilder im Film," *Die Kinotechnik* 1 (September 1919): 12-17.

Guido Seeber, *Kamera-Kurzweil: Allerlei interessante Möglichkeiten beim Knipsen und Kubeln* (Stuttgart: Union Deutsche Verlagsgesellschaft, 1930).

Harro Seegeberg, ed. *Mediengeschichte des Films* (Munich: Wilhelm Fink Verlag, 1996-2012).

R. Seligmann, "Kinematograph und Traum," *Frankfurter Zeitung und Handelsblatt* 289 (18 October 1911): 2.

Adolf Sellmann, *Kino und Volksbildung* (M. Gladbach: Gemeinnützige Volksbibliothek, 1914).

Ted Shawn, *Every Little Movement. A Book About François Delsarte, the Man and His Philosophy, His Science and Applied Aesthetics, the Application of This Science* (New York, NY: Dance Horizons, 1968).

Victor Shklovsky, "Art as Technique [1916]," in *Russian Formalist Criticism: Four Essays*, eds. Lee T. Lemon and Marion J. Reiss, 3-24 (Lincoln, NE: University of Nebraska Press, 1965).

Zoltán Simon, *The Double-Edged Sword: The Technological Sublime in American Novels Between 1900 and 1940* (Budapest: Akadémiai Kiadó, 2003).

Ben Singer, "The Ambimodernity of Early Cinema: Problems and Paradoxes in the Film-and-Modernity Discourse," in *Film 1900: Technology, Perception, Culture*, eds. Annemone Ligensa and Klaus Kreimeier, 38-51 (New Barnet, Herts: John Libbey Publishing, 2009).

David J. Skal, *Hollywood Gothic: The Tangled Web of Dracula From Novel to Stage to Screen* (New York, NY: Norton, 1990).

Vivian Sobchack, *Screening Space: The American Science Fiction Film* (New Brunswick, NJ: Rutgers University Press, 1987).

Society of Motion Picture Engineers Progress Committee, "Progress in the Motion Picture Industry," *Journal of the Society of Motion Picture Engineers* 2 (August 1932): 123.

"Somlyo Joins Famous Players," *Motion Picture News* (4 September 1926): 830.

Spektator, "Autorenkünstler [sic] und Riesenfilms," *Der Kinematograph* 349 (3 September 1913).

Oswald Spengler, *The Decline of the West* [1918/1922], trans. Charles Francis Atkinson (New York, NY: Knopf, 1966).

Ludwig Spitzer, "Fritz Lang über den Film der Zukunft," *Die Filmtechnik* 2 (15 July 1925): 34-36.

Genevieve Stebbins, *Delsarte System of Expression* [1886] (New York, NY: Dance Horizons, 1977).

Rudolf Steiner, "Blut ist ein ganz besonderer Saft. Berlin, 25. Oktober 1906," in *Die Erkenntnis des Übersinnlichen in unserer Zeit und deren Bedeutung für das heutige Leben*, 35-65 (Dornach: Rudolf Steiner Verlag, 1983).

Philipp Stiasny, *Das Kino und der Krieg: Deutschland 1914-1929* (Munich: edition text + kritik, 2009).

Adalbert Stifter Briefe, ed. Hans Schumacher (Zurich: Manesse-Verlag, 1947).

Stiftung Deutsche Kinemathek, ed. *Das wandernde Bild: Der Filmpionier Guido Seeber* (Berlin: Stiftung Deutsche Kinemathek, 1979).

Georg Otto Stindt, "Die entfesselte Kamera," *Filmtechnik* 25 (1926): 491-494.

Bram Stoker, *Dracula* [1897] (New York, NY: Barnes and Noble, 2003).

Karl Struss, "Recollections of Karl Struss, Interviewed by John Dorr," in *An Oral History of the Motion Picture in America*, Appendix II: *Sunrise* (Los Angeles, CA: University of California, 1969).

Karsten R. Stueber, *Rediscovering Empathy. Agency, Folk Psychology and the Human Science*s (Cambridge, MA: MIT Press, 2006).

William Stull, "Development of Mobile Camera Carriages and Cranes," *American Cinematographer* (19 May 1933): 12-13; 36-37.

Leon Surette, *The Birth of Modernism: Ezra Pound, T.S. Eliot, W.B. Yeats, and the Occult* (Montreal & Kingston: McGill-Queen's University Press, 1993).

Joseph Tabbi, *Postmodern Sublime: Technology and American Writing from Mailer to Cyberpunk* (Ithaca, NY: Cornell University Press, 1995).

Frederick A. Talbot, *Moving Pictures: How They Are Made and Worked* (London: William Heinemann, 1912).

Herbert Tannenbaum, "Film Advertising and the Advertising Film [1920]," in *The Promise of Cinema: German Film Theory, 1907-1933*, eds. Anton Kaes, Nicholas Baer, and Michael Cowan, trans. Tara Hottman, 179-182 (Berkeley, CA: University of California Press, 2016).

Herbert Tannenbaum, *Kino und Theater* (Munich: Steinebach, 1912).

Herbert Tannenbaum, "Problems of the Film Drama [1913]," in *The Promise of Cinema: German Film Theory, 1907-1933*, eds. Anton Kaes, Nicholas Baer, and Michael Cowan, trans. Alex H. Bush, 192-196 (Berkeley, CA: University of California Press, 2016).

Maria Tatar, *The Hard Facts of the Grimms' Fairy Tales* (Princeton, NJ: Princeton University Press, 2003).

Nicholas Taylor, "The Awful Sublimity of the Victorian City: Its Aesthetic and Architectural Origins," in *The Victorian City: Images and Realities*, eds. H. J. Dyos and Michael Wolff, 431-447 (London: Routledge & Kegan Paul Books, 1973).

Sheldon Teitelbaum, "Special Effects: Aliens," *Cinefantastique* 16 (October 1986).

"The American Film Institute and American Society of Cinematographers Seminar with Hal Mohr, ASC, Held March 24, 1973" (Beverly Hills, CA: Center for Advanced Film Studies, 1978).

"The German Kinematograph and Foreign Countries," *Der Kinematograph* 38 (18 September 1907).

"The Picture Parade," *Motion Picture Magazine* 6 (January 1927): 61.
Walter Thielemann, "Der neue Weg im Film," *Film und Brettl* 1 (January 1922).
Kristin Thompson, *Herr Lubitsch Goes to Hollywood: German and American Film after World War I* (Amsterdam: Amsterdam University Press, 2005).
Kristin Thompson, "Im Anfang war...: Some Links between the German Fantasy Films of the Teens and the Twenties," in *Before Caligari: German Cinema 1895-1920*, eds. Paolo Cherchi Usai and Lorenzo Codelli, 138-161 (Pordenone: Edizioni Biblioteca dell'Immagine, 1990).
Walther Thorner, *Verfahren und Vorrichtung zur Herstellung kinematographischer Kombinationsaufnahmen*. Deutsches Reichspatent 598,712, filed 19 May 1932, and issued 18 June 1934.
Pamela Thurschwell, *Literature, Technology, and Magical Thinking, 1880-1920* (Cambridge, UK: Cambridge University Press, 2001).
Tsvetan Todorov, *The Fantastic: A Structural Approach to a Literary Genre*, trans. Richard Howard (Ithaca, NY: Cornell University Press, 1975).
Ernst Toller, "The German Hinkemann [1923]," *Plays Two*, ed. and trans. Alan Raphael Pearlman (London: Oberon Books, 2019).
Ernst Toller, "The Machine Breakers [1922]," *Plays Two*, ed. and trans. Alan Raphael Pearlman (London: Oberon Books, 2019).
Ernst Toller, "Transformation [1919]," *Plays One*, ed. and trans. Alan Raphael Pearlman (London: Oberon Books, 2000).
Corinna Treitel, *A Science for the Soul: Occultism and the Genesis of the German Modern* (Baltimore, MD: Johns Hopkins University Press, 2004).
Hermann Treuner, "Carl Hoffmann," in *Filmkünstler: Wir über uns selbst* (Berlin: Sibyllen-Verlag, 1928).
François Truffaut, *Hitchcock* (New York, NY: Simon & Schuster, 1985).
Yuri Tsivian, "Cyberspace and its Precursors: Lintsbach, Warburg, Eisenstein," in *Transmedia Frictions: The Digital, the Arts, and the Humanities*, eds. Marsha Kinder and Tara McPherson, 80-99 (Berkeley, CA: University of California Press, 2014).
Julie Turnock, "Patient Research on the Slapstick Lots: From Trick Men to Special Effects Artists in Silent Hollywood," *Early Popular Visual Culture* 2 (May 2015): 152-173.
Julie Turnock, *Plastic Reality: Special Effects, Technology and the Emergence of 1970s Blockbuster Aesthetics* (New York, NY: Columbia University Press, 2015).
Malcom Turvey, *Doubting Vision: Film and the Revelationist Tradition* (Oxford: Oxford University Press, 2008).
Casper Tybjerg, "The Faces of Stellan Rye," in *A Second Life: German Cinema's First Decades*, eds. Thomas Elsaesser and Michael Wedel, 151-159 (Amsterdam: Amsterdam University Press, 1996).

King Vidor, *A Tree is a Tree* (Hollywood, CA: S. French, 1989).
Berthold Viertel, "In the Cinematographic Theater [1910]," in *The Promise of Cinema: German Film Theory, 1907-1933*, eds. Anton Kaes, Nicholas Baer, and Michael Cowan, trans. Eric Ames, 77-78 (Berkeley, CA: University of California Press, 2016).
Alfred de Vigny, "Paris [1831]," in *Oeuvres complètes. Texte présenté et commenté par Ferdinand Baldensperger*, vol. 1, 164-165 (Paris: Gallimard, 1948).
Auguste Villiers de l'Isle-Adam, *Tomorrow's Eve* [1886], trans. Robert Martin Adams (Chicago, IL: University of Illinois Press, 2001).
Christiane Voss, "Fiktionale Immersion," *montage A/V* (February 2008): 69-86.
Richard Wagner, "The Art-work of the Future [1850]," in *The Art-work of the Future and Other Works*, trans. William Ashton Ellis, 69-214 (Lincoln, NE: University of Nebraska Press, 1993).
Chris Wahl, *Multiple Language Versions Made in Babelsberg. Ufa's International Strategy, 1929-1939* (Amsterdam: Amsterdam University Press, 2016).
Willi Warstat, "Zwischen Theater und Kino," *Die Grenzboten* 23 (1912): 483-488.
Richard Watts, "Some Additional Remarks on that New German Photoplay," *New York Herald Tribune* (11 July 1926): E3.
Max Weber, "Science as Vocation [1918]," in *The Vocation Lectures*, eds. David Owen and Tracy B. Strong, trans. Rodney Livingstone, 1-32 (Indiapolis, IN: Hackett Publishing Company, 2004).
Michael Wedel, *Der deutsche Musikfilm: Archäologie eines Genres 1914-1945* (Munich: edition text + kritik, 2007).
Paul Wegener, "On the Artistic Possibilities of the Motion Picture [1916]," in *The Promise of Cinema: German Film Theory, 1907-1933*, eds. Anton Kaes, Nicholas Baer, and Michael Cowan, trans. Alex H. Bush, 206-208 (Berkeley, CA: University of California Press, 2016).
Paul Wegener, "Warum ich für den Film spiele," *Berliner Börsen-Courier* 321 (12 July 1914).
W…f, "Zum ersten Male Schüfftansches Patent," *Die Kinotechnik* 18 (25 September 1925): 456.
"What's a Camera for?" *Motion Picture News* 7 (14 August 1926): 565.
Kristen Whissel, *Spectacular Digital Effects: CGI and Contemporary Cinema* (Durham, NC: Duke University Press, 2014).
Daniel Wiegand, *Gebannte Bewegung: Tableaux vivants und früher Film in der Kultur der Moderne* (Marburg: Schüren-Verlag, 2016).
"Wie *Nosferatu* entstand," *Film BZ* (5 March 1922).
Heidi Wiese and Henri Alekan, "Man muß die Technik überwinden, um zur Kunst zu gelangen: Ein Gespräch mit dem Kameramann Henri Alekan," in *Die Metaphysik des Lichts: Der Kameramann Henri Alekan*, ed. Heidi Wiese, 22-57 (Marburg: Schüren-Verlag 1996).

Oscar Wilde, *The Soul of Man under Socialism* [1891] (New York, NY: Harper & Row, 1970).
Leo Wiltin, "Die Technik," *Film-Kurier* 24 (19 November 1926).
K. W. Wolf-Czapek, "Deutsche Films – technisch bewertet," *Lichtbild-Bühne* 21 (25 May 1912): 22-23.
Robin Wood, "Murnau's Midnight and Sunrise," *Film Comment* 3 (May/June 1976): 4-19.
Frederic Wynne-Jones, "When the Camera's Eye Lies for Entertainment," *The New York Times* (2 May 1926): X5.
Fedor von Zobeltitz, "Film-Literatur," *Das literarische Echo* 13:15 (1 May 1911): 1103.
Werner Zurbuch, "Eugen Schüfftan – Meister der Filmtechnik," *Filmtechnikum* (June 1962): 179-180.

Index

For Product Safety Concerns and Information please contact our EU representative GPSR@taylorandfrancis.com
Taylor & Francis Verlag GmbH, Kaufingerstraße 24, 80331 München, Germany

www.ingramcontent.com/pod-product-compliance
Ingram Content Group UK Ltd.
Pitfield, Milton Keynes, MK11 3LW, UK
UKHW020018200926
475702UK00014B/1948

* 9 7 8 1 0 4 1 1 8 6 5 2 6 *